River Deltas Research - Recent Advances

Edited by Andrew J. Manning

Published in London, United Kingdom

IntechOpen

Supporting open minds since 2005

River Deltas Research – Recent Advances
http://dx.doi.org/10.5772/intechopen.78857
Edited by Andrew J. Manning

Contributors
Fikrat M. Hassan, Abdul Hameed M. Al Obaidy, Rana Riyadh R. Al-Ani, María Del Refugio Castañeda Chávez, Fabiola Lango-Reynoso, Punarbasu Chaudhuri, Subhamita Chaudhuri, Raktima Ghosh, Igor Cretescu, Liliana Lazar, Liliana Teodorof, Adrian Burada, Zsofia Kovacs, Madalina Sbarcea, Gabriela Soreanu, Dan Padure, Charles Nyanga, Loi To Thi Bich, Beatrice Njeri Obegi, Mathieu Le Meur, Vo Le Phu, Nicolas Gratiot, Nde Samuel Che, Sammy Bett, Manny Mathuthu, Enyioma Chimaijem Okpara, Peter Oluwadamilare Olagbaju, Omolola Esther Fayemi, Matteo Postacchini, Maurizio Brocchini, Lorenzo Melito, Eleonora Perugini, Andrew J. Manning, Joseph P. Smith, Joseph Calantoni, Leiping Ye, Tian-Jian Hsu, James Holyoke, Jorge A. Penaloza-Giraldo

Notice
Statements and opinions expressed in the chapters are these of the individual contributors and not necessarily those of the editors or publisher. No responsibility is accepted for the accuracy of information contained in the published chapters. The publisher assumes no responsibility for any damage or injury to persons or property arising out of the use of any materials, instructions, methods or ideas contained in the book.

First published in London, United Kingdom, 2022 by IntechOpen
IntechOpen is the global imprint of INTECHOPEN LIMITED, registered in England and Wales, registration number: 11086078, 5 Princes Gate Court, London, SW7 2QJ, United Kingdom
Printed in Croatia

British Library Cataloguing-in-Publication Data
A catalogue record for this book is available from the British Library

Additional hard and PDF copies can be obtained from orders@intechopen.com

River Deltas Research – Recent Advances
Edited by Andrew J. Manning
p. cm.
Print ISBN 978-1-78985-670-5
Online ISBN 978-1-83880-165-6
eBook (PDF) ISBN 978-1-83880-681-1

We are IntechOpen,
the world's leading publisher of
Open Access books
Built by scientists, for scientists

6,000+

Open access books available

148,000+

International authors and editors

185M+

Downloads

Our authors are among the

156

Countries delivered to

Top 1%

most cited scientists

12.2%

Contributors from top 500 universities

Interested in publishing with us?
Contact book.department@intechopen.com

Numbers displayed above are based on latest data collected.
For more information visit www.intechopen.com

Meet the editor

Professor Andrew J. Manning is a principal scientist (rank grade 9) in the Coasts and Oceans Group at HR Wallingford, UK, who has more than twenty-three years of scientific research experience in both industry and academia examining natural turbulent flow dynamics, fine-grained sediment transport processes, and assessing how these interact. He also lectures in Coastal and Shelf Physical Oceanography at the University of Plymouth, UK. Internationally, Professor Manning has been appointed visiting/adjunct/guest professor at five universities and is a highly published and world-renowned scientist in depositional sedimentary flocculation processes. He is a fellow of the Royal Geographical Society and recipient of a 2007 UoP Vice Chancellor's Research Fellowship and 2015 Exemplary Act Award from the United States Department of the Interior and US Geological Survey. He has contributed to more than 100 peer-reviewed publications in marine science, of which more than 50 have been published in international scientific journals. He also has more than 140 articles in refereed international conference proceedings to his credit. Professor Manning supervises graduate, postgraduate, and doctoral students on a range of research topics in marine science. He has led numerous research projects investigating sediment dynamics in aquatic environments around the world, including estuaries, tidal lagoons, river deltas, salt marshes, intertidal, coastal waters, and shelf seas.

Contents

Preface

Typically bordering between land and sea, river deltas are among the most diversely utilized environments on the planet. River deltas are extremely important environments socioeconomically, and their usage places ever-increasing stresses on these very sensitive aquatic regions.

The effective management of predominantly shallow aquatic environments requires a detailed scientific understanding of the various contributary natural processes. This has both environmental and economic implications, especially where there is any anthropogenic involvement. Numerical models are often the tool used for predicting the trends and patterns in these situations, as they can estimate the various spatial and temporal changes. However, the processes (e.g., physical, biological, and chemical) can vary quite considerably depending upon local conditions. Thus, for more than half a century, scientists, engineers, hydrologists, and mathematicians have been researching the many aspects that influence river delta environments. These issues range from processes such as water quality, pollution, land reclamation, and tidal dynamics to how geomorphodynamics, water column structure, and ecosystem habitats can be applied within river delta environmental frameworks.

River Deltas Research - Recent Advances draws on international scientific research to examine the following river delta-related issues: water quality and contamination, environmental disasters, climate change, land reclamation, geomorphology, biodiversity, sediment dynamics, classification, circulation hydrodynamics, ecosystems, and anthropogenic stresses. These key topics are supported by case study examples of river deltas from around the world. The research included herein was carried out by researchers who specialise in shallow water processes and related issues.

The book includes nine chapters written by an international group of research scientists who specialise in areas such as geomorphology, sediment dynamics, water quality, hydrology, and numerical modelling. Most of the chapters are concerned with natural process-related issues in river delta environments. They discuss water management in the Danube Delta, anthropogenic activities in the Mekong River Delta, and land use and land cover changes in the Crocodile River catchment, South Africa. Other chapters discuss wave-forced dynamics, oil-contaminated sediments, physico-chemical parameters, river-sea interaction, mangroves, and delta sinking issues.

This book is an excellent source of information on recent research on river delta environments from an interdisciplinary perspective. I would like to thank all the authors for their contributions, and I highly recommend this textbook to both scientists and engineers who deal with related issues.

Andrew J. Manning
HR Wallingford Ltd,
Coasts and Oceans Group,
UK

University of Hull,
UK

University of Delaware,
USA

University of Florida,
USA

Stanford University,
USA

Technical University Delft,
Netherlands

University of Plymouth,
UK

Chapter 1

What Is the Future of the Lower Mekong Basin Struggling against Human Activities? A Review

Mathieu Le Meur, Vo Le Phu and Nicolas Gratiot

Abstract

The Mekong River (MR) is recognized the 12th biggest rivers in the world. The Mekong watershed is the biggest one in Southeast Asia (795,000 km^2), is densely populated (70 million people), is considered as the most productive one in Southeast Asia and is economically essential to the region. However, nowadays, the Lower Mekong River (LMR) and its delta are facing several emerging and critical anthropogenic stressors (dams construction, climate change, water poor quality, delta sinking). This review attempts to: (i) present the Mekong regional characteristics (geography, topological settings, climatic conditions, hydrology, demographic features and the anthropogenic activities), (ii) present the different factors that endanger the LMR, including the dam's impacts, the climate change, the delta subsidence, and the degradation of the water quality, (iii) make comparison with different big rivers around the world and (iv) promote future decisions in order to minimize the negative impacts and seek for a trajectory that assures well-being and sustainability. International consultation and cooperation leading to sustainable management is now of a pivotal importance to try to avoid the deterioration of the LMR and its delta.

Keywords: Mekong River, Mekong delta, dams, climate-change, delta sinking

1. Introduction

Rivers and deltas are highly populated areas and their ecosystems are highly vulnerable. Most of the deltas on earth are nowadays severely degraded due to anthropogenic activities [1]. Man-made delta degradation, in consequence, affects the livelihood of millions of people [2].

The Mekong River (MR) is known as one of the 32 largest river on earth [3], flowing through the largest watershed in Southeast Asia (795,000 km^2) and crossing six different countries (China, Lao PDR, Myanmar, Thailand, Lao PDR, Cambodia and Vietnam) (Mekong River Commission (MRC) [4]). This basin is densely populated with around 72 million inhabitants and it is about 52,000 km^2 [5]. The economy of the region and its population have been highly growing in the last decades, with an exponential need for electricity in China, Thailand and in Vietnam [6]. The Mekong watershed is very productive with a lot of aquatic economic activities [7], providing livelihoods for most of its population [8]. Along the Mekong Basin (MB), more than 60 million people are dependent of the natural resources [9]. This important productivity is mainly due to the natural hydrological cycle with an important flood pulse occurring during the monsoon season (from July to September). The large Cambodian floodplain and particularly the Tonle Sap Lake (TSL) is a highly productive ecosystem because of the sediments brought by the important flood pulse during the monsoon [7] and a reversed flow during the dry season.

However, the Mekong River is now threatened by different anthropogenic stressors and some considerable changes deeply affected the region over the last decades. Land use change, water infrastructures and regional climate changes are ongoing factors that affect the natural functioning of the basin and its population [10]. At the downstream ending of the Mekong River, the 20 million inhabitants are now threatened by the surrounding sea level rise (rising sea level and land subsidence) [11–13].

This chapter aims to answer the following questions: (i) What are the different stressors that threaten the MB? How these stressors evolve spatially and temporally? (iii) What are the future directions needed to minimize the different damages?

To answer the different research questions, the chapter will be articulated in different sections:

- Introduction of the Mekong regional characteristics

- Presentation and discussion of the different issues that the MR and its delta are facing to

- Comparison of the different stressors with different big rivers in the world

- identification of some future directions that should be taken in order to minimize damages even if some changes will not be recoverable (tipping point) for a sustainable water management and the research procedure that could adequately accompany these scenarios.

2. Mekong regional settings

2.1 Physical features

The Mekong Basin is the largest one in Southeast Asia (795,000 km^2), covering six countries (China, Laos, Myanmar, Thailand, Cambodia and Vietnam) [14]. **Figure 1** and **Table 1** of this paper present the main features of the Mekong Basin. The MR is 4800 km long (12th longest in the world), discharges 475 km^3 yr.$^{-1}$ of water (8th largest), and discharges 160 * 106 tons yr.$^{-1}$ of sediments (10th largest sediment load in the world) [16]. The Tibetan Plateau signs the start of the Mekong's journey, at an altitude of 4970 meters. There, the river flows through the Chinese provinces (Qinghai and Yunnan) which are a very steep topography. Across 2000 km, the average slope is 2 m km^{-1}. Then, the river marks the border between Myanmar and Lao PDR. In Laos and Thailand, less mountainous regions tuck the river. The Mekong Basin is usually divided into upper and lower parts where the geographical boundary is at Chiang Sean in Thailand [17]. Downstream Chiang Sean, the slope is moderated with an average of 0.25 m km^{-1}. The Khone waterfalls in Laos' Champasak province, the only waterfalls of the Mekong River mark the beginning of the Mekong's plains. Finally, the Mekong spills in the Cambodian and Vietnamese floodplains and ends its course into the East Sea [4].

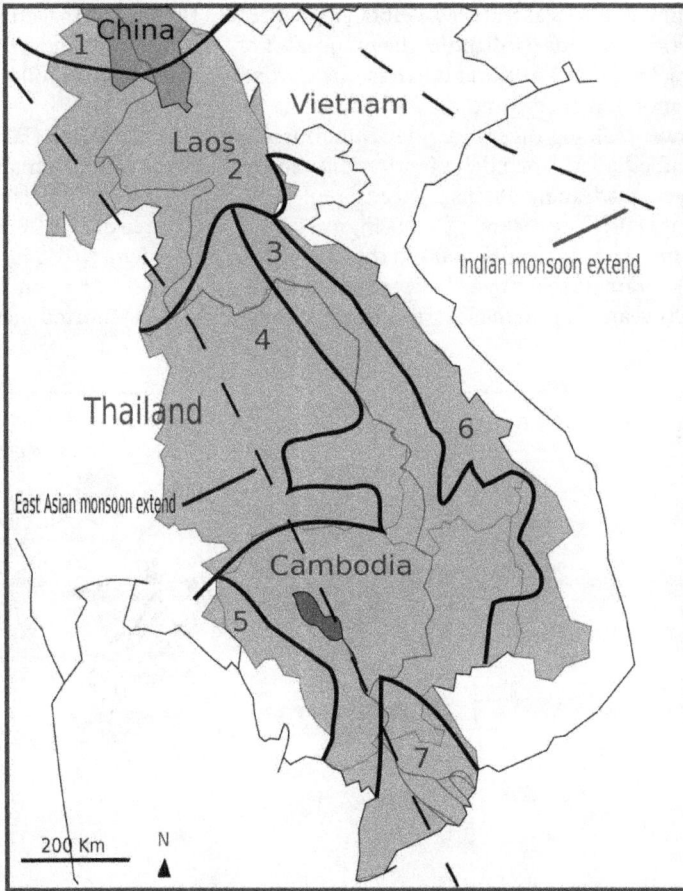

Figure 1.
Map of the Mekong River basin (MRB) showing the different countries, the upper Mekong Basin (red), the lower Mekong Basin (gray), the seven physical separations: the mountainous panhandle, the mountains of northern Laos and Thailand, the Mekong lowland, The Korat uplands, the Cardamon and Elephant hills, the Annamite mountain range and the Mekong Delta *(from [15]), the monsoon extends.*

Mekong River characteristics	
River length (km)	4800
Countries along the river	Cambodia, China, Lao People's Democratic Republic, Myanmar, Thailand, Vietnam
River basin (km^2)	795
Population (10^6)	58
Discharge (Km^3.yr^{-1})	475
Discharge sediments (tons yr^{-1})	160 ˙ 106
Annual suspended sediment load (10^6 tonnes)	150
Planned number of dams	123
Mean erosion rate (m yr-1)	12

Table 1.
Mekong basin characteristics.

According to physical features, Gupta [15] separated the MB into seven different units: the mountainous panhandle, the mountains of northern Laos and Thailand, the Mekong lowland, The Korat uplands, the Cardamon and Elephant hills, the Annamite mountain range and the Mekong Delta (MD) (**Figure 1**).

The Lower Mekong River is divided into four subsystems: Tonle Sap (lake and river), Cambodian Mekong delta (entire MD except the Tonle Sap), Vietnamese MD (Long Xuyen quadrangle, Plains of Reeds, region between Tien and Hau rivers) and coastal zones [18]. The Mekong Delta (**Figure 2**) has a total area of 52,000 km^2 [19]. The last 6 km (late Holocene) marked the fast growth of the delta [19]. The average progradation rate was 30 m yr. $^{-1}$ over the past 700 years [20] and 16 m yr.$^{-1}$ over the past 300 years [21]. In the last few decades, several authors reported instead of

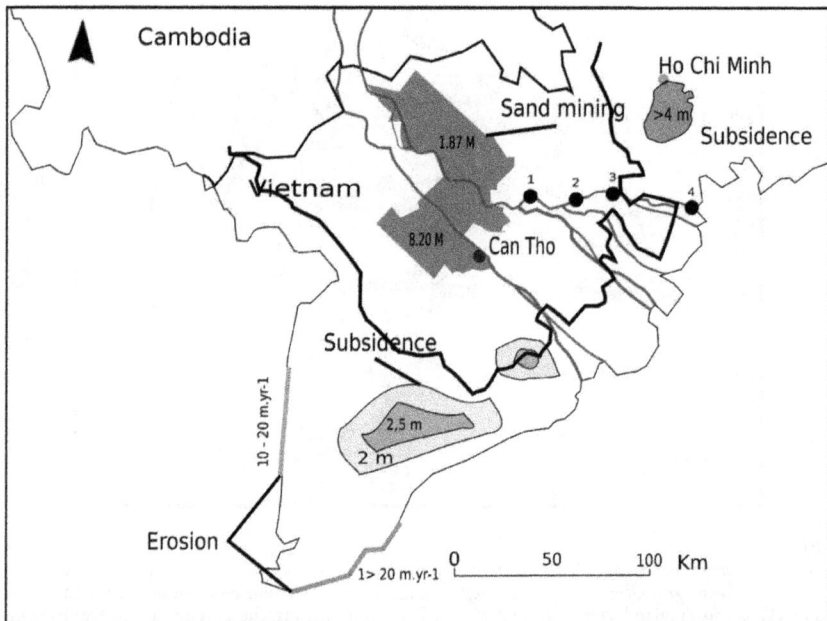

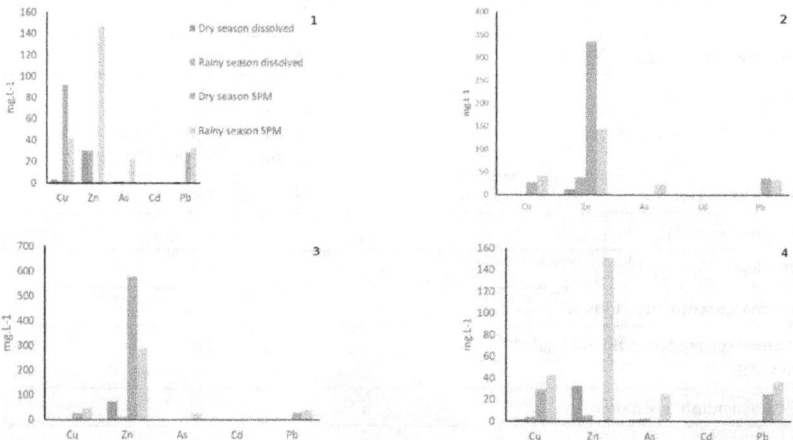

Figure 2.
Map of the Mekong delta showing the different issues in the region. Coastal erosion (m.yr.⁻¹), subsidence (m), sand mining (m³), trace metal concentration in dissolved and SPM fractions (mg.L⁻¹) salt intrusion contour in 2010 (5 ppt).

gradual progradation, some serious shoreline erosion [22, 23]. The mean erosion rate was estimated to be at 12 m yr.$^{-1}$ [22] with some sub regions where erosion rates and driven factors are different [24]. The delta is constituted by two main geomorphological types: sandy beach-ridge and mud-dominated coasts westwards [10, 22]. The growth of the MD resulted in the formation of nine tributaries, so-called "Nine Dragon" and a dense man-made canal network, through the Tien River and Hau (Bassac) River, including: the Tieu, Dai, Ba Lai, Ham Luong, Co Chien, Cung Hau, Dinh An, Bat Xac and Tran De estuaries [25].

2.2 Climate and hydrology

In the basin, the climate varies from a cool to tropical weather. The Tibetan Plateau, where the MR takes its source, is mainly composed by high mountains, where the temperature is below 0 degrees during winter. There are also several glaciers with a total surface of 221 km^2 [26]. The summer precipitations are abundant. The lower basin is composed by tropical savanna and monsoon climate zones where the climate is seasonal. From November to February, the Northern monsoon brings dryness and cool temperatures whereas the hot wet season which lasts from June to September is dominated by the Southwest monsoon. The start of the wet season is marked by the beginning of the South West monsoon and later (August, September, October) during the wet season; the tropical cyclones also bring moisture to the region from the East Sea [4, 27].

Due to this dual climate variation, the hydrological regime is characterized by a monomodal flood pulse [28]. This annual flood pulse is the answer to the western North-Pacific monsoon and to the Indian monsoon and is a key hydrological characteristic of the basin. During the dry season, the water level is low whereas during the wet season, large areas of the MD are overflooded. The tributaries in the central and southern Laos contribute mostly to the Mekong flow in the lower basin, the mountainous topography permits the creation and the development of many rivers that pour into the Mekong River. In fact, 82% of the Mekong annual water discharge comes from four different sources: the mountains of northern Laos, the southern mountains, the Mun Chi system and the outflow of the Tonle Sap [15]. The upper catchment (China, Myanmar) represents 24% of the total catchment area and contributes in 18% of the total runoff [29]. However, this part of the basin is of a crucial importance for the lower basin, especially during the dry season. Indeed, from April to May, the upper basin represents 35% of the total runoff [30]. The Cambodian floodplains and the MD receive more than 90% of their available water resources from upstream [30]. In Vietnam, 95% of the Mekong water's discharge depends of upstream sources [31].

2.3 Socio-economic features

The Lower Mekong basin is the livelihood region of over 70 million people [27] and this number is probably going to reach 90 million by 2025 [10, 32]. Half of the total population lives within fifteen kilometers of the river banks [33, 34] and the delta hosts a population of nearly 20 million inhabitants [8]. The river remained relatively unexploited during the 20th century due to the wars that occurred in Indochina and the poor development of the Chinese provinces near the MB. The peace and the recent and fast development of China has led to the exponential economic development of the basin. The high demand for electricity and the mountainous landscape led to the construction of dams throughout the basin [35]. The agricultural sector is also well represented and a large proportion of the Laotian population relies on agricultural practices for their livelihood. Consequently, the Mekong River is intensively used for irrigation. The inventory of the diversions for agriculture is difficult

Figure 3.
Floating fish cages in the Tien River (Mekong River).

to make, but most of them are small diversions, directly connected in river channels and tributaries located in the Khorat Plateau of Thailand [36]. The MD is very dynamic and is the first area for agriculture and aquaculture production in South East Asia [37]. During the wet season, large areas of the lower basin are naturally flooded due to the rain and regulated for agriculture optimization [38]. These floods are making the lower basin highly fertile and create numerous ecosystems rich in biodiversity [39]. 781 species of freshwater fishes are known for living in the MR [40]. The MD detains the highest range of fish species (484) whereas the headwaters detains the lowest (24 species). Many people in the lower basin are dependent of the Mekong resources and especially fishes. In Cambodia, 80% of the population eats fish and it is the first source of proteins [41]. The lower MB has a big fishery industry (**Figure 3**) with 2.6 metric tons yr.$^{-1}$ and a total income of 7 billion USD yr.$^{-1}$ [8].

3. Damming the Mekong Basin

3.1 General information on Mekong dam's

Prior 1990, hydro-power was undeveloped on the Lower Mekong River. The rapid development of the Southeast Asian region has increased the use of natural resources and is now accompanied by high a demand in electricity. The different governments of the Mekong Basin chose large-scale hydro-power projects because the region was so far limited in term of dam development infrastructures [42]. Consequently, the development of massive water infrastructures is now booming [35, 43]. The first dam directly built on the Lancang River (Chinese name of the Mekong) was the Manwan Dam, in 1986 and was fully operational in 1996 [44]. Other dams in the Chinese region were built rapidly: Dachaoshan dam (2001); Jinghong dam (2008); Xiaowan dam (2009); Gongguoqiao dam (2011) and Nuozhadu dam (2012) and some other dams are under now construction or being planned [44]. Lower on the MR, the Xayaburi dam (Laos) is almost only used for commerce. In total, 176 dams are now being built or planned through the MRB [16]. The MR has a total hydro-power capacity of ~60 GW [45], and about 20 GW is already being exploited [16]. These dams have considerable impacts on the ecosystem and on human life.

3.2 Sediment load changes

One of the firsts impact of dams is the changes of the sediment load as a dam traps up to 95% of sediments and consequently, reduces the amount of sediment downstream. The sediment discharge in the pre dam area was 160 million tons.yr.$^{-1}$ [46] and about 50% of the sediment load was provided by the Lancang River basin [16, 45]. The reduction of total suspended solid concentration was firstly observed in 1993 after the construction of Manwan dam in China [30]. The estimated sediment trapping for the entire cascade of the eight dams in the middle and lower Lancang River may reach 78–81% [44, 47]. However, incertitude between the data concerning the post dam sediment discharge has been observed (from 67 to 145 million tons) [14, 48–50].

Downstream the Lancang River and especially in the lower MB, the data concerning the sediment discharge remains uncertain and their improvement would require further efforts [50]. Actually, the dam's impact on sediment discharge can be noised by other effects such as climate change and land use [14]. Manh et al. [18] used numerical model to unravel the effects of dams and climate change. They showed that damming the Mekong exacerbates the sediment dynamics in the MD. Suif et al. [51] examined through modeling, scenarios analysis of 19 dams (existing, under construction and planned dams) and indicated that the annual suspended sediment load is largely reduced ranging from 20 to 33%, 41–62% and up to 71–81% for existing, construction and planned dams respectively.

The reduction of sediment load in the delta clearly depends on the position and quantity of dams in the mainstream and tributaries and hence, on political decisions [43]. The construction of all planned dams would have consequences on sediment transportation (95% being trapped), leading to a reduction of the amount of sediment in the delta at 9Mt.yr.$^{-1}$ [43]. Direct measurements on site revealed that from 2009 to 2016, severe drop of suspended sediment concentration have been observed in the MD [52]. Koehnken [48] suggested that the sediment load that enter the LMB was reduced from an average of 84.7 Mt. yr.$^{-1}$ to 10.8 Mt. yr.$^{-1}$ at Chiang Saen (1960–2002) and from 147 Mt. yr.$^{-1}$ to 66 Mt. yr.$^{-1}$ at Pakse. The dam's portfolio already in action upstream seems to impact the sediment load with their high trapping ability [53]. Concerning the Vietnamese delta, the future seems dramatic with a total sediment load into the delta about 16–40 Mt./yr., that is around 10–25% of the pre-dam sediment load of 160 Mt./yr. (MONRE VN, 2016).

3.3 Hydrological changes

Dam's impacts on the LMR hydrology have been studied in the literature using both statistical analyses from measured data in site and hydrological modeling [27, 54]. The results showed that the hydro-power operations have already altered the natural hydrological cycle with lower flood peaks and higher dry season flows. However, the different studies used different models, leading to the change of magnitude [55]. At Kratie the dry season flows are expected to be higher (25–160%)and the flood peaks lower (5–24%) if the projected dam's portfolio is realized [56].

Other studies showed that the presence of upstream dams did not affect considerably the flood hydrology or low flow hydrology and that the low flow water levels were firstly attributed to poor rainfall during the period of studies and deforestation in the catchment [50, 57, 58]. Downstream the Chinese dams, the monsoon water flows are now lower than those in the dry season due to the water storage in the hydro-power reservoirs [45]. Toan et al. [59] expect an increase of the number of years with a small flood, a decrease of the years with an important flood and a slightly decrease of years with a medium flood due to the hydro-power dam's development.

4. Climate change

4.1 Different models used to simulate climate change

Climate change is another stressor that could impact the basin. The IPCC studies projected an increase of the average temperatures in the MB by the mid and late 21st century but the precipitations projection are still unclear due to numerous uncertainties [60]. Climate change studies focused on the MRB are scarce and mostly use modeling to study the effects on hydrology [54, 56, 61–63]. Some studies only use one General Circulation Models (GCM) [54, 63] and projected that the Mekong's discharge will be impacted by climate change. However, Different GCMs have different sensibility, especially concerning the precipitations. It is then essential to use multiple GCMs to investigate the different climate scenarios that could occur and the various hydrological impacts.

Kingston et al. [62] used several GCMs down scaled to the MB and large uncertainties were observed (from −16% to +55% of monthly river discharge). Lauri et al. [56] used five GCMs and two emissions scenarios to project the hydro-logical response to climate change and the hydro-power development for the period of 2032–2042. They found that dams are expected to have higher impacts on MR hydrology than the projected climate changes. Hoang et al. [64] also expected a reduction in the frequency and magnitude of extremely low flows. Thilakarathe and Sridhar [65] showed an increase in drought risks in the Lower MR Basin.

4.2 El Nino southern oscillation and extreme climatic events

El Nino Southern Oscillation (ENSO) is a climatic event that connects the ocean and the atmosphere in the tropical Pacific Ocean [66]. The Mekong Region is known as being influenced by three monsoon systems, including: the East Asian Monsoon, the South West Asian Monsoon and also the Western North Pacific Monsoon [67]. This ENSO also impacts the precipitations and discharges in the watershed [68]. A study of the Mekong discharge data showed that the combination of the ENSO and the runoff may have increased during the 1993–2005 period in comparison to the 1950–1993 period [69]. Rasanen et al. [27] also studied the relation between ENSO and the Mekong hydrology. They observed that the ENSO dependence was the highest in the Southern part of the basin. They also concluded that the precipitation and discharges decreased and the annual flow was shorter during El Nino. Ha et al. [52] also showed a decrease in water and sediment supply during El Nino.

Extreme climatic events such as tropical cyclones can impact the precipitation regime and lead to more sediment transport through the watershed. Darby et al. [58] combined, suspended sediment load data with hydrological model simulations in order to better understand the role of tropical cyclones on precipitations and sediment load. They found that 32% of the suspended sediments that reach the delta are generated by tropical cyclone rainfalls.

4.3 Regional Sea level rise

Regional sea level rise (RSLR) is currently about 4.0 mm yr.$^{-1}$ along the Mekong coast [70]. Vu et al. [71] estimated SLR using the RCP 6.0 emission scenario. They projected an increase of SLR between 25 to 30 cm by 2050. Relative Sea Level Rise associated with delta subsidence will have inundation consequences in the MD which has an average elevation of less than 2 m [11, 12]. This RSLR will affect rice production yield, up to 50–60 km into the river and approximately 30,000 ha of agricultural area may be affected. Currently, up to 40 km of the river mouth is affected by the intrusion of saline water [72]. RSLR due to climate change is one of the consequences of the delta sinking. Subsidence is another consequence of the delta sinking and is 10 times higher than the sinking due to RSLR.

5. Delta sinking

With 20 million inhabitants, the MD is recognized as one of the largest populous deltas in the world [8] and is considered as the "rice bowl" of South East Asia, providing 50% of the Vietnam's food. It is a very active area in term of agricultural development and is the second most important rice exporter (Vietnam [73]).

5.1 History of the Mekong Delta

5300 to 3500 years ago, the Mekong estuary increased rapidly due to high sediment supply transforming it into a delta. However, since the past few years (2005 seems to be the transition year), the delta is experiencing high riverbank and coastal erosion (**Figures 2** and **4**). Anthony et al. [22] quantified the shoreline erosion using high-resolution satellite images. They found that between 2003 and 2012, the erosion affected over than 50% of the 600 km long delta shoreline. However, the erosion is not the same for all the delta shoreline and three sections are observed: the sand dominated delta distributing mouths, the mud-dominated South China Sea coast and the mud-dominated Gulf of Thailand coast. Recently, Marchesiello et al. [24] undertook a comprehensive numerical exercise, coupling field surveys, laboratory experiences and remote sensing data, to assess natural against man-induced coastal erosion processes. While sediment trapping by dams and sand mining for construction is clearly identified as factors of risks enhancing coastal erosion, this later could also be largely attributed to other processes, such as shore sediment redistribution by the ocean forces (waves, currents and tides), mangrove squeeze and/or subsidence.

5.2 Incision of the Mekong Delta

Brunier et al. [19] studied the channel changes morphology in the MD. Computed data from 10 years comparison period showed that the channels changed with significant incision, expansion and deepening of numerous pools. The authors explained the morphology channels changes by the large extraction of construction materials. The regional exponential socio-economic development engendered a high increase in the demand for sand mining [74]. The United Nations Comtrade Database (UN [75]) revealed that Singapore demands a high amount of sand with 80 Mt. bought from Cambodia and 71 Mt. from Vietnam for the 2000–2016 periods (**Figure 5**). Bravard et al. [76] estimated sand extraction to be approximately 54 My. yr.$^{-1}$ which is around one third of the natural flux of sediments. This extraction causes serious impacts on the environment and on human's livelihoods, regularly reported in national newspapers ([74] among others). Le et al. [77] pointed out that a critical value of 0.4 g.L^{-1} of suspended sediment concentration must be maintained in the Mekong estuary for sediment processes. If a SSC beyond this value is measured, the deposition rate will be strongly reduced and the erosion rate will probably increase.

(a)

(b)

(c)

Figure 4.
(a) Sand extraction in the Mekong Delta, used as construction material in Ho Chi Minh City (photo from the Saigoneer, published on Monday, 19 March 2018). (b) Photo of Sim Chi Yin in National Geographic showing the erosion effects in the Mekong Delta (photo from the Saigoneer, published on Monday, 19 March 2018). (c) Photo showing the erosion effects in the Mekong Delta (photo by Vo Le Phu).

Figure 5.
Photo showing sand extraction in the Mekong delta (photo from the Saigoneer, published on Monday, 07, August 2017).

5.3 Dikes implementation and delta subsidence

The MD has the capacity to produce three rice crops per year, and even to supply the whole region. This important production is possible because of the implementation of irrigation, drainage channels and flood protection infrastructures [38, 78, 79]. However, these infrastructures also change flows and sediment transports. Triet et al. [80] observed that high-dykes permitted to decrease local flood risks but, on the other side, increased downstream flood risks. In addition, high dikes removed high fertility sediments delivered by the floodwaters balance [81].

The recent literature [11, 82, 83] clearly highlighted that subsidence exacerbates the delta sink and associated risks of coastal erosion, extended duration of flooding and salinization. This salinization will have impacts on the MD fisheries. The *Pangasius* needs freshwater to survive. An increase of 0.73 m of the sea level, combined with a decrease of 29% of the Mekong River flow, a scenario expected in Vietnam by 2100, will impact the area dedicated to the *Pangasius* farming by 11% [84]. Erban et al. [82] estimated the subsidence rate at 1.6 cm yr.$^{-1}$. This subsidence is due to over exploitation of the groundwater with a decline of 0.3 m yr.$^{-1}$ of groundwater level. A continuation of this pumping at the present rates, due to population increase, might result in a 0.88 m of land subsidence by 2050. Minderhoud et al. [12] alarm on the possible drown of the Mekong delta if water extraction is allowed to increase continuously. However, a limitation of the water extraction could result in limiting future elevation loss.

6. Water quality degradation

6.1 General information

Compared to temperate watersheds, the Mekong River is not well monitored, despite a clear warning on threats [31]. In the early 1990, the mainstreams of the delta were already too contaminated for drinking [85]. However, in the MD, surface water is used by the population for irrigation, aquaculture and for drinking [86]. In the MD, more than 50,000 Km of man-made canals were reported [87].

Studies made by Wilbers et al. [86] showed that pH, turbidity, ammonium, arsenic, barium, chromium, mercury, manganese, aluminum, iron, and *Escherichia coli* in the canals exceed the thresholds set by Vietnamese quality guidelines for domestic purposes [88]. Strady et al. [89] analyzed trace metal concentrations in water, SPM and surface sediments in the Tien River. They showed relatively low concentration ranges in the different phases.

6.2 Groundwater

Groundwater is a main water supply source for domestic purposes and is tapped wherever the salinity level is not too high [90]. Groundwater resource quality has been exacerbated by trace metals and other pollutants, in which Arsenic contamination became an emerging concern. Stanger et al. [91] was the first study to alarm about arsenic (As) groundwater contamination in the MD and showed that As concentrations from 100 to 120 m deep was extremely high and exceeded WHO standards (10 µg/L) [92]. As groundwater was recorded in the Kandal province (Cambodia) and also in Vietnam [93]. The high concentration of As in water leads to serious health issues in the population [94]. The highest As concentration recorded in the MD was 1610 $µg.L^{-1}$ (100 times higher than the WHO standards) [95], which is about one-hundred times higher than the WHO drinking water standard. Furthermore, during a groundwater sampling campaign, Buschmann et al. [96] observed that 37% of the samples exceeded the WHO standards and 26% of the samples had an As concentration over 50 $µg.L^{-1}$. However, Guédron et al. [97] reported, for the Laotian capital (Vientiane), low total Hg concentration in the MR close to the city, which is due to the low indus- trialization of the city.

6.3 Dams impact on water quality

Dams also lead to eutrophication and pollution of the water resources locally and downstream, as a result of the regulated water flux and water stagnation [98]. The closure of Manwan dam led to the decrease of heavy metal concentrations at downstream parts. The nutrients in the reservoir showed variations of concentra- tions between 1993 and 2004, and the total amount of phosphorus consistently increased [44]. The water temperature was also recorded and the overall mean surface temperature in the reservoir increased due to the dam closure. Downstream the dam, the temperature is now cooler during summer and warmer during winter [44]. Chanta and Sok [99] studied the impacts of a Dam on the Sesan River, a major tributary of the Mekong River, located in central Vietnam and north-east Cambodia. They showed a decrease of water quality downstream the dam with an increase of the suspended particulate matter concentration, total phosphorus, nitrogen and ammonium.

6.4 Water pollution due to human activities

As previously above-stated, the population in the MB is rapidly growing and the estimations report over 100 million inhabitants by 2050 [100]. This growing population already has impacts on the land use with a deep transformation of agri- culture over the last 15–20 years [101]. Water pollution in the Mekong due to human activities is relatively low compared to other tropical catchments [102]. However, the increasing amount of industrial, agricultural and urban (Vientiane, Phnom Penh) wastewater could have potential concern in the future [102]. In the lower MB, contamination could become the main issue through irrigation, wastewater, fuel or coal combustion [103]. A recent report of the World Bank, stated that water pollution is a great threat that could cost Vietnam up to 3.5 percent of GDP each yearby 2035 [31]. Nutrients, pathogens and pharmaceuticals can be encountered as waste from livestock. In Vietnam, two third of the 84.5 million tons of this livestock waste is not treated [104].

Spatial water quality variability studies can be done using self-organizing map (SOM). With this technique, Chea et al. [105] classified more than 117 monitoring stations. They found that the Laos, Thailand and Cambodia had good water quality with low nutrient and high dissolved oxygen levels. Two other clusters including the MD (Vietnam), Northwest of Thailand, Tonle Sap Lake and Vientiane urban center in Laos showed moderate to poor water quality indexes, with high nutrient load and low dissolved oxygen levels. They explained that these differences are due to human activities and especially to population growth and agricultural development in the man-made canals in the lower delta in Vietnam.

7. Comparison with other world big rivers

The Mekong River is ranked 12th biggest river in the world and the Mekong watershed is the biggest in southeast Asia. Other rivers and watersheds on earth also struggle against human activities. The objective of this section it to make an inter-comparison between the MR and other different big rivers around the word.

7.1 Dams in the world

A new hydroelectric portfolio has been developed on different earth big rivers since 2000 with a craze for the construction of megadams (dams with a height superior to 15 m). Worldwide, hydropower capacity has increased by 55% from 2000 to 2015 [106, 107]. The construction of megadams is the response of the world electric demand. However, this high level of construction can lead to several risks such as: downstream sediment and nutrient reductions, changes of the annual flood pulse and reservoir siltation [107, 108].

The Amazon River is one of the most dammed river in the world. Effectively, 62 dams are localized in the Andes, 76 dams on the cratonic rivers of the Amazon basin, and 2 on the Madeira River. Additionally, up to 286 dams are planned in the basin [109]. The Dam Environmental Vulnerability Index (DEVI) is a method permitting to assess the impacts of damming, using different parameters. The results for the Amazon basin reveals high vulnerability for several sub-basins with large potential changes.

Another example of negative dam's impacts can be illustrated by the Haung He (Wellow) River in China. This river was characterized by the highest total sediment flux of any river on earth (1.6 Gt year^{-1}) [110] due to deforestation and agricultural development. The implementation of dams resulted in a decrease of the sediment deposition in the lower Huang He River. Between 2000 and 2002, the sediment deposition passed from 111×106 m^3 yr.$^{-1}$ to a channel incision with a net erosion of up to 361×106 m^3 yr.$^{-1}$ [111].

7.2 Climate change

As we detailed in the previous section, the prediction of the impacts of the climate change on a big river such as the MR is difficult. In addition to the sediment mining, and the damming of the basin, variation of the climate will have conse-quences on the future delivery of sediments to the Mekong Delta.

The modeling of climate change and the impacts on rivers is difficult to predict and depends on the characteristics of the basins [112]. Some studies show that the intensification of the hydrological cycle is mainly due to the increase of extreme events during short time [113]. In addition, the increase in the magnitude and return period of floods is observed in different rivers such as the Amazon, Congo, Niger and also big rivers in Southeast Asia. Decrease in flood magnitude and return period is also predicted in the Nile, tigris-Euphrate, Danube Volga, Ob and parts of the Mississippi basin [114].

Atmospheric rivers are narrow ribbons of large moisture flux from the tropics to mid latitude [115]. These rivers are susceptible to see their flow regime change with the global climate change. Some atmospheric rivers will have large influence in high flow contribution (Zhuiang, Volga, Tigris-Euphates rivers) whereas other rivers will not be influent (Amazon, Congo, Magdalena, Nile rivers) [116].

The rivers where the flow is based on the monsoon rainfall and snowmelt could in the future suffer from the diminution of their flow in consequence to the reduction of the glacier size such as the Indus and Brahmaputra [117, 118]. Other rivers such as the Huang he River, where the meltwater is a small percentage, the increase of precipitations will enhance water availability [118].

The russia's great Artic rivers (Ob Yenisey and Lena) flood hydrograph is controlled by the snow and ice melt [119]. A warmer climate will increase ice melting resulting in an increase of the flow regime in the polar zone. In addition, the permafrost, logically permanently frozen will start to melt, leading to a change in the water pathway with an increase of the groundwater proportion in the river flow [120].

7.3 Water pollution, sand mining, shoreline erosion

Water pollution is an omnipresent subject when dealing with big Rivers. This review stated that the MR water is relatively preserved compared to other rivers. Indeed, excepted some Rivers, such as the Danube, where better management practices during the last two decades [121], resulted in better water quality, most of big rivers are still suffering large pollution. Best [3] stated that 80% of the world's transboundary rivers are severely polluted by nutrients (nitrogen, phosphorus) that can lead to eutrophication and/or wastewater (human waste) (UNEP-DHI, 2016). One of the most polluted river, The Gange River, India, faces to several problems such as untreated fecal waste, pesticides and heavy metal pollution [122]. Macro and micro plastic pollution is another problem encountered in big Rivers. Seventy-four per cent of the world plastic pollution is coming from large Chinese rivers such as the Changjiang, Indus and Huang He. Eighty-six per cent of the global plastic waste is coming from Asian rivers [123].

Sediment mining is a big issue for the Mekong River that has been documented in this review. The MR sediment extraction is 55 Mt. yr.$^{-1}$, representing 47–95% of the total annual suspended sediment load [124]. Other rivers in the world are also impacted by this activity that can, *in fine*, considerably decrease downstream sediment fluxes, change channel morphology, increase salt intrusion. The Zhujiang river, China is affected by this phenomenon with the removal of 60 Mt. yr.$^{-1}$ of sand, equivalent to the annual sediment load, engendering large channel incision [125]. The Changjiang River is also impacted with the removal of 40 Mt. yr.$^{-1}$ representing 17% of the total annual suspended sediment flux [126].

This paper revealed shoreline erosion of the Mekong delta (**Figure 2**). Other tropical rivers also show erosion of their coastal shorelines. The Irrawaddy River is considered as a major tropical river in the world [127]. The Irrawaddy basin covers 60% Myanmar's country [128]. Its delta is highly populous with about 15 million people living in an area of 35.000 km2 [129]. Chen et al. [130] showed that the delta's front accreted by 10.4 m. $yr.^{-1}$ between 1974 and 2018. In addition, 42% of its shoreline was subjected to erosion. However, the repartition of the erosion is disparate with a predominance of the erosion on western coastline and an accretion on the Yangon lobe. However, the mainstream of the Irrawaddy became increasingly straightened since 1974 suggesting the influence of dams and sediment extraction reducing sediment supply.

8. Future directions

As reported in this review, the Mekong River is currently facing profound transformations that lead to several issues, some emerging and some already critical (**Figure 6**). This basin is unique, with an exceptional productivity and a dense population depending on it. It is hence necessary to face these issues in order to reduce impacts, with the goal of sustaining the Lower Mekong river resources and services [5].

8.1 Local and regional strategies

The issues are multi-level and need strategy at both local and regional scales like, for example, groundwater resources. In 2007, Ho Chi Minh City encouraged the diminution of the groundwater pumping. This regulation permitted to decrease the subsidence [131]. Groundwater pumping could also be attenuated by connecting the Vietnamese

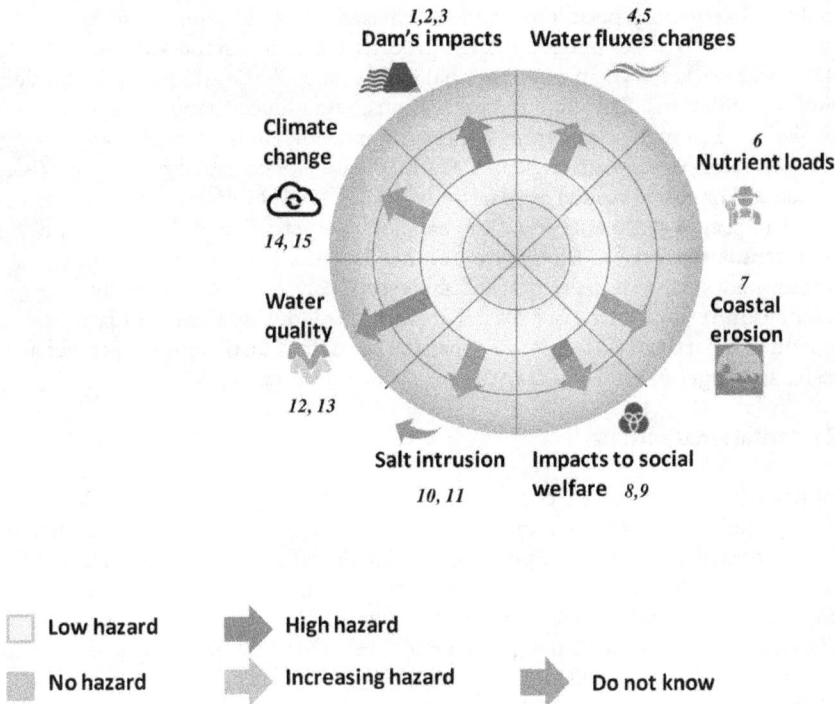

Figure 6.
Sketch representing the hazard level of the different issues that the Mekong River is facing to. Numbers are referring to the publications in the review: 1. Schmitt et al. [43]; 2. Suif et al. [51]; 3. Manh et al. [18]; 4. Koehnken [48]; 5. Pokhrel et al. [55]; 6. Toan et al. [59]; 7. Chanta and Sok [99]; 8. Minderhoud [11, 12]; 9. Trieu and Phong [84]; 10. Nguyen The Hinh [104]; 11. Chea et al. [105]; 12. Minderhoud [11]; 13. Trieu and Phong [84]; 14. Buschmann et al. [90, 96]; 15. Chanta and Sok [99]; 16. Hoang et al. [64]; 17. Darby et al. [58].

population to safe public water [132] or by accompanying farmers to less demanding freshwater production modes and agricultural activities, such as brackish aquaculture, and mangrove associated aquaculture [133]. The subsidence in the delta could also be reversed by using organic residues from rice production [134] or by restoring the mangrove to prevent coastline erosion [135, 136].

Among the other major issues introduced in this review, most of them need political agreements between the different neighbor countries so they can be tackled. For instance, the ongoing development of dams, at regional scale, seems to be irreversible, but there is a clear benefit to develop a culture of dam's portofolio for the strategy of implementation, as recently proposed by Schmitt et al. [43]. While dams have and will have impacts in the basin in the next decades; the best efficient development will probably make the difference, in terms of cost/benefits balance. The data concerning the construction of megadams is clearly insufficient [137]. Too little data is nowadays available concerning big rivers the flux of sediments, nutrients and water to guide decisions. Sediment flux modeling evolves rapidly and the advances made can help decision makers to find the best site for future infrastructure as dam's position is of a pivotal importance [138]. A strategic dam position for the environmental flow of the Mekong is a position with little impact on the sediment budget and fish habitats [139]. Some studies showed that the presence of smaller dams in cascade with equivalent energy production is better than very large dam in term of sediment budget [140].

The use of satellites in order to precisely measure the water level could permit to understand better hydrological dam's impacts. In 2020, the Surface Water Ocean Topography (SWOT) mission will launch a satellite to pursue this work [141]. Satellites observations permit to provide unbiased, spatially explicit and repeated data permiting to better understand the processes in action in the MR and delta [142]. Sand extraction is another regional issue that could be attenuated by implementing regulations. Banning sand extraction could reduce change of channel morphology and reduce long term sediment starvation [143]. Downstream, the lack of sediment is regularly reported as a main issue for coastal dynamic and risk of erosion. The recent results of Marchesiello et al. (2019) unfortunately showed that the impact of sediment reduction may not have yet affected the shore, and may accentuate risks of erosion in the coming decades. The reforestation of the mangrove is a simple measure that should be seriously considered to ensure the preservation/restoration of the coastal areas. Indeed, during tidal inundations, the sediment particles flocculate and form larger flocs. The mangroves just act as a passive scavenger of mud and trap the suspended sediments [144, 145].

8.2 International actions

International actions also come from international organizations. The Mekong River Commission (MRC) was created in 1995 and aims to share scientific data, promote the sustainable development of the basin and promote the communication between Laos, Thailand, Cambodia and Vietnam. The MRC reviews projects in relation with potential development impacts. Other big rivers in the world, such as Nile, has an international powerful management [146]. This kind of management could be copied to improve the Mekong's management.

The MR is currently confronted to several anthropogenic stressors that impact the environment, with some feedback effects on the population and its well-being. Some of these impacts could be irreversible in the next years or decades, even if sustainable solutions are being settled up now. The time to initiate a network of collaborations, as stated by the Vienna declaration on the Status and Future of the World's Large Rivers is over and it is now time to act. Sustainable development can still reduce the impacts. This development must integrate different actors from scientists to decision makers in order to take into account the scientific information in the future decisions.

Acknowledgements

The authors would like thanks the Faculty of Environment and Natural Resources - Ho Chi Minh City University of Technology (HCMUT) – VNU HCM and CARE at HCMUT for their support and assistance in providing and sharing data on the lower MB and the MD.

Author details

Mathieu Le Meur[1*], Vo Le Phu[2] and Nicolas Gratiot[3]

1 EA 4592 G&E, Bordeaux INP - Université Bordeaux Montaigne - Carnot ISIFoR, 1 allée F. Daguin, 33607 Pessac, France

2 Faculty of Environment and Natural Resources, Ho Chi Minh City University of Technology – VNU HCM, 268 Ly Thuong Kiet Street, District 10, Ho Chi Minh City, Vietnam

3 IRD, 268 Ly Thuong Kiet Street, Ward 14, District 10, Ho Chi Minh City, Vietnam

*Address all correspondence to: m.lemeur@hotmail.fr; mathieu.le_meur@bordeaux-inp.fr

IntechOpen

References

[1] Syvitski, J.P.M., A.J. Kettner, I. Overeem, E.W.H. Hutton, M.T. Hannon, G.R. Brakenridge, J. Day, C. Vörösmarty, Y. Saito, L. Giosan, R.J. Nicholls. 2009. Sinking deltas due to human activities. Nature Geoscience 2 (10), 681-686.

[2] Vörösmarty, C.J., J. Syvitski, J. Day, A. De Sherbinin, L. Giosan, C. Paola. 2009. Battling to save the world's river deltas. The Bulletin of the Atomic Scientists 65 (2), 31-43.

[3] Best, J. 2019. Anthropogenic stresses on the world's big rivers. *Nature Geoscience* 12: 7-21. doi.org/10.1038/s41561-018-0262-x.

[4] Mekong River Commission 2005. Overview of the Hydrology of the Mekong Basin, Vientiane, Lao PDR.

[5] Campbell, I.C. 2009. Development scenarios and Mekong river flows. In *The Mekong-Biophysical Environment of an International River Basin*, ed. I.C. Campbell, 389-402. New York: Elsevier.

[6] Electricity Generating Authority of Thailand (EGAT) 2008. Thailand power development plan 2007-2021: Revision 1, systems planning division, Thailand.

[7] Lamberts, D. 2006. The Tonle Sap lake as a productive ecosystem. International Journal of Water Resources Development 22: 481-495.

[8] Mekong River Commission 2010. State of the basin report 2010, Vientiane, Lao PDR.

[9] Bui, T.K.L., L.C. Do-Hong, T.S. Dao, and T.C. Hoang. 2016. Copper toxicity and the influence of water quality of Dongnai River and Mekong River waters on copper bioavailability and toxicity to three tropical species. Chemosphere 144: 872-878.

[10] Li, X., J.P. Liu, Y. Saito, V.L. Nguyen. 2017. Recent evolution of the Mekong Delta and the impacts of dams. Earth Science Reviews 125: 1-17.

[11] Minderhoud, P.S.J., L. Coumou, G. Erkens, H. Middelkoop, and E. Stouthamer. 2019a. Mekong delta much lower than previously assumed in sea-level rise impact assessments. Nature Communications 10 (1), 3847. https://doi.org/10.1038/s41467-019-11602-1

[12] Minderhoud, P.S.J., L. Coumou, G. Erkens, H. Middelkoop, and E. Stouthamer. 2019b. Digital elevation model of the Vietnamese Mekong delta based on elevation points from a national topographical map. Pangaea https://doi.org/10.1594/PANGAEA.902136

[13] Minderhoud, P.S.J., L., Coumou, L.E., Erban, H., Middelkoop, E., Stouthamer, E.A., Addink. 2018. The relation between land use and subsidence in the Vietnamese Mekong delta. Science of the Total Environment 634, 715-726. doi:10.1016/j.scitotenv.2018.03.372

[14] Wang, J.J., X.X. Lu, and M. Kummu. 2011. Sediment load estimates and variations in the lower Mekong River. River Research and Applications 27: 33-46.

[15] Gupta, A. 2009. Geology and landforms of the Mekong basin. In *The Mekong Biophysical Environment of an International River Basin*, ed. I.C. Campbell, 29-52. New York: Elsevier.

[16] Nhan N.H., and N.B. Cao. 2019. Chapter 19 – Damming the Mekong: Impacts in Vietnam and Solutions. *Coasts and Estuaries: The Future*. 321-340.

[17] Raju, K.S., and D.N. Kumar. 2018. Impact of Climate Change on Water Resources. Springer Science and Business Media LLC.

[18] Manh, N.V., N.V. Dung, N.N. Hung, M. Kummu, B. Merz, and H. Apel. 2015. Future sediment dynamics in the Mekong delta floodplains: Impacts of hydropower development, climate change and sea level rise. Global and Planetary Change 127: 22-33.

[19] Brunier, G., E.J. Anthony, M. Goichot, M. Provansal, and P. Dussouillez. 2014. Recent morphological changes in the Mekong and Bassac river channels, Mekong delta: The marked impact of river-bed mining and implications for delta destabilization. Geomorphology 224: 177-191.

[20] Liu, J.P., D.J. DeMaster, C.A. Nittrouer, E.F. Eidam, and T.T. Nguyen. 2017. A seismic study of the Mekong subaqueous delta: Proximal versus distal accumulation. Continental Shelf Research 147: 197-212.

[21] Xue, Z., J.P. Liu, D. DeMaster, V.L. Nguyen, and T.K.O. Ta. 2010. Late Holocene evolution of the Mekong subaqueous delta, southern Vietnam. Marine Geology 269: 46-60.

[22] Anthony, E.J., G. Brunier, M. Besset, M. Goichot, P. Dussouillez, and V.L. Nguyen. 2015. Linking rapid erosion of the Mekong River delta to human activities. Scientific Reports 5: 14745.

[23] Besset, M., E.J. Anthony, G. Brunier, and P. Dussouillez. 2016. Shoreline change of the Mekong River delta along the southern part of the South China Sea coast using satellite image analysis (1973-2014). Geomorphologie Relief Processus Environnement 22: 137-146.

[24] Marchesiello, P., N.M. Nguyen, N. Gratiot, H. Loisel, E.J. Anthony, and T. Nguyen. 2019. Erosion of the coastal Mekong delta: Assessing natural against man induced processes. Continental Shelf Research 181: 72-89.

[25] W. Szczuciński, R. Jagodzinski, T.J.J. Hanebuth, K. Stattegger, A. Wetzel, M. Mitrega, D. Unverricht, and P.V. Phung. 2013. Modern sedimentation and sediment dispersal pattern on the continental shelf off the Mekong River delta, South China Sea. Global Planetary Change: 110: 195-213.

[26] Brun, F., E. Berthier, P. Wagnon, A. Kaab, and D. Treichler. 2017. A spatially resolved estimate of High Mountain Asia glacier mass balances from 2000 to 2016. Nature Geoscience 10 (9): 668-673.

[27] Rasanen, T.A., J. Koponen, H. Lauri, and M. Kummu. 2012. Downstream hydrological impacts of hydropower development in the upper Mekong Basin. Water Resources Management 26: 3495-3513.

[28] Junk, W.J. et al. 2006. The comparative biodiversity of seven globally important wetlands: A synthesis. Aquatic Sciences 68: 400-414.

[29] Mekong River Commission 2003. State of the Basin Report 2003, Phnom Penh, Cambodia.

[30] Kummu, M., and O. Varis. 2007. Sediment-related impacts due to upstream reservoir trapping, the lower Mekong River. Geomorphology 85: 275-293.

[31] World Bank. 2019. "Vietnam: Toward a Safe, Clean, and Resilient Water System." World Bank, Washington, DC.

[32] Lu, X.X., and R.Y. Siew. 2006. Water discharge and sediment flux changes over the past decades in the lower Mekong River: Possible impacts of the Chinese dams. Hydrology and Earth Systems sciences 10: 181-195.

[33] Gratiot, N., A. Bildstein, T.T. Anh, H. Thoss, H. Denis, H. Michallet, H. Apel. 2017. Sediment flocculation in the Mekong River estuary, Vietnam, an important driver of geomorphological

changes. Compte Rendu Géoscience. 349: 260-268.

[34] Mekong River Commission 2011. Technical Paper: Flood Situation Report 2011, Vientiane, Lao PDR.

[35] Grumbine, R.E., J. Dore, and J. Xu. 2012. Mekong hydropower. Drivers of change and governance challenges. Frontiers in Ecological Environments 10: 91-98.

[36] Hoanh, C.T., N. Phong, J. Gowing, T. Tuong, N. Ngoc, and N. Hien. 2009. Hydraulic and water quality modeling: A tool for managing land use conflicts in inland coastal zones. Water Policy 11: 106-120.

[37] Smajgl, A., T.Q. Toan, D.K. Nhan, J. Ward, N.H. Trung, L.Q. Tri, V.P.D. Tri, and P.T. Vu. 2015. Responding to rising sea levels in the Mekong Delta. Nature Climate Change 5:167-174, https://doi.org/10.1038/nclimate2469.

[38] Aires, F., J.P. Venot, S. Massuel, N. Gratiot, P. Binh, and C. Prigent. 2020. Surface water evolution (2001-2017) at the Cambodia/Vietnam border in the upper Mekong Delta using satellite MODIS observations. Remote Sensing. 12: 800 doi:10.3390/rs12050800.

[39] Winemiller, K.O., et al. 2016. Balancing hydropower and biodiversity in the Amazon, Congo and Mekong. Science 351 (6269): 128-129.

[40] Vaidyanathan, G. 2011. Remaking the Mekong. Scientists are hoping to stall plans to erect a string of dams along the Mekong River. *Nature* 478: 305-307.

[41] Will, G. 2010. Der Mekong: Ungelöste Probleme regionaler Kooperation. SWP-Studies S7, Stiftung Wissenschaft und Politik. Deutsches Institut für Internationale Politik und Sicherheit, Berlin.

[42] Kuenzer, C., I. Campbell, M. Roch, P. Leinenkugel, V. Quoc Tuan, and S.

Dech. 2013. Understanding the impact of hydropower developments in the context of upstream-downstream relations in the Mekong river basin. Sustainable Science 8: 565-584.

[43] Schmitt, R.J.P., S. Bizzi, A. Castelletti, J.J. Opperman, G.M. Kondolf. 2019. Planning dam portfolios for low sediment trapping shows limits for sustainable hydropower in the Mekong. Science Advances 5: 1-12.

[44] Fan, H., D. He, and H. Wang. 2015. Environmental consequences of damming the mainstream Lancang-Mekong River: A review. Earth-Science Reviews 146: 77-91.

[45] Mekong River Commission 2016. Integrated Water Resources Management-Based Basin Development Strategy 2016-2020. Vientiane, Lao PDR.

[46] Milliman, J.D., and J.P.M. Syvitski. 1992. Geomorphic/tectonic control of sediment discharge to the ocean: The importance of small mountainous rivers. Journal of Geology 100: 525-544.

[47] Kummu, M., X.X. Lu, J.J. Wang, and O. Varis. 2010. Basin-wide sediment trapping efficiency of emerging reservoirs along the Mekong. Geomorphology 119: 181-197.

[48] Koehnken, L. 2014. Discharge Sediment Monitoring Project 2009-2013 Summary and Analysis of Results. Final report. Mekong River Commission, Phom Penh, Cambodia.

[49] Liu, C., Y. He, E. Des Walling, and J.J. Wang. 2013. Changes in the sediment load of the Lancang-Mekong River over the period 1965-2003. Science China Technological Science 56: 843-852.

[50] Lu, X.X., M. Kummu, and C. Oeurng. 2014. Reappraisal of sediment dynamics in the lower Mekong River, Cambodia. Earth Surface Processes Landforms 39: 1855-1865.

[51] Suif, Z., A. Fleifle, C. Yoshimura, and O. Saavedra. 2016. Spatio temporal patterns of soil erosion and suspended sediments in the Mekong River basin. Science of the Total Environment 568: 933-945.

[52] Ha, T.P., C. Dieperink, H.S. Otter, and P. Hoekstra. 2018. Governance conditions for adaptive freshwater management in the Vietnamese Mekong Delta. Journal of Hydrology 557: 116-127.

[53] Zhai, H.J., B. Hu, X.Y. Luo, L. Qiu, W.J. Tank, and M. Jiang. 2016. Spatial and temporal changes in runoff and sediment loads of the Lancang River over the last 50 years. Agricultural Water Management 174: 74-81.

[54] Hoanh, C.T., K. Jirayoot, G. Lacombe, and V. Srinetr. 2010. Impacts of climate change and development on Mekong flow regimes. First assessment – 2009. MRC Technical Paper No 29, Mekong River Commission, Vientiane, Lao PDR.

[55] Pokhrel Y., S. Shin, Z. Lin, D. Yamakasi, and J. Qi. 2018. Potential disruption of flood dynamics in the lower Mekong River basin due to upstream flow regulation. Scientific Reports. 8: 17767|DOI:10.1038/s41598-018-35823-4

[56] Lauri, H., H. de Moel, P.J. Ward, T.A. Räsänen, M. Keskinen, and M. Kummu. 2012. Future changes in Mekong River hydrology: Impact of climate change and reservoir operation on discharge. Hydrology and Earth System Sciences 16: 4603-4619.

[57] Adamson, P.T., I.D. Rutherford, M.C. Peel, and I.A. Conlan. 2009. The hydrology of the Mekong River, In *The Mekong: Biophysical Environment of an International River*, ed. I.M. Campbel, 53-76. Amsterdam: Elsevier.

[58] Darby, S.E., C.R. Hackney, D.R. Parsons, J.L. Best, A.P. Nicholas, and R. Aalto. 2016. Fluvial sediment supply to a mega-delta reduced by shifting tropical- cyclone activity. Nature 539:276-279.

[59] Toan T.Q. et al., 2016. Synthesis report on science and technology results: Study for assessing the impacts of hydropower dam ladders on the Mekong downstream mainstream to waterflow, environment and socio economics in the Mekong Delta and proposing mitigation measures. State level project of code: KC08.13/11-15 (In Vietnamese).

[60] IPCC, 2018. Global warming of 1.5 C. an IPCC special report on the impacts of global warming of 1.5 C above pre-industrial levels and related global greenhouse gas emission pathways, in the context of strengthening the global response to the threat of climate change, sustainable development, and efforts to eradicate poverty.

[61] Eastham, J., F.M. Mpelasoka, C. Mainuddin, P. Ticehurst, G. Dyce, R. Hodgson, R. Ali, and M. Kirby. 2008. Mekong River basin water resources assessment: Impacts of climate change, CSIRO, water for a healthy country national research flagship report.

[62] Kingston, D., J.R. Thomson, and G. Kite. 2011. Uncertainty in climate change projections of discharge for the Mekong river basin. Hydrological Earth Systems Science 15: 1459.

[63] Västilä, K., M. Kummu, C. Sangmanee, and S. Chinvanno. 2010. Modelling climate change impacts on the flood pulse in the lower Mekong floodplains. Journal of Water Climate Change 1: 67-86.

[64] Hoang, L.P., H. Lauri, M. Kummu, J. Koponen, M.T.H. van Vliet, I. Supit, R. Leemans, P. Kabat, and F. Ludwig. 2016. Mekong River flow and hydrological extremes under climate change. Hydrology and Earth System Sciences 20 (7): 3027-3041.

[65] Thilakarathne, M., and V. Sridhar. 2017. Characterization of future drought conditions in the lower Mekong river basin. Weather and Climate Extremes 17: 47-58.

[66] Alexander, M.A. et al. 2002. The atmospheric bridge: The influence of enso teleconnections on air–sea interaction over the global oceans. Journal of Climatology 15: 2205-2231.

[67] Holmes, J.A., E.R. Cook, and B. Yang. 2009. Climate changes over the past 2000 years in Western China. Quaternary International 194: 91-107.

[68] Kiem, A., H. Hapuarachchi, H. Ishidaira, J. Magome, and K. Takeuchi. 2004. Uncertainty in Hydrological Predictions Due to Inadequate Representation of Climate Variability Impacts. University of Yamanashi, Japan.

[69] Xue, Z., J.P. Liu, and Q. Ge. 2011. Changes in hydrology and sediment delivery of the Mekong River in the last 50 years: Connection to damming, monsoon, and ENSO. Earth Surface Processes and Landforms 36: 296-308.

[70] Church, J.A., et al. 2013. Sea level change. In *Climate Change: The Physical Science Basis. Contribution of Working Group I to the Fifth Assessment Report of the Intergovernmental Panel on Climate*, ed. Change. T.F.

[71] Vu, D.T., T. Yamada, and H. Ishidaira. 2018. Assessing the impact of sea level rise due to climate change on seawater intrusion in Mekong Delta, Vietnam. Water Science and Technology 77: 1632-1639.

[72] Nowacki, D., A.S. Ogston, C.A. Nittrouer, A.T. Fricke, and P. Van. 2015. Sediment dynamics in the lower Mekong River: Transition from tidal river to estuary. Journal of Geophysical Research 120: 363-383.

[73] Vietnam News, 2012. http://www.gso.gov.vn/default_en.aspx?tabid=491. Accessed on April 3rd 2019.

[74] Saigoneer 2018. In the Mekong delta, excessive sand mining is destroying local homes. Retried on 7th April 2019, URL: https://saigoneer.com/vietnam-news/12854-in-the-mekong-delta,-excessive-sand-mining-is-destroying-local-homes>.

[75] UN Comtrade, 2016. Natural sand except sand for mineral extraction URL https://comtrade.un.org/db/mr/daCommoditiesResults.aspx?&px=&cc=2505 (accessed 7.24.19).

[76] Bravard, J.-P., M. Goichot, S. Gaillot. 2013a. Geography of sand and gravel mining in the lower Mekong River. EchoGéo 26, 13659.

[77] Le, H.A., N. Gratiot, W. Santini, O. Ribolzi, D. Tran, X. Meriaux, E. Deleersnijder, and S. Soares Frazao. 2020. Suspended sediment properties in the lower Mekong River (LMR), from fluvial to estuarine environments. *Estuarine Coastal and Shelf Science* doi: 10.1016/j.ecss.2019.106522.

[78] Dang, T.D., T.A. Cochrane, M.E. Arias, P.D.T. Van, and T.T. de Vries. 2016. Hydrological alterations from water infrastructure development in the Mekong floodplains. Hydrological Processes 30: 3824-3838. doi:10.1002/hyp.10894

[79] Kondolf, G.M., et al. 2018. Changing sediment budget of the Mekong: Cumulative threats and management strategies for a large river basin. Science of the Total Environment. 625: 114-134.

[80] Triet, N.V.K., N.V. Dung, H. Fujii, M. Kummu, B. Merz, and H. Apel. 2017. Has dyke development in the Vietnamese Mekong Delta shifted flood hazard downstream. Hydrological and Earth System Science 21: 3991-4010.

[81] Chapman, A.D., S.E. Darby, H.M. Hồng, E.L. Tompkins, and T.P.D. Van. 2016. Adaptation and development trade-offs: Fluvial sediment deposition and the sustainability of rice cropping in an Giang Province, Mekong Delta. Climatic Change 137: 593-608. doi:10.1007/s10584-016-1684-3.

[82] Erban, L.E., S.M. Gorelick, and H.A. Zebker. 2014. Groundwater extraction, land subsidence, and sea-level rise in the Mekong Delta, Vietnam. *Environmental Research Letters* 9 (8): 084010. doi10.1088/1748 9326/9/8/084010.

[83] Erkens, G., T. Bucx, R. Dam, G. de Lange, and J. Lambert. 2015. Sinking coastal cities. Proceedings of the International Association of Hydrological Sciences 372: 189-198.

[84] Trieu, T.T.N., and N.T. Phong. 2014. The impact of climate change in salinity intrusion and Pangasius (Pangasianodon Hyphopthalmus) farming in the Mekong Delta, Vietnam. Aquaculture International 23: 523-534.

[85] Fedra, K., L. Winkelbauer, and V.R. Pantulu. 1991. Expert systems for environmental screening: an application in the lower Mekong basin, RR-91-19.

[86] Wilbers, G.J., M. Becker, L.T. Nga, Z. Sebesvari, and F.G. Renaud. 2014. Spatial and temporal variability of surface water pollution in the Mekong Delta, Vietnam. Science of the Total Environment 485: 653-665.

[87] Truong, T.V. 2006. Flood identification, forecast and control in Cuu Long River Delta. 459 agriculture publication.

[88] Lam, S., G. Pham, H. Nguyen-Viet. 2018. Emerging health risks from agricultural intensification in Southeast Asia: a systematic review. *International Journal of Occupational and Environmental Health*. 250-260.

[89] Strady, E., Q.T. Dinh, J. Némery, T.N. Nguyen, S. Guédron, N.S. Nguyen, H. Denis, P.D. Nguyen. 2017. Spatial variation and risk assessment of trace metals in water and sediment of the Mekong Delta. Chemosphere doi:10.1016/j.chemosphere.2017.03.105

[90] Buschmann, J., M. Berg, C. Stengel, L. Winkel, M. Sampson, P.T.K. Tran, and P.H. Viet. 2008. Contamination of drinking water resources in the Mekong delta floodplains: Arsenic and other trace metals pose serious health risks to population. Environment International 34: 756-764.

[91] Stanger, G., T.V. Truong, L.T.M. Ngoc, T.V. Luyen, and T. Tran. 2005. Arsenic in groundwaters of the lower Mekong. Environmental Geochemistry and Health 27 (4): 341-357.

[92] World Health Organization (WHO) 2011. Guideline for Drinking Water Quality. 4th ed. WHO press, Geneva, Switzerland.

[93] Berg, M., C. Stengel, P.T.K. Trang, P.H. Viet, M.L. Sampson, M. Leng, S. Samreth, and D. Fredericks. 2007. Magnitude of arsenic pollution in the Mekong and Red River deltas-Cambodia and Vietnam, Science of the Total Environment 372: 413-425.

[94] Agusa, T., T.K.T. Pham, M.L. Vi, H.A. Duong, S. Tanabe, H.V. Pham and M. Berg. 2014. Human exposure to arsenic from drinking water in Vietnam. Science of the Total Environment 488-489: 562-569.

[95] Phan T. H. Van, T. Bonnet, S. Garambois, D. Tisserand, F. Bardelli, R. Bernier-Latmani, and L. Charlet. 2017. Arsenic in shallow aquifers linked to the electrical ground conductivity: The Mekong Delta source example. Geoscience Research 2(3): 180-195.

[96] Buschmann, J., M. Berg, C. Stengel, and M. Sampson. 2007. Arsenic

and manganese contamination of drinking water resources in Cambodia: Coincidence of risk areas with low relief topography. Environmental Science and Technology 41: 2146-2152.

[97] Guédron, S., D. Tisserand, S. Garambois, L. Spadini, F. Molton, B. Bounvilay, L. Charlet, and D.A. Polya. 2014. Baseline investigation of (methyl) mercury in waters, soils, sediments and key foodstuffs in the lower Mekong Basin: The rapidly developing city of Vientiane (Lao PDR). Journal of Geochemical Exploration 143: 96-102.

[98] Li, J., S. Dong, S. Liu, Z. Yang, M. Peng, and C. Zhao. 2013. Effects of cascading hydropower dams on the composition, biomass and biological integrity of phytoplankton assemblages in the middle Lancang-Mekong River. Ecological Engineering 60: 316-324.

[99] Chanta, O., and Sok T. 2020. Assessing changes in flow and water quality emerging from hydropower development and operation in the Sesan River basin of the lower Mekong region. Sustainable Water Resources Management 6: 27-39.

[100] Varis, O., M. Kummu, and A. Salmivaara. 2012. Ten major river basins in monsoon Asia-Pacific: An assessment of vulnerability. Applied Geography 32: 441-454.

[101] Valentin, C., F. Agus, R. Alamban, A. Boosaner, J.P. Bricquet, V. Chaplot, T. de Guzman, A. de Rouw, J.L. Janeau, D. Orange, K. Phachomphonh, Do Duy Phai, P., Podwojewski, O., Ribolzi, N., Silvera, K., Subagyono, J.P., T. Tran Duc Toan, T. Vadari. 2008. Runoff and sediment losses from 27 upland catchments in Southeast Asia: Impact of rapid land use changes and conservation practices. Agriculture, Ecosystems and Environment 128: 225-238.

[102] Snidvongs, A., and S. Teng. 2006. Global international waters assessment: Mekong River, GIWA regional assessment 55 (University of Kalmar, Sweden, on behalf of United Nations Environment Programme).

[103] Swain, E.B., et al. 2007. Socioeconomic consequences of mercury use and pollution. Ambio 36 (1): 45-61.

[104] Nguyen The Hinh, Thực trạng xử lý môi trường chăn nuôi và đề xuất giải pháp quản lý, 2017. Available at: http://tapchimoitruong.vn/pages/a r t i c l e.

[105] Chea, R., G. Grenouillet, and S. Lek. 2016. Evidence of water quality degradation in lower Mekong basin revealed by self-organizing map. PLoS One 11: e0145527.

[106] Gross, M. A. 2016. global megadam mania. Current Biology 26, R779–R782.

[107] Henning, T., D. Magee. 2017. Comment on 'an index-based framework for assessing patterns and trends in river fragmentation and flow regulation by global dams at multiple scales'. Environmental Research Letters 12, 038001.

[108] Henning, T. 2016. Damming the transnational Ayeyarwady basin. Hydropower and the water-energy nexus. Renewable Sustainable Energy Reviews 65, 1232-1246.

[109] Latrubesse, E. M. et al. 2017. Damming the rivers of the Amazon Basin. Nature 546, 363-369.

[110] Chen, Y. et al. 2015. Balancing green and grain trade. Nature Geoscience 8, 739-741.

[111] Kong, D. et al. 2017. Environmental impact assessments of the Xiaolangdi reservoir on the most hyperconcentrated river, Yellow River, China. *Environ*. Science and Pollution Research 24, 4337-4352.

[112] Palmer, M. A. et al. 2008. Climate change and the world's river basins: Anticipating management options. Frontiers in Ecological Environments 6, 81-89.

[113] Alfieri, L. et al. 2017. Global projections of river flood risk in a warmer world. Earth's Future 5, 171-182.

[114] Arnell, N.W., S.N. Gosling. 2016. The impacts of climate change on river flood risk at the global scale. Climate Change 134, 387-401.

[115] Dacre, H.F., P.A. Clark, O. Martinez-Alvarado, M.A. Stringer, D.A. Lavers. 2015. How do atmospheric rivers form? Bulletin of the American Meteorological Society 96, 1243-1255.

[116] Paltan, H. et al. 2017. Global floods and water availability driven by atmospheric rivers. Geophysical Research Letter 44, 10387-10395.

[117] Bookhagen, B., D.W. Burbank, 2010. Toward a complete Himalayan hydrological budget: Spatiotemporal distribution of snowmelt and rainfall and their impact on river discharge. Journal of Geophysical Research 115, F03019.

[118] Immerzeel, W.W., L.P.H. van Beek, M.F.P. Bierkens. 2010. Climate change will affect the Asian water towers. Science 328, 1382-1385.

[119] Ye, B., D. Yang, D.L. Kane. 2003. Changes in Lena River streamflow hydrology: Human impacts versus natural variations. Water Resources Research 39,1200.

[120] Smith, L.C., T.M. Pavelsky, G.M. MacDonald, A.I. Shiklomanov, R.B. Lammers. 2007. Rising minimum daily flows in northern Eurasian rivers: a growing influence of groundwater in the high-latitude hydrologic cycle. *Journal of Geophysical Research* 112, G04S47.

[121] Schiemer, F., C. Baumgartner, K. Tockner. 1999. Restoration of floodplain rivers: The 'Danube restoration project'. Regul. Riv. Res. Manag. 15, 231-244.

[122] Singh, S. K. J.P.N. Rai. 2003. Pollution studies on river ganga in Allahabad District. Pollution Research 22, 469-472.

[123] Lebreton, L. C. M. et al. 2017. River plastic emissions to the world's oceans. Nature Communications 8, 15611.

[124] Bravard, J.P., M. Goichot, and H. Tronchère. 2013b. An assessment of sediment-transport processes in the lower Mekong River based on deposit grain sizes, the CM technique and flow-energy data. Geomorphology 207: 174-189.

[125] Lu, X.X., S.R. Zhang, S.P. Xie, P.K. Ma. 2007. Rapid incision of the lower Pearl River (China) since the 1990s as a consequence of sediment depletion. Hydrology and Earth System Sciences 11, 1897-1906.

[126] Xiqing, C., Qiaoju, Z., Erfeng, Z. 2006. In-channel sand extraction from the mid-lower Yangtze channels and its management: Problems and challenges. Journal of Environmental Planning and Management 49, 309-320.

[127] Syvitski, J.P.M., S. Cohen, A.J. Kettner, G.R. Brakenridge. 2014. How important and different are tropical rivers? — An overview. Geomorphology 227, 5-17.

[128] Furuichi, T., Z. Win, R.J. Wasson. 2009. Discharge and suspended sediment transport in the Ayeyarwady River, Myanmar: Centennial and decadal changes. Hydrological Processes 23 (11), 1631-1641.

[129] Brakenridge, G.R., J.P.M. Syvitski, E. Niebuhr, I. Overeem, S.A. Higgins, A.J. Kettner, L. Prades. 2017. Design with nature: Causation and avoidance of

catastrophic flooding, Myanmar. Earth Science Reviews 165, 81-109.

[130] Chen, D., X. Li, Y. Saito, J.P. Liu, Y. Duan, L. Zhang. 2020. Recent evolution of the Irrawaddy (Ayeyarwady) Delta and the impacts of anthropogenic activities: A review and remote sensing survey. Geomorphology, 107231.

[131] Minderhood, P.S.J., G. Erkens, V.H. Pham, V.T. Bui, L. Erban, H. Kooi, and E. Stouthamer. 2017. Impacts of 25 years of groundwater extraction on subsidence in the Mekong delta, Vietnam. Environmental Research Letter 12 (6), 064006

[132] Cheesman, J., J. Bennett, and T.V.H. Son. 2008. Estimating household water demand using revealed and contingent behaviors: Evidence from Vietnam. Water Resources Research 44: 11. doi:10.1029/2007WR006265

[133] CCAFS-SEA, 2016. The Drought and Salinity Intrusion in the Mekong River Delta of Vietnam. CGIAR Research Program on Climate Change, Agriculture and Food Security-Southeast Asia (CCAFS-SEA), Hanoi, Vietnam.

[134] Wakeham, S.G., and E.A. Canuel. 2016. The nature of organic carbon in density-fractionated sediments in the Sacramento-San Joaquin River Delta (California). Biogeosciences 13, 567-582. doi:10.5194/bg-13-567-2016

[135] Thu, P.M., and J. Populus. 2007. Status and changes of mangrove forest in Mekong Delta: Case study in Tra Vinh, Vietnam. Estuary and Coastal Shelf Science, Sedimentological and ecohydrological processes of Asian deltas: The Yangtze and the Mekong 71, 98-109. doi:10.1016/j.ecss.2006.08.007.

[136] Veettil, B.K., D.R. Ward, N.X. Quang, N.T. Thu Trang, and T.H. Giang. 2019. Mangroves of Vietnam: Historical development, current state of research and future threats. Estuarine, Coastal and Shelf Science. 218: 212-236.

[137] Grill, G. et al. 2015. An index-based framework for assessing patterns and trends in river fragmentation and flow regulation by global dams at multiple scales. Environmental Research Letter 10, 015001.

[138] Kondolf, G.M., Z.K. Rubin, and J.T. Minear. 2014. Dams on the Mekong: Cumulative sediment starvation. Water Resources Research 50: 5158-5169. doi:10.1002/2013WR014651

[139] Schmitt, R.J.P., 2016. CASCADE - a Framework for Modeling Fluvial Sediment Connectivity and its Application for Designing Low Impact Hydropower Portfolios. Politecnico di Milano, Milan, Italy.

[140] Wild, T.B., D.P. Loucks, G.W. Annandale, and P. Kaini. 2016. Maintaining sediment flows through hydropower dams in the Mekong River basin. Journal of Water Resources Planning and Management 142:1. doi:10.1061/(ASCE)WR.1943-5452.0000560.

[141] Blancamaria, S., D.P. Lettenmaier, and T.M. Pavelsky. 2016. The SWOT Mission and its capabilities for land hydrology. Surveys in Geophysics 37: 307-337.

[142] Kuenzer, C. et al. 2020. Profiling resilience and adaptation in mega deltas: A comparative assessment of the Mekong, yellow, Yangtze, and Rhine deltas. Ocean & Coastal Management, *198*, 105362.

[143] Jordan, C., J. Tiede, O. Lojek, J. Visscher, H. Apel, H.Q. Nguyen, C.N.X. Quang, and T. Schlurmann. 2019. Sand mining in the Mekong Delta revisited - current scales of local sediment deficits. *Scientific Report* 9 :17823. doi.org/10.1038/s41598-019-53804-z.

[144] Besset M., N. Gratiot, E.J. Anthony, F. Bouchette, M. Goichot and P. Marchesiello. 2019. Mangroves and shoreline erosion in the Mekong River delta, Viet Nam. Estuarine Coastal and Shelf Science 226. 106263. ISSN 0272-7714

[145] Gratiot, N., and E.J. Anthony. 2016. Role of flocculation and settling processes in development of the mangrove-colonized, Amazon-influenced mud-bank coastof South America. Marine Geology 373: 1-10.

[146] Transboundary River Basins: Status and rends (UNEP-DHI, UNEP, TWAP, 2016) http://geftwap.org/publications/river-basins-technical-report

Chapter 2

An Assessment of Land Use and Land Cover Changes and Its Impact on the Surface Water Quality of the Crocodile River Catchment, South Africa

Nde Samuel Che, Sammy Bett, Enyioma Chimaijem Okpara, Peter Oluwadamilare Olagbaju, Omolola Esther Fayemi and Manny Mathuthu

Abstract

The degradation of surface water by anthropogenic activities is a global phenomenon. Surface water in the upper Crocodile River has been deteriorating over the past few decades by increased anthropogenic land use and land cover changes as areas of non-point sources of contamination. This study aimed to assess the spatial variation of physicochemical parameters and potentially toxic elements (PTEs) contamination in the Crocodile River influenced by land use and land cover change. 12 surface water samplings were collected every quarter from April 2017 to July 2018 and were analyzed by inductive coupled plasma spectrometry-mass spectrometry (ICP-MS). Landsat and Spot images for the period of 1999–2009 - 2018 were used for land use and land cover change detection for the upper Crocodile River catchment. Supervised approach with maximum likelihood classifier was used for the classification and generation of LULC maps for the selected periods. The results of the surface water concentrations of PTEs in the river are presented in order of abundance from Mn in October 2017 (0.34 mg/L), followed by Cu in July 2017 (0,21 mg/L), Fe in April 2017 (0,07 mg/L), Al in July 2017 (0.07 mg/L), while Zn in April 2017, October 2017 and April 2018 (0.05 mg/L). The concentrations of PTEs from water analysis reveal that Al, (0.04 mg/L), Mn (0.19 mg/L) and Fe (0.14 mg/L) exceeded the stipulated permissible threshold limit of DWAF (< 0.005 mg/L, 0.18 mg/L and 0.1 mg/L) respectively for aquatic environments. The values for Mn (0.19 mg/L) exceeded the permissible threshold limit of the US-EPA of 0.05 compromising the water quality trait expected to be good. Seasonal analysis of the PTEs concentrations in the river was significant ($p > 0.05$) between the wet season and the dry season. The spatial distribution of physicochemical parameters and PTEs were strongly correlated ($p > 0.05$) being influenced by different land use type along the river. Analysis of change detection suggests that; grassland, cropland and water bodies exhibited an increase of 26 612, 17 578 and 1 411 ha respectively, with land cover change of 23.42%, 15.05% and 1.18% respectively spanning from 1999 to 2018. Bare land and built-up declined from 1999 to

2018, with a net change of - 42 938 and − 2 663 ha respectively witnessing a land cover change of −36.81% and − 2.29% respectively from 1999 to 2018. In terms of the area under each land use and land cover change category observed within the chosen period, most significant annual change was observed in cropland (2.2%) between 1999 to 2009. Water bodies also increased by 0.1% between 1999 to 2009 and 2009 to 2018 respectively. Built-up and grassland witness an annual change rate in land use and land cover change category only between 2009 to 2018 of 0.1% and 2.7% respectively. This underscores a massive transformation driven by anthropogenic activities given rise to environmental issues in the Crocodile River catchment.

Keywords: water quality, potential toxic element contamination, land use and land cover (LULC) change, electrochemical detection

1. Introduction

The availability of clean water sources is essential for the survival of any living species. Rivers play a significant role in maintaining human health and has been recognized as the fundamental right of all living beings [1]. Improved access to clean water contributes towards achieving the 2030 agenda for sustainable development goals (SDGs) particularly SDG 6.1 and 6.2 [2]. However, river deterioration due to anthropogenic activities remains one of the contemporary challenges faced by river basin management both at regional and global scale [3–5]. Anthropogenic activities have been exacerbated over the past decades by socio-economic drivers such as the intensification and expansion of irrigation systems for agricultural purposes, increase in population and pressure on existing freshwater usage, climate variability through uneven distribution of precipitation, floodgate constructions, and untreated wastewater disposal into receiving waters bodies [6, 7]. Because of the misuse of river water resources driven by the need to sustain our economies, water resources are one of the most rapidly declining and degrading in our environment [8]. Thus, recognizing the devastating effects of river pollution on human health demands that the main cause of the problem be identified, managed effectively and efficiently [9, 10].

Globally, is estimated that 2 million tons of sewage, industrial, and agricultural wastewater is discharged into rivers leading to the death of at least 1.8 million people with diseases related to unsafe water [11, 12]. In 2012, it was estimated that 842 000 people died of diarrhea due to directly or indirectly consuming poor water quality, of which 43% of the mortality case reported were children. According to Dube, Shoko [13], 29.9% of global freshwater is reserved underground, being a critical source of water supply and a buffer against drought in rural communities, where surface water is limited especially in developing countries [14]. However, most of the rural communities in developing countries are now at threat and vulnerable from the effect of climate change which affects people's daily water availability and consumption. For instance, it is estimated that the daily intake of drinking water by a human being is 7% of the body weight which is essential for the person's healthy growth and existence [15].

Opportunities to address outstanding water issues in Africa have been undercut by intense and prevalent poverty hampering many cities and communities' capacity to make available services for sanitation and potable water, adequate for economic activities, and further forestall deterioration of water quality [16]. These factors, including finance and poor water management, and lack of proper coordination, has further deepened the water crises in Sub-Sahara Africa, thereby undermining any hope of making potable water available in the near future for the populace [17, 18]. This situation is further compounded by several environmental issues arising in the 21st century including climate change, eutrophication, salinization, toxic metal contamination, *E*-coli, phosphate, nitrate, amongst others [19].

The impact of water pollution in different parts of the world can be grouped under two broad themes according to published literature; Increase public health awareness of the negative impact of river pollution from different governmental and non-governmental agencies through education and mass sensitization. Secondly, through the development of sustainable management practices and models to mitigate the impact of river pollution [20]. Surprisingly, all these measures have yielded less results most probably because of the point and non-point sources of pollutants and also because developing and implementing sustainable mitigation measures requires a sound knowledge of the linkages between the different types and diffuse sources of pollutants, conveyor and sinks. Correspondingly also, the need for constant, effective, low cost and outdoor assessment of any available water in circulation in the ecosystem has emerged as a crucial concern for economic development and biological survival [21, 22].

1.1 Fate of African water bodies

In particular, "Africa is the fastest urbanizing continent on the planet and the demand for water and sanitation is outstripping supply in cities" quoted Joan Clos, Executive Director of UN-HABITAT [23]. Northern Africa and Sub-Saharan Africa although in the same continent, have attained different degrees of progress towards the Millennium Development Goal (MDG) on water. With ninety-two percent coverage, North Africa was already on the way to achieve their stipulated ninety-four percent target prior to 2015 [24, 25]. On the contrary, the experience of Sub-Saharan Africa is a dissimilar situation with forty percent of the 783 million people, not having access to better sources of drinking water in the whole region. Sub-Saharan Africa, operates far below the MDG on the water with only sixty-one percentage coverage, and consequently may not have attained the seventy-five percent regional coverage target following their current pace. Available data from 35 countries in Sub-Saharan Africa, which covers a swooping eighty-four percent of the population of the region, reflects high discrimination between the poorest and the richest twenty percentage of the populace in both rural and urban areas. More than ninety percent of the richest quintile (twenty percent) in urban places have access to better water supply, and more than sixty percent have piped water in the environs. Meanwhile, forty percent of the poorest in the rural areas do not have piped water network in their premises and not up to half of the population make do with any form of an improved water source.

Another concern is poor sanitation that overwhelms the safety of our usable water. African was and likely, is one of the two main continents with the least performance in fulfilling the MDG on sanitation as at 2015. This calls for serious concern sequel to the concomitant health challenge, a lot of people who do not have fundamental sanitation orientation indulge in unhealthy sanitary activities such as, indiscriminate disposal of solid waste and wastewater, and open defection [26]. Additionally, Africa's increasing population is driving more the need for water and expediting the depletion of available water sources. Amidst the regions still developing, Sub-Saharan Africa has a projected highest commonness of urban slums and it is likely to double to around 400 million by this year (2020) [27]. Notwithstanding the attempts by some Sub-Saharan African countries and cities, to broaden fundamental services and make reasonable urban housing conditions improvements. Precipitous and unplanned growth of housing, at the urban areas, has heightened the figure of settlements on uneven, floodable, and high-risk zones where natural incidents such as landslides, rains, and earthquakes have demoralizing after-effects. Settlers at such dysfunctional environment resort to any available water supply for both domestic and possibly drinking uses.

Furthermore, need for constant, effective, low cost and outdoor assessment of any available water in circulation in the ecosystem, has emerged a crucial concern for both economic development and biological survival [21, 22].

As part of remediation measures to this overwhelming challenge, in recent times, there has been a strong interest in investigating the impact of land use and land cover (LULC) on water quality [28, 29]. This is because land use and land cover is an integral component of the global environmental changes that affects ecosystems processes at various levels such as hydrological dynamics, sustainability of water bodies to mankind, increasing demand for agricultural cultivated products, shift in grassland to urban and agricultural land [6, 30]. LULC changes provide first-hand information on the transformation of the natural environment due to anthropogenic activities [31]. A range of studies has investigated the association of land use and land cover change that affect water quality in different environments [19, 28, 32–34]. This has been made possible by emerging developments in the use of spatial data acquisition technologies where different attributes of the landscape configuration can be analyzed more effectively by acquiring satellite imagery [35]. This has enabled land use planners to better interpret and to explain the interaction between hydrological components and land uses activities in a catchment and allow better water conservation strategies to be formulated. However, the perusal of literature suggests that the LULC impact of change has not been previously investigated in the upper Crocodile River catchment thus a study of this kind is necessary.

1.2 PTEs in water and adverse health effects

Generally, most elements are classified as been potentially toxic. These elements are grouped into transition metals, metalloids, lanthanides and actinides. Most of these metals occur naturally in soils, and their concentrations are highly dependent on the parent material through weathering processes, while others are included in the environment through anthropogenic activities [36]. The presence of toxic elements in water typically compromises the quality traits expected to be good for drinking, industrial processing and for biodiversity purposes [37]. However, human-induced activities have modified the natural level, biochemical balance and geochemical cycling of PTEs in the environment [38]. A good number of the metals associated with biodegradable organic and inorganic contaminants are themselves not biodegradable and hence cannot be removed or deactivated through naturally occurring processes [39, 40]. Hence, once exposed to the environment, these metals can stay for decades or centuries due to the fact they are not biodegradable [36]. Although the presence of some of these metals is essential to the ecosystem and are still needed in organisms and human body, beyond which level referred to as maximum concentration limit (MCL), they pose a threat to human health and the environs.

Nickel surpassing its necessary level could cause critical kidney and lung problems, besides distress in the gastrointestinal, skin dermatitis and pulmonary fibrosis [41–44]. Zinc as a trace element, is important for human health. It is essential for the physiological functioning of living tissues and many biochemical processes depend on it for regulation. However, beyond the MCL, zinc can pose serious threat to health like stomach cramps, vomiting, nausea, skin irritation and anemia [45, 46]. Copper is crucial to animal metabolism. Nonetheless, excessive exposure could cause serious toxicological threats like convulsions, vomiting cramps and can be in some severe cases lethal [47]. On the other hand, some metals like lead (Pb), cadmium (Cd), arsenic (As), chromium (Cr) are highly toxic even in minute amount and could critically affect the process of biological degradation of organic matters and severely harm humans [36]. Pb could cause pathological alterations in the endocrine system and kidney that lead to failure in reproduction [48]. With the exception of passage through urine, which is usually extremely slow, there is no other means of eliminating the lead in humans [49]. The irrevocable tubular damage in kidney, caused by exposure to increased level of Cd in the body can no longer be denied. The stability of genes could be negatively impacted by the inhibition in the repair of damaged DNA leading to increased chances of mutations [50]. In the disruption of the endocrine, precisely affects the reproductive system of men, thereby reducing semen quality [51, 52].

The exposure to Cd occupationally, not even involving changes proven to be clinically pathogenic, also threatens to result in visual motor function impairment, promoting changes in emotional balances and causes loss of concentration [53]. Hence these metals including Cd, Pb, As, and Cr are seen as the "Environmental health hazards" having a ranking of the first ten on the list from "Agency for Toxic Substances and Disease Registry Priority List of Hazardous Substances", relative substance toxicity and possible exposure to infested soil, air and water [54–56]. Various global agencies such as Joint Food and Agricultural Organization (FAO)/WHO Expert Committee on Food Additives (JECFA), and International Agency for Research on Cancer (IARC), Centre for Disease Control (CDC) and World Health Organization (WHO), United Nations Environmental Protection Agency (US-EPA,) have been actively involved in the control of its pollution in the environment.

1.3 PTEs sources in African water bodies: an overview

The water bodies in Africa are increasingly at the risk of PTEs exposure [57], a sequel to the growing human population leading to broadening settlement, urbanization and concomitant industrialization [58–60]. The general result is commonly the increasing discharge of completely untreated or poorly treated domestic and industrial effluent, responsible for the largest origin of heavy metal contamination and consequently, generate a continuous rise in metallic contamination in water bodies in most of the globe [59, 61]. In particular, sources of heavy metal pollution are either natural or anthropogenic [59], which are distributed across settlements. The greatest source of heavy metal pollution in the rural settlements are natural while that of the urban areas are fundamentally anthropogenic [59, 60]. However, 'bossy' and at times illegal mining activities, in some of the rural areas can also contribute to heavy mining pollution of some fresh water bodies [62].

Natural Sources of toxic elements in most rural African countries include weathering of mineral deposits, bush burning and windblown dust, comets, leachate, wet and dry fallout of atmospheric particulate matter, and volcanic eruptions [59, 62]. The anthropogenic sources on the other hand seem, to be as large as the development of the societies in most African countries where environmental protection, waste management, and disposal are still poorly managed. These include activities directly or indirectly connected with, industrial effluents, fossil fuel and coal combustion, mining and metal processing, solid waste disposal, fertilizers, battery and paint manufacturing, petroleum refining, cement and ceramic production, and steel production [62]. Others include mineral exploitation, ore transportation, smelting and refining, disposal of the tailings and waste waters around mines, weathering of rocks, and heaped waste materials in mining sites [63, 64]. The list goes on to include draining of sewerage, dumping of hospital wastes, recreational activities, shipping, mining, breweries, tanning, fishing, and agro-processing factories [64, 65]. Further activities include urban storm water runoff, atmospheric sources, boating, biocides runoff, nutrients and pathogens from agricultural lands, urban areas and informal settlements [60], metal fabrication and scraping industries, and indiscriminate use of heavy metal-containing fertilizer and pesticides in agricultural fields [65]. For instance, Reza and co-worker [65] reported that mine water, run-off from abandoned watersheds and associated industrial discharges are the major source of heavy metal contamination, total dissolved solids (TDS) and low pH of streams in the mining area [66–69]. The rivers in urban areas have also been associated with water quality problems. This is due to the practice of discharging of untreated domestic and small scale industries into the water bodies, which leads to the increase in the level of metals concentration in river water [70–74]. It may hence, not be an overstatement to assert that the risk of toxic metals pollution is to the degree, of the number of any chemical process going on in the society, especially in the Sub-Saharan region [75, 76]. The list appears intimidating and further strengthens the need for constant environmental monitoring the presence of the heavy metal in our water bodies.

1.4 Aim and objectives of the study

The upper Crocodile River catchment has witnessed an increase land use and land cover change mainly because of the increased population, increase agricultural practices along the Crocodile River, increase in private resort accommodation and other developmental projects over the past few decades. Regarding the worsening situation on site, the National Environmental Act (Act of 108 of 1998) governs the overall conservation, correct utilization of natural resource and management of natural resource, promote sustainable development, and prohibit activities that will affect the environment. In this regards it requires an Integrated Water Resource Management (IWRM) geared towards maximizing water resource in a sustainable manner, which vital for ecosystems conservation. The key question to be asked is; *is water and other conditions in the Crocodile River have been altered by human activities? What are the sources of the potentially toxic element in the river?* Rustenburg is one of the fastest-growing towns in the North-West Province in South Africa and hosts most of the country operating mining and agricultural activities. Due to the ongoing anthropogenic activities bringing about changes in land use pattern (mining and intensive cultivation), irrigation from the Crocodile River, resultant dynamics stable river system will be distinctively different from what would be present under natural setting in the catchment. However, estimated changes in land use and land cover has not been reported to assess the overall impact on the surface water quality of the Crocodile River. Hence the knowledge of LULC dynamics is thus necessary to safeguard the health of the riverine population and to inform management of appropriate measures where mitigation action is necessary. This study aims to: [1] Assess the spatial distributions of physicochemical parameters and PTEs concentrations in the Crocodile River, [2] To evaluate LULC change in the catchment for the period of 1999–2018 using geographical information system (GIS) techniques.

2. Material and methods

2.1 Study area

The upper Crocodile River catchment is situated in Rustenburg, the economic hub of the North-West Province, South Africa (**Figure 1**). The area hosts a number

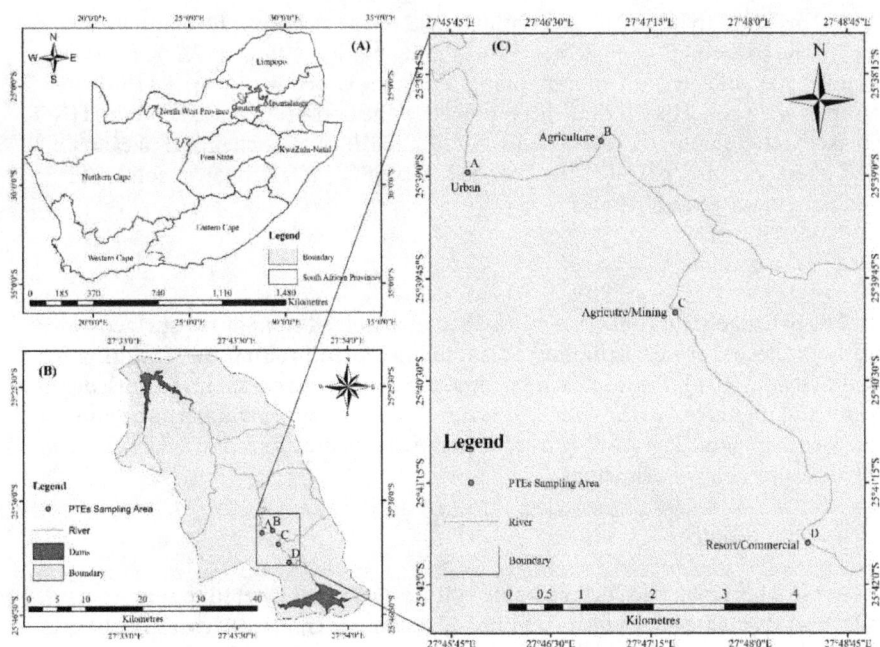

Figure 1.
Study area.

of manufacturing industries, steel and iron smelting, mining and intensive commercial and subsustence agriculture along the Crocodile River. The sub-catchment has two major dams (Roodekopjes and Hartbeespoort) with scattered dams throughout the catchment (**Figure 1B**). These dams act as a source of water supply to the cultivated farmlands and additional water supply is sourced from the borehole and artificial dams. An increasing number of resort accommodations located close these major dams and along the Crocodile River for local and international tourists. The resultant effects of these anthropogenic activities in the marine environment have been reported to be a regular occurrence of filamentous cyanobacteria also known as blue-green algae with several highly toxic biologically active compounds [77–79].

2.2 Surface water sampling

Surface water pollution has been reported as the direct consequence of anthropogenic activity [80–82], and has significantly contributed to the deterioration the Crocodile River [37]. The surface water sampling framework along the Crocodile River was developed based on two considerations; firstly, a proper understanding of the contributing sources as the river traverses the different land uses [20]. Secondly the duration of the sampling framework should be long enough to account for seasonal variation physicochemical parameters and PTEs concentrations in the river. Thus, a longitudinal transect was adopted based on the different land uses within the vicinity of the river. Four sampling point were chosen along the Crocodile River during the field survey to ensured that each of the sampling points was within the vicinity of the different land uses (**Figure 1C**) as prescribed by Chetty and Pillay [83]. Those land uses which overlay each other were considered as areas of nonpoint sources contributing to the contamination of the river [37]. From the stratified sampling sites, surface water was collected on a quarterly basis for 15 months from April 2017 to July 2018. A handheld GPS (Garmin E-Trex 12 channel) was to record the coordinates for each of the sampling points. A total of 72 surface water samples was collected at different points along the Crocodile River. All the water samples were collected in three litter polyethylene bottles, pre-washed with HNO_3. Surface water quality was analyzed according to the physicochemical parameters that is temperature, pH, electrical conductivity (EC), total dissolves solids (TDS) and potentially toxic elements.

2.2.1 In situ and laboratory analysis

The pH, electrical conductivity (EC) total dissolved solid of the surface water freshly collected at each sampling sites were measured in situ using a multi-meter (CRISON MM40+). Prior to each reading, the meter probe was rinsed with distilled water and immersed in the collected water sample for approximately one minute to reach equilibrium. The reading of each parameter was recorded in a data sheet when the measurement was constant.

2.2.2 ICP-MS analysis

In the laboratory, the surface water samples were first filtered to remove all solid and impurities through a (number 42) filter paper. For each sample, 10 mL of nitric acid was added to a 50 mL of water samples as prescribed by [37] and was analyzed using the inductively coupled plasma spectrometry-mass spectrometry (ICP-MS) (Perkin-Elmer Nixon 300Q) for the following elements; copper (Cu), lead (Pb), cadmium (Cd), zinc (Zn), arsenic (As), chromium (Cr), aluminum (Al), manganese (Mn) and iron (Fe). The instrument was calibrated using a standard calibration solution as the atomic spectrometric standard of the mass calibration stability measured using 10 mg/L multi-element standards solution Al, Ba, Ce, Co, Cu, In, Li, Mg, Mn, Ni, Pb, Tb, U and Zn. The instrument was set to run a blank and a standard check for ten samples for quality control for each measurement. Based on three times the standard deviation of the blank using three second integration time and peak hopping at 1-point per mass. The detection limit (mg/l (ppb)) of the selected metals; Ni (< 0.5), Fe (< 1.5), Cu (0.5) and As (< 0.25) and were then converted to mg/L.

2.2.3 Statistical analysis

The statistical analysis was employed using Microsoft Excel (version 2016) and Stata (version 13). Significant relationships between the physicochemical parameters and PTEs was performed using the person's correlation matrix at 95% confidence level ($p > 0.05$).

2.3 Remote sensing data collection

In order to monitor the LULC change, data sets spanning from two time periods for comparison is needed [84]. Suitable images for the following years, 1999, 2009, and 2018 of the study area was acquired from the South African National Space Agency (SANSA) archive. In order to quantify the LULC changes in the study area, remote sensing approach was employed as it involves the usage of satellite images of multiple dates [84]. Landsat and Spot imagery are readily and freely available in South Africa. However, SPOT images were preferred due to high spatial resolution and to ensure consistency in the cover classes and phenology dates of imagery were selected between May and July for all the three images.

2.3.1 Image processing and analysis

ERDAS Imagine 2020 software package was used for image analysis and processing. A subset of the images corresponding to the study area was created after converting all images to a common format. Subsequently, a pre-processing procedure was necessary to make comparable satellite images obtained from different sensors (SPOT) with different radiometric characteristics and acquisition conditions. Moreover, much of the pre-processing, radiometric, and geometric corrections were accomplished using ERDAS Imagine 2020. Additionally, due to the differences in radiometric resolution, the technique adopted to fit this purpose involved the calibration of the digital numbers (DN's) were converted in the image data from to at-sensor radiance (L_{SAT}) units (W m^{-2} sr^{-1} μm^{-1}).

2.3.2 Geometric corrections and image segmentation

Since the images had different spatial resolution, it became necessary for the images to be geometrically corrected [85]. In order to bring the pixel sizes to a common value, due to differences in date, the Root Mean-Square Error (RMSE) was used. The reason is to avoid registration errors to be interpreted as LULC change which can lead to an overestimation of actual change. Because Landsat data series is characterized by spectral bands which are very sensitive to both vegetation and other earth related features, this was central to the study in mapping the LULC changes [29]. To accurately measure the LULC change, a topographic map with a scale of 1:50,000 produced in 1996 was used for geometric correction using GCP (Ground Control Points) to geocode the image of 2009. The image was then used to register the image of 1999 and 2018, using a nearest-neighbor algorithm. From the three images, the RMSE was less than 0.4 pixel which is acceptable [86]. Image segmentation was conducted using the multiresolution segmentation algorithm [87]. The algorithm requires the specification of the weights of the band, the shape (and its mutual color), the scale parameter and the compactness (and its mutual smoothness), which are expounded by Benz and co-workers [88].

2.3.3 LULC cover change classification and accuracy assessment

In order to investigate changes that would have occurred in the study area, the maximum likelihood classifier (MLC) was used. This method provides an effective and robust supervised classification method. This method has widely been used by different scholars as it evaluates both the variance and covariance of spectral response pattern whereby each pixel is assigned to the class for which it has the highest possibility of association and is considered to be most accurate classifier [29, 84]. MLC assumes that spectral values of the pixels are statistically distributed according to a multivariate normal probability density. Accuracy assessment used an error (confusion) matrix, in which producer's accuracy (PA, %), user's accuracy (UA, %), the Kappa coefficient (K), and overall accuracy (OA, %) were computed [29]. Using ground checkpoints and digital topographic maps of the study area, supervised classification was made use of. The area was classified into five main classes: water bodies, cropland, grassland, bare land, and built-up, as presented in **Table 1** with the description of the land cover classes given therein. To represent different land cover classes of the study area, the assessment of 200 random points was generated for the MLC of the study area per image date using the random stratified method. The "create precision points" function in ERDAS Imagine 2020 was used on the MLC classified images to generate a set of random points.

Class	Description
Water bodies	An area containing open bodies of water, which includes brackish, streams, rivers, dams, and natural ponds as well as artificial ponds.
Cropland	Areas cultivated with annual crops, vegetables, or fruit. These crops are irrigated mainly from the water of the Crocodile river and/or groundwater. Most of the cultivated area is newly reclaimed.
Grassland	For the study area, the plants can be classified into nine life forms such as evergreen non-succulent perennial sub-shrubs, evergreen succulent perennial sub-shrubs, annuals perennial grasses, perennial herbs, evergreen succulent perennial shrubs, evergreen non-succulent perennial shrubs, deciduous perennial shrubs and partially deciduous perennial sub-shrubs.
Bare land	Land areas of exposed soil surface as influenced by human impacts and/or natural causes as well as changes in topsoil that comprises areas with active excavation and quarries and opencast mines. These areas contain sparse vegetation with very low plant cover value as a result of overgrazing, woodcutting, etc.
Built-up	Includes construction activities of all kinds in the study area such as apartment buildings, single houses, shacks, shopping centres, industrial and commercial facilities as well as highways and major streets be it tarred or gravel.

Table 1.
Description of different land cover classes in the study area.

The reference data against which to judge the correctness of classification were obtained from 10 m resolution images on Google Earth® of dates close to the SPOT images. Ancillary data and the result of visual interpretation was integrated with the classification result using GIS in order to increase the accuracy of land cover mapping of the three images and improve the classification accuracy of the classified imagery.

2.4 Quality control/quality assurance

This study has established a sound quality control/quality assurance over a similar study and is references therein [37].

3. Results and discussion

3.1 Spatial variation PTEs in the Crocodile River and its implication to water quality from 2017 to 2018

The results of the trend analysis of the PTEs concentrations in the Crocodile River are presented in order of abundance of Mn in October 2017 (0.34 mg/L), < Cu in July 2017 (0.21 mg/L), < Fe in April 2017 (0.07 mg/L), < Al in July 2017 (0.07 mg/L), and < Zn in April 2017, October 2017 and April 2018 (0.05 mg/L) respectively (**Figure 2**). Similar findings was also reported by Marara and Palamuleni [89] in which Mn, Fe and Zn were amongst the most abundant element in the Klip river in South Africa. This results shows an increase in metal concentrations during the first quarter in the sampling months, owing to low rainfall intensities and runoff [81]. Non-point sources of PTEs in the river might be attributed to dust blown into the river from the cultivated field and mining areas, runoff, iron smelting and exhaust automobile [90]. During the second quarter of the sampling months, changes in rainfall pattern might have influenced the PTEs concentrations in the river due to the diluting effect from the different land uses. Usually, the rainfall season begins in October and peaks in intensity from October to February.

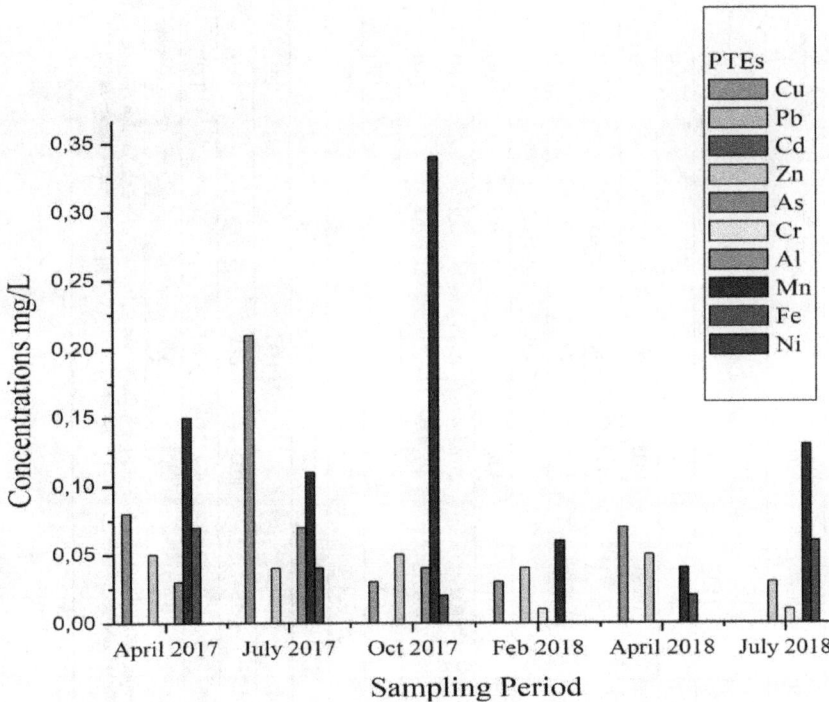

Figure 2.
Trend analysis of PTEs concentrations during the sampling periods.

The concentrations of PTEs during this sampling month might likely have had some diluting effect in the river metals concentration. A similar study by du Preez and co-workers [91] asserts that a reduction in nutrients in the Crocodile River could be attributed to the diluting effect, especially during periods of high current flow.

Further, Ogoyi and co-workers [92] examined the content of PTEs in water, sediment and microalgae from Lake Victoria, which is the largest tropical fresh water lake in the world [93], representing an exceptional ecosystem with the largest fresh water fishery in the continent [92]. It is located in East Africa and surrounded by Uganda on the North West, Kenya on the North East, Rwanda on the far West and Tanzania on the South–South [94, 95]. They collected water samples from two different points namely from Winam and Mwanza gulf and using atomic absorption spectrophotometry (AAS) examined the level of heavy metal pollution of lead, cadmium, chromium, mercury and zinc. The analysis of the water sample as summarized in **Table 2** indicates that the presence of lead, cadmium and chromium at the Mwanza gulf point (LVEA-MGP) were 2.2, 2.3 and 1.4 times respectively higher than the recommended permissible threshold standard by WHO (**Table 2**), while the mercury and zinc were within the recommended limit for safe water. At the Winam gulf point (LVEA-WGP), the level of PTEs concentration for lead and chromium was 82.3 and 3.56 times respectively higher than the recommended permissible threshold limit by WHO while the rest were within safe limits. They argued that there is a link between PTEs pollution and anthropogenic activities like waste disposal and mining in the environs [75]. They concluded that the PTEs pollution at these points of the lake was relatively low, but emphasized the need for continuous monitoring of the PTEs pollution in the lake [65]. Chief Albert Luthuli Local Municipality is situated on the eastern scarp of Mpumalanga Province of Republic of South Africa. The Municipality covers a land area of nearly 5.560km^2,

PTES/water Source	$\frac{W_{Pb}}{WHO_{Pb}}$	$\frac{W_{Cd}}{WHO_{Cd}}$	$\frac{W_{Hg}}{WHO_{Hg}}$	$\frac{W_{Cr}}{WHO_{Cr}}$	$\frac{W_{Ca}}{WHO_{Ca}}$	$\frac{W_{Zn}}{WHO_{Zn}}$	$\frac{W_{Fe}}{WHO_{Fe}}$	$\frac{W_{Mn}}{WHO_{Mn}}$	$\frac{W_{As}}{WHO_{As}}$	References
LVEA-MGP	2.2	2.3	0	1.4	—	0.006	—	—	—	[92]
LVEA-WGP	82.3	0	0	3.56	—	0.017	—	—	—	[92]
BHMPRSAJ	>8	—	—	>0.2	>0.01	>0.03	—	>1	—	[64]
DZindi River	3	—	—	—	0.025	0.033	4.433	1.5	—	[62]
AAP	—	1.7	2	—	—	0.043	2.233	8.54	0.3	[94]
OAP	—	1.7	3	—	—	0.00167	3.11	7.92	0.3	[94]
US	—	1.7	2	—	—	0.00167	4	7.23	0.3	[94]
MS	—	5666.7	9	—	—	0.0097	4.74	5.77	0.2	[94]
DS	—	1.7	154	—	—	0.0547	22.2	8.72	0.2	[94]
AB	—	2.7	2	—	—	0.0013	5.9	1.13	0.4	[94]
LWW	102	183.3	—	22	0.255	1	—	—	—	[98]
PBW	279	476.7	—	100	1.51	4	45.4	175	—	[99]
BH	455	463.3	—	326.8	0.58	4.923	77.5	991.4	—	[99]
STREAM	29	746.7	—	402.2	1.755	2.063	20	113.2	—	[99]
RIVER	669	1546.7	—	101.2	1.075	1.547	105.8	61.2	—	[99]
HDW	401	320	—	2275.6	23.175	49.1	47.1	936.5	—	[99]
Crocodile River	2	—	—	—	0.01	0.017	—	1.9	—	Present work

LVEA-WGP: Lake Victoria East Africa-Winam Gulf Point; LVEA-MGP: Lake Victoria East Africa-Mwanza Gulf Point. BHMPRSAJ: Borehole at Mpumalanga South Africa, ELH: East London Harbor; PEH: Port Elizabeth Harbor; Accra Abandoned Pit (AAP), OAP: Obuasi Abandoned Pit, AB: Accra Borehole, US: Up stream, MS: Main Stream, DS: Down Stream; LWW: Lagoon Waste water; BH: Borehole, HDW: Hand dug well; PBW: Pipe Borne Water.

Table 2.
Heavy metal pollution level in some selected African water bodies.

and a report from the Stats SA 2016 Community Survey, indicates its home to some 187,630 people, which have increased. The Municipality is made up of various communities confronted with a society that faces sundry economic, social, environmental, and governmental challenges. Approximately 80% of the populace live in the rural areas concentrated in the east of the area; the two main service centres of Emanzana and Carolina provide a home for 15% of the people while the remaining population are found in the forestry and farming areas of the Municipality [96]. Nthunya and co-workers [64] investigated the source of toxic metals in drinking water in this Chief Albert Luthuli Local Municipality in Mpumalanga, South Africa. Their work was so detailed and captured five different points over four seasons of the year, winter, spring (August 2014), summer (November 2014), autumn (February 2015). The sampling points included a drinking water treatment plant in Eerstehoek bout 5 km from Lochiel, a 50 m deep open well used largely by the community and the students of a nearby school designated as well 1; an open shallow well located in the upper part of Lochiel and used by the residents designated as well 2; Tanks 1 and 2 located in the Lochiel Primary school premises and the community respectively. The latter of the two tanks is being used by the larger part of the community and finally a borehole in Masakhane primary school supplying water to the school tank and taps. Using ICP-OES spectrometer suited with iTEVA software for measurements of all the analytes at maximum wavelength, they investigated the presence of nine heavy metal pollutants in the drinking water which are namely: cadmium, chromium, copper, cobalt, iron, manganese, nickel, lead and zinc (**Figure 3**).

Figure 3 represents the physical properties of the various water samples and the concentration of toxic metals in ppm. Their results indicate that the concentration of toxic metals varied across the seasons and sources. The lead concentration was found to be above WHO limit for drinking water in well 1 & 2, Tanks 1 & 2, surprisingly in both raw and treated water in February 2015, and bore hole for all seasons considered. In autumn, the level of Manganese rose above the WHO limit in the untreated water. Cobalt for most of the periods of the year considered remained above WHO limits for safe and potable water. The rest of the metals were largely within the WHO drinking water limit. The borehole is ground water mainly used by a greater percentage of African populace as already established in the earlier part of the review [97]. **Table 2** indicates that the borehole water taken in July designated as BHMPRSAJ has a lead and cobalt concentration that is greater than the WHO limit by a factor greater than 8 and 1 respectively as at 2014/2015. They argued that the source of these toxic metal accumulation in this locality is both natural and anthropogenic, which include weathering of mineral rich rocks and indiscriminate disposal of metal rich wastes at the landfills. In conclusion, they underscored that long-term exposure to the toxic heavy metal can be fatal and hence, the need to further purify and monitor the quality of drinking water regularly.

Another detailed work was done on the assessment of heavy metals in drinking water, at Datuku in the Talensi-Nabdam District in the Upper East region of Ghana by Cobbina and co-workers [94]. They aimed to evaluate the impact of small scale gold mining on the drinking water quality in that community. Samples were collected from six sources namely: Accra abandoned pit (AAP), Obuasi abandoned pit (OAP), mainstream (MS), upper stream (US), Accra borehole (AB) and down stream (DS). Using the Shimadzu model AA 6300, they evaluated the trace concentration of Zn, As, Cd, Fe, Mn and Hg in these five places. Their results show that Cd, Fe, Hg and Mn level was higher than the standard for safe water by WHO, while As and Zn were within the limit safety for all the sources (**Table 2**). The level of Cd concentration on the mainstream source (MS) was 5666.7 times higher than the WHO standard for safe water. The level of Fe contamination was taken with

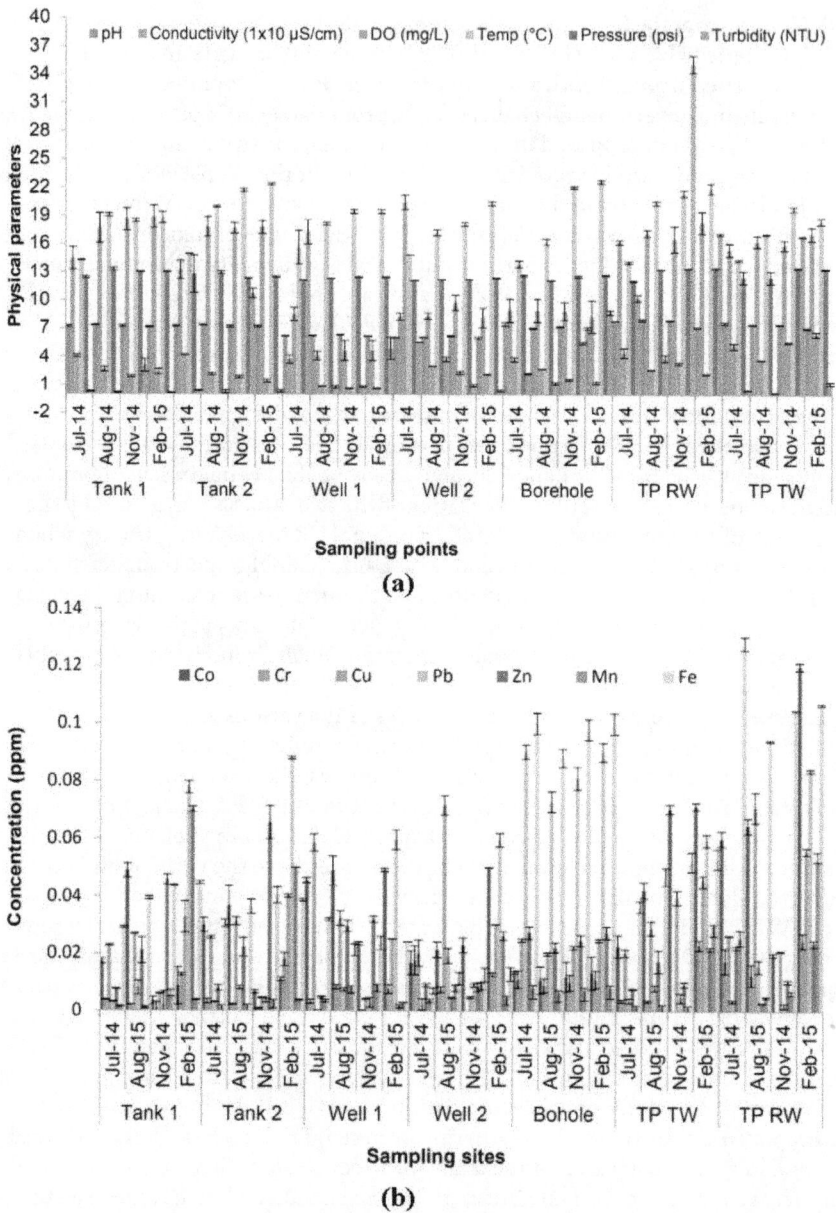

Figure 3.
(a) The physical properties of the various water sources under consideration (b) the number of toxic metals present from the various water sources in winter, spring (August 2014), summer (November 2014), autumn (February 2015). TP TW: Treated Plant Treated Water and TP RW: Treated Plant Raw Water [64].

reference to US-EPA and was also found to be higher than the accepted limit by a factor greater than 2 for all the sources of the water. They opined that cadmium pollution could be as a result of seepage from the parent rock, use of cadmium containing products such as batteries, plastics and mining tools.

Orata and Birgen [98] studied the uptake of heavy metals by different fishes and their tissues in a lagoon waste water (LWW). They proposed that their study would provide a useful tool for envisaging human exposure to PTEs through consuming fish under different contamination scenarios. The lagoon wastewater body had heavy metals concentration of the most lethal class of lead, cadmium, and chromium in an amount that is 102, 183.3 and 22 times respectively higher than the WHO accepted standard of safe water (**Table 3**) and other environmental agencies (**Table 4**). They hence concluded that various species of fishes studied in this scenario were unsafe for consumption sequel to the uptake of heavy metals in various parts of their bodies. Another similar and detailed work was done to inspect the physicochemical properties and heavy metal content of water sources in Ife North Local Government Area of Osun State, Nigeria by Oluyemi and co-workers [99, 100]. While they concluded that the physical parameters of the water collected from pipe borne water (PBW), borehole (BH), stream, river and hand-dug well (HDW) were within limits for potable and household water, the AAS results of heavy metal concentration of Pb, Cd, Cu, Cr, Fe, Mn and Zn is a far cry from safe limits for drinking water (**Table 2**). Pd and Cd levels were 279 and 476.6 times in pipe borne water and 455 and 463.3 times in borehole (BH) above the WHO standard for safe domestic and drinking water. These two sources of water have been validated as the most common sources of water for Africans in rural settings. They opined that such high concentration of these PTEs cannot be disconnected from mining activities, leaching of metals from wastes site to the ground water plus rural and urban water run-off, and possible wearing of lead from metal pipes into the water during the distribution.

In this study, the following elements Cd, As, and Ni, concentrations in all the sampling points were below the detection limits except for Cr (0.1 mg/L) in point C (Agriculture/Mining) and Pb (0.02 mg/L) in point D (Resort/Commercial).

Metal	WHO	EPA	ECE	FTP-CDW	PCRWR	ADWG	NOM-127	DWAF
Aluminum	N/A	N/A	N/A	N/A	N/A	N/A	N/A	< 0.005
Nickel	0.07	0.04	0.020	N/A	0.020	0.020	N/A	< 1
Copper	2	1.3	0.200	0.100	0.200	0.200	0.200	< 2
Zinc	3	5	N/A	5.000	5.000	3.000	5.000	5
Cadmium	0.003	0.005	0.005	0.005	0.010	0.002	0.005	0.003
Lead	0.01	0.015	0.01	0.01	0.05	0.01	0.01	0.01
Mercury	0.001	0.002	0.001	0.001	0.001	0.001	0.001	
Arsenic	0.010	0.010	0.010	0.010	0.050	0.010	0.025	0.01
Antimony	0.020	0.006	0.005	0.006	0.005	0.003	N/A	
Iron	N/A	0.300	0.200	0.300	N/A	0.300	0.300	0.1
Uranium	0.030	0.030	N/A	0.020	N/A	0.017	N/A	
Manganese	0.10	0.500	0.500	0.500	0.500	0.500	0.150	0.18
Thallium	N/A	0.002	N/A	N/A	N/A	N/A	N/A	
Silver	N/A	0.100	N/A	N/A	N/A	0.100	N/A	
Chromium	0.050	0.100	0.050	0.050	0.050	0.050	0.050	0.05

Key: DWAF* = Department of Water Affairs and Forestry, South Africa. EPA* = US- Environmental Protection Agency(2011). WHO* = World Health Organization (2011), N/A* = Not reported or Not available and BDL* = Below detection limits, ECE: European Commission Environment (1998), FTP-CDW: Federal-Provincial-Territorial Committee on Drinking Water, Health Canada (2010), PCRWR: Pakistan Council of Research in Water (2008), ADWG: Australian Drinking Water Guidelines (2011), NOM-127: Norma Official Mexicana NOM-127-SSA1–1994 (1994).

Table 3.
Standards and guidelines for heavy metals in drinking water (mg/L), recommended by the Environmental Protection Agency (EPA) and world health organizations (WHO) for drinking water that is based on data of toxicity and scientific findings.

Metals	Seawater($mg \cdot L^{-1}$)		Sediment ($mg \cdot L^{-1}$)	
	EEC	ANZECC	CEPA	PSAG
Cd	2.5	2	2	10
Cu	5	5	8	500
Fe	—	—	—	—
Pb	15	5	22	500
Mn	—	—	—	—
Zn	40	50	40	750

Table 4.
Guidelines for metals in seawater and sediment by EEC: European Commission environment; ANZECC: Austrialian and new Zeland environmental conservation council; CEPA: Cannadian Environmental Protection Agency; PSAG: Proposed South African guidelines.

Site ID	Cu	Pb	Cd	Zn	As	Cr	Al	Mn	Fe	Ni
A	0.02	BDL*	BDL*	0.04	BDL*	BDL*	0.02	0.13	0.02	BDL*
B	0.10	BDL*	BDL*	0.05	BDL*	BDL*	0.03	0.22	0.04	BDL*
C	0.16	BDL*	BDL*	0.05	BDL*	0.01	0.02	0.15	0.14	BDL*
D	0.02	0.02	BDL*	0.05	BDL*	BDL*	0.04	0.19	0.09	BDL*

Table 5.
Average PTEs concentrations (mg/L).

The spatial distribution of Mn, Cu, Fe, Al and Zn along the different land uses in the Crocodile River is presented in **Table 5** and **Figure 4**. The concentration of Mn is quite variable along the different land uses and the highest value of 0.22 mg/L was recorded in point B (Agriculture) while the least value (0.13 mg/L) in point A (Urban). The concentration of Cu also varied spatially along the river with the highest value in point C (Agriculture/Mining) while point A (Urban) and D (Resort/Commercial) had the lowest values of 0.02 mg/L respectively (**Table 2; Figure 4**). Fe had the highest concentration in point C, while point A had the lowest concentration value. The concentrations of Al along the different land uses were slightly different from each sampling and point A and C had the lowest concentrations of 0.02 mg/L respectively. The average concentration of Zn in the river indicates that point A, B and C all had the same concentration value of 0.05 mg/L, respectively but with a slight drop in the concentration value of point A (0.04 mg/L) (**Table 2; Figure 4**).

Three water quality guidelines permissible threshold values were used to gauge the level of PTEs concentrations in the river (**Table 5**). The results indicate that most of the elements were within the DWAF (Department of Water Affairs and Forestry, South Africa, 1997, and 1997b), stipulated guideline for aquatic environments except for Al, Mn and Fe exhibiting high concentration values above the permissible threshold limit of DWAF of <0.005, 0.18 and 0.1 mg/L respectively (**Table 3**). Similarly, the value of Mn in the Crocodile River exceeded the recommended threshold guideline for EPA of 0.05 mg/L. The concentration of Al in the river exceeded the DWAF guideline in all the sampling points, while Mn concentration exceeded the recommended threshold value by EPA, for all the sampling points, and also that of DWAF at point B (Agriculture) and D (Resort & Commercial). In contrast, the values of Fe exceeded the permissible limit of DWAF

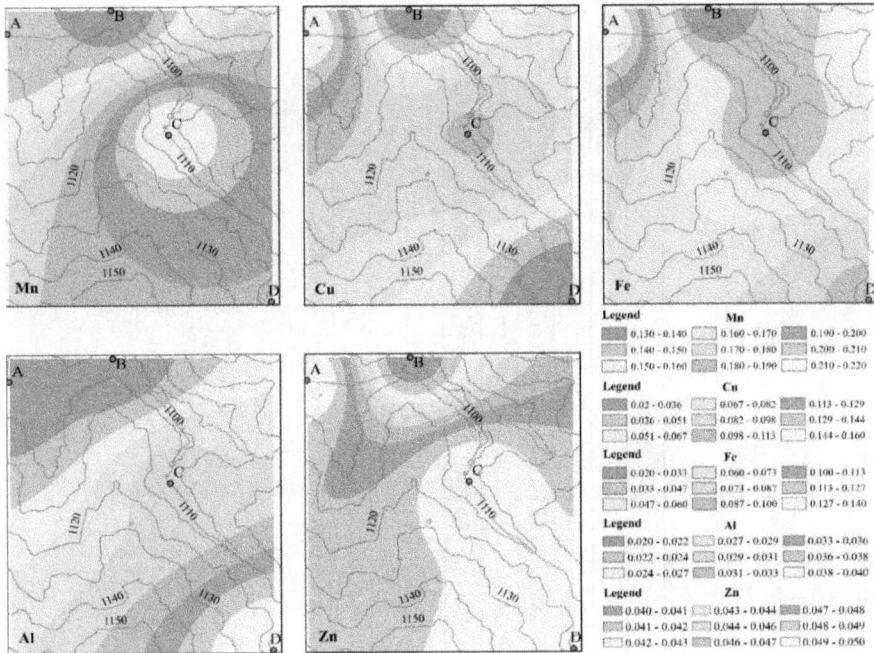

Figure 4.
Summary of the average concentrations (mg/L) of PTEs in the Crocodile River.

at point C (Agriculture/Mining). Cd, As and Ni concentrations in the river were below the detection limit or were not present in the water.

Although the following elements As, Ni, Cd, Pb and Cr analyzed exhibited low concentrations values; however, it cannot be concluded that the river is not contaminated. For instance, Pb concentration in point D (0.02 mg/L) exceeded all the water quality guidelines (DWAF, WHO and EPA) as seen in **Table 3** and a plausible explanation could be attributed to point-source contamination. This is an indication that the river might eventually be polluted in the future if proper mitigation measure is not put in place due to the diverse anthropogenic activities within the vicinity of the river [36]. These changes might also be due to the spatial–temporal input from agricultural areas, surface runoff from different mining areas, untreated wastes disposal from resort accommodation, catchment sensitivity, and settlement dumpsites close to the river [62].

3.2 Seasonal variation of physicochemical parameters and PTEs in the Crocodile River

3.2.1 Physicochemical parameters in the Crocodile River

Studies by Okonkwo and Mothiba [101] and Somerset and co-workers [102] have reported changes in physicochemical and heavy metal concentrations in South African rivers due to changes in the seasons. The results of the physicochemical parameters between the different seasons are shown in **Table 6**. The analysis of the water temperature at the different sampling points was slightly different but was distinctively different between the wet (summer) and the dry season (winter) (**Table 6**). At the time of the water collection, the wet season had a maximum temperature of 28.6°C ± 0.35 while the dry season (winter) had a minimum

Sampling Points	Site ID	Temp(°C)		EC (µs/cm)		pH		TDS (mg/L)	
		Wet	Dry	Wet	Dry	Wet	Dry	Wet	Dry
Point A	Urban	28 ± 0.14	19.8 ± 1	520.0 ± 4.24	561.5 ± 51.84	8.2 ± 0.01	7.6 ± 0.29	355.5 ± 34.6	393.0 ± 95.64
Point B	Agriculture	28.6 ± 0.35	20.0 ± 0.91	517.0 ± 0	544.3 ± 37.87	8.5 ± 0.55	7.5 ± 0.41	381.5 ± 71.4	396.5 ± 75.39
Point C	Agriculture/Mining	28.4 ± 0.35	19.9 ± 0.84	509.0 ± 4.24	568.0 ± 25.36	8.4 ± 0.26	7.4 ± 0.57	321.5 ± 0.7	391.0 ± 88.74
Point D	Resort/Commercial	28.3 ± 0	19.6 ± 1.06	533.0 ± 14.14	563.8 ± 54.03	8.1 ± 0.02	7.7 ± 0.41	355.0 ± 0	412.0 ± 97.97
DWAF*		N/A*	N/A*	400–900		5.0–9.5		450–900	
WHO*		N/A*	N/A*	N/A*		7.0–8.5		N/A*	
EPA*		N/A*	N/A*	N/A*		$6.5 \leq pH \leq 8.5$		500	

N/A* = Not Available. DWAF* = Department of Water Affairs and Forestry, South Africa. EPA* = (US- Environmental Protection Agency). WHO* = World Health Organization.

Table 6.
Seasonal variation in the average concentrations of the physiochemical parameters in the crocodile river.

temperature of 19.8°C ± 1. The EC values from across each of the sampling points ranged from 509 µs/cm to 533 µs/cm in the wet season while during the dry season the readings ranged from 544.3 µs/cm to 568 µs/cm.

The pH concentration in the river varied slightly between each sampling points with a maximum value of 8.5 for the wet season and 7.7 for the dry season (**Table 6**). According to du Preez and co-workers [91], an increase in pH concentrations might have a negative impact on water quality and its suitability in watering crops and animals. Although the pH values were generally lower, its value, however, indicates that the water is slightly alkaline in most of the sampling points for drinking water which is deleterious for the animals and human in the catchment. Evidence from the field visit also suggests that the water from the Crocodile River is abstracted and irrigated for agricultural purpose. Bouaroudj and co-workers [103] report that the continuous irrigation of crops with saline waters may lead to a gradual or rapid increase in soil salinity. The concentration of TDS (mg/L) in the river from the different sampling point varied from a 321.5 ± 0.7 to 381.5 ± 71.4 for the wet season while for the dry season it varied from 391.0 ± 88.74 to 412.0 ± 97.97 (**Table 6**). A study by du Preez and co-workers [91], attributed an increase in EC and pH in the Crocodile to anthropogenic activities likely from runoff caused by agricultural activity while Wongsasuluk and co-workers [104] attributed an increase in EC due to seasonal variation, thus confirming the role of seasonal variations in physicochemical parameters.

3.2.2 Seasonal variations in PTE concentrations in surface water

The assessment of PTEs in the Crocodile River suggests there is a significant variation ($p > 0.05$) of each element between seasons (**Figures 5** and **6**). The average concentration of Cu in the dry season ranged from 0.01 to 0.018 mg/L while those for the wet season ranged from 0.03–0.04 mg/L signifying an elevated concentration during the wet season. Although the value of Cu between the two seasons was within the safe permissible limit stipulated by DWAF (< 0.2 mg/L), WHO (< 0.2 mg/L) and EPA (US) (0.3 mg/L). However, a study by Ahmad and Goni [105] states that Cu concentration at 0.01 to 0.02 mg/L might be toxic because of the presence of salts (chlorides and litigates). Analysis of Al in the river for the

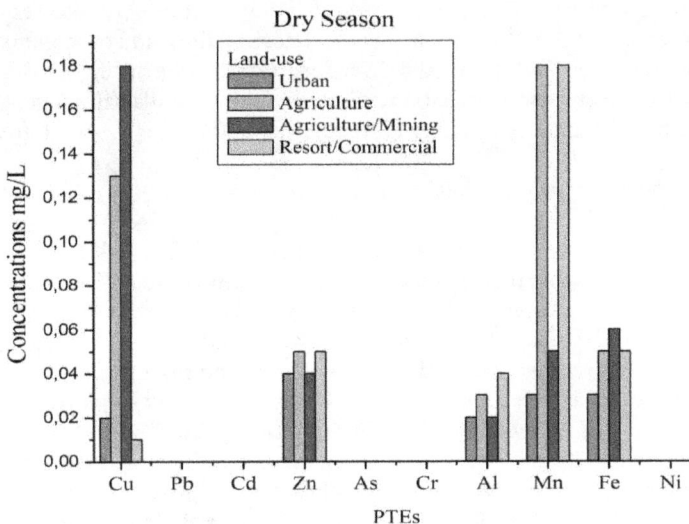

Figure 5.
Average PTEs concentration (mg/L) in surface water.

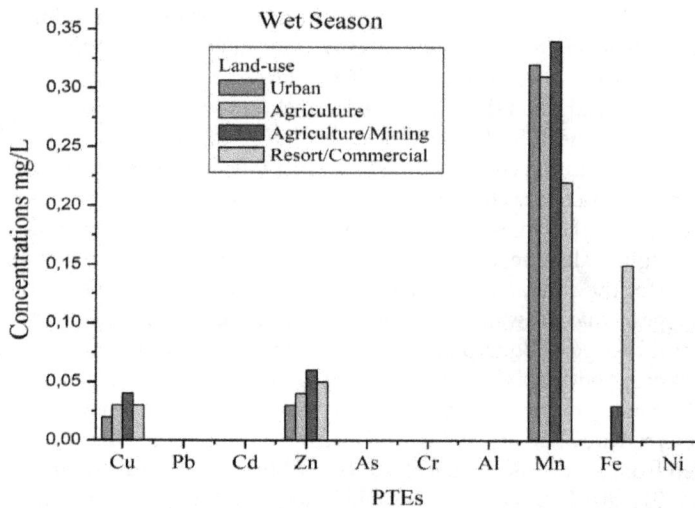

Figure 6.
Average PTEs concentration in the Crocodile River.

dry seasons ranged from 0.02–0.04 mg/L whereas during the wet season it was not detected in the water samples. A plausible reason why Al was found in the water during the dry season might be due to the discharge of waste effluent from nearby private resort accommodation, agricultural surface runoff and commercial waste dumping directly into the river. Marara and Palamuleni [89] reported an increase in toxic element in the Klip river, South Africa to high evaporation rates and low flow rates of water during the season which is similar to the findings of this research.

The average concentration of Mn in the river ranged from 0.03–0.18 mg/L (dry season) while for the wet season it ranged from 0.22–0.34 mg/L. The findings of this study is in line with those reported by Li and Zhang [106], whereby an increase concentration of Mn during the wet seasons in the Upper Han River in China. Fe concentration for the dry season ranged from 0.03–0.05 mg/L while in the wet season, it ranged from 0.03–0.15 mg/L recorded only at points C and D respectively, whereas point A and B were below the detection limit and or might not be available in the water. The observed high concentration of Fe during the dry season compared to the wet season might be attributed to significant anthropogenic disturbance dominated primarily by physical weathering in the river as source areas [89].

3.2.3 Correlation matrix of PTEs in the water samples

The results of the Pearson's correlation coefficients (r) $(p > 0.05)$ between the PTEs and the physicochemical parameters are shown in **Table 7**. The results of the physicochemical parameters of the water showed a highly significant positive correlation with each of the parameters with temperature and EC $(r = 0.96)$, temperature and pH $(r = 0.99)$, temperature and, TDS, $(r = 0.98)$ and pH and EC, $(r = 0.99)$, TDS and EC $(r = 1)$ as indicated in **Table** 7.10. Also, the pH was significantly positively correlated with all the PTEs, thus indicating that the pH influences the concentration of PTEs in the Crocodile River. Similarly, the temperature correlation with the PTEs showed positive to significantly strong positive with Zn, Mn and Cu $(r = 0.63, r = 0.64$ and $r = 0.76)$ respectively. The correlation between the PTEs showed a significant positive correlation between Mn and Zn $(r = 0.70)$, Mn and Al $(r = 0.71)$ while Fe and Cu, Fe and Zn showed strong positive correlation **Table** 7.

	Tem	EC	pH	TDS	Cu	Zn	Al	Mn	Fe
Tem	1								
EC	0,96	1							
pH	0,99	0,99	1						
TDS	0,98	1,00	1,00	1					
Cu	0,76	0,77	0,77	0,77	1				
Zn	0,63	0,39	0,52	0,46	0,54	1			
Al	−0,05	−0,31	−0,17	−0,24	−0,43	0,52	1		
Mn	0,64	0,44	0,55	0,50	0,08	0,70	0,71	1	
Fe	0,16	0,03	0,10	0,06	0,62	0,65	0,02	−0,08	1

Correlation is significant at the p > 0.05 level. (2-tailed).

Table 7.
Pearson correlation coefficient matrix of the physiochemical parameters and PTEs in the river.

3.3 Land use and land cover change detection in the catchment

The Crocodile River catchment witnessed a considerable change in land use and land cover during the two decades. The results from the observed changes of the land use and land cover in the study area during the selected periods (1999–2009-2018) are illustrated in **Tables 8–10** and **Figure 7**. **Table 8** shows the results of the accuracy assessment for the study area. Thematic map of the study area shows the overall accuracy classification of 77% with an overall kappa statistic of 0.7579 in 1999. Cropland and grassland user accuracy yielded 73% and 70% respectively. Bare land was correctly classified at 75% user accuracy, and built-up land yielded a classified user accuracy of 78% as per the actual representation on the ground. Water bodies were correctly classified at 88%, thus making it the highest user's accuracy. Classification of 2009 had an overall accuracy of 84% ($\hat{K}$ = 0.8341), slightly better than the 1999 image. Water bodies had the highest user's accuracy, at 100%. Bare land had a user's accuracy of 73%, built-up area had user's accuracy of 85%, while cropland and grassland had user's accuracy of 75% and 88% respectively. On the other hand, the 2018 image produced an overall kappa statistic of 0.7832 with an overall accuracy of 79%. The built-up class had a user's accuracy of 80%, while the cropland area had 73%. The bare land class, as well as the grassland, were both classified with a user's accuracy of 75%, while the water bodies' class produced a user's accuracy of 93%, which was the highest out of all the five classes.

3.3.1 Change detection in the study area

Tables 8, 9 and **Figure 7** shows all the major land use classes in the area. It was noted that between 1999 to 2009, cropland increased by 25 462 ha and with a land cover change of 21.8% but decreased from 2009 to 2018 by −7 884 ha and with a − 5.5% changes in land cover. However, from 1999 to 2018, cropland witness 15.05% general change land cover and 1.44% change rate. The observed change can be attributed to a number of natural factors such as climate changes, and anthropogenic factors such as loss in soil fertility, changes in land use pattern/management, bush encroachment amongst others. It is also possible that climate change has played a leading role to the loss of cropland from 2009 to 2018. Grassland decreased by −2 159 ha between 1999 to 2009 with a land cover change of −1.9% but increased

Accuracy assessment for study area 1999 MLC classified

Classification	Cropland	Grassland	Bare land	Built-up	Water bodies	Row total	User 'accuracy
Cropland	29	3	3	3	2	40	73%
Grassland	4	28	3	2	3	40	70%
Bare Land	2	4	30	4	0	40	75%
Built-up	3	3	3	31	0	40	78%
Water	2	2	1	0	35	40	88%
Column total	40	40	40	40	40	200	—
Producer's accuracy	73%	70%	75%	78%	88%	—	77%
Overall Kappa ($\hat{K}$) = 0.7579							
Accuracy assessment for study area 2009 MLC classified							
Cropland	30	2	7	1	0	40	75%
Grassland	3	35	1	1	0	40	88%
Bare Land	5	2	29	4	0	40	73%
Built-up	2	1	3	34	0	40	85%
Water	0	0	0	0	0	40	100%
Column total	40	40	40	40	40	200	—
Producer's accuracy	75%	88%	73%	85%	100%	—	84%
Overall Kappa ($\hat{K}$) = 0.8341							
Accuracy assessment for study area 2018 MLC classified							
Cropland	29	4	3	2	2	40	73%
Grassland	6	30	2	1	1	40	75%
Bare Land	3	3	30	4	0	40	75%
Built-up	2	2	4	32	0	40	80%
Water	2	2	1	0	35	40	88%
Column total	40	40	40	40	40	200	—
Producer's accuracy	73%	70%	75%	78%	88%	—	77%
Overall Kappa ($\hat{K}$) = 0.8341							

Table 8.
Accuracy of LULC obtained from satellite data for the selected periods.

from 2009 to 2018 by 28 771 ha having a land cover change of 25.8%. Also, from 1999 to 2018 grassland witness an overall increase of 23.42% change in land cover with an annual change rate of 2.76%. This could be attributed to increased conservation in protected areas for game hunting as the number of privately owned resort accommodation increased for ecotourism [29].

Similarly, between 1999 and 2009 and from 2009 to 2018, bare land decreased from −22 163 ha to −20 775 ha respectively, with an annual negligible land cover

Land cover categories	1999–2009		2009–2018		1999–2018
	Area (ha)	Percentage Change	Area (ha)	Percentage change	Percentage change
Cropland	+25 462	+21.8	−7 884	−5.5	15.05
Grassland	−2 159	−1.9	+28 771	+25.8	23.42
Bare Land	−22 163	−19	−20 775	−22	−36.81
Built-up	−1 978	−1.7	−685	−0.6	−2.29
Water bodies	+838	+0.7	+573	+0.5	1.18

Table 9.
Trend changes in study area land cover categories.

Land cover categories	1999–2009		2009–2018		1999–2018
	Area (ha)	Percentage Change	Area (ha)	Percentage change	Percentage change
Cropland	+2 546.2	+2.2	−876	−0.7	1.44
Grassland	−215.9	−0.2	+3 196.78	+3	2.76
Bare Land	−2 216.3	−1.9	−2 308.33	−2	−3.88
Built-up	−197.8	−0.2	−76.11	−0.1	−0.28
Water bodies	+83.8	+0.1	+63.67	+0.1	0.18

Table 10.
Annual rate of change in land cover categories for study area.

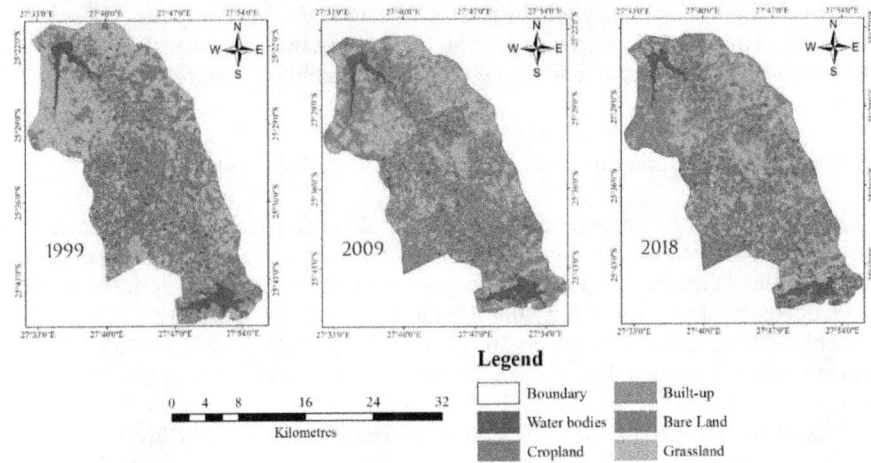

Figure 7.
Land use and land cover map of the upper crocodile river catchment from 1999 to 2018.

change of 1.9% and − 22% respectively and with an overall land cover change of −36.81% and annual change rate of −3.88 spanning from 1999 to 2018. The decrease in bare land suggests that other land uses such as grassland are slowly occupying the bare land. Similarly, Built-up also decrease from 1999 to 2009 by −1 978 ha and between 2009 to 2018 by −685 ha witnessing a change of -2.29% in land cover change and − 0.28% annual change rate from 1999 to 2018. Similar explanation as for bare land hold true with the exception that built-up areas are highly influence by man reconfiguring the environment.

Water bodies increased from +838 ha from 1999 to 2009 and with a slight increment of 0.7% and from 2009 to 2018, it further increased by 573 ha with an overall land cover change of 1.18% and annual change rate of 0.18 spanning from 1999 to 2018. This increase could be attributed to the construction of artificial dams used for irrigation water of crops in the area. The area is known for large scale intensive cultivation of both perishable crops (vegetables) fruits and grains (corn and wheat). Also, river environments are pristine and fragile, thus the restriction of human on these environments critical for sustainability [29]. The reconfiguration of the environments and the land use and land cover change may have had a negative effect on the river most probably influenced by the increased numbers and concentration of privately own accommodation along the river. A similar study by Namugize and co-workers [107], also attributed the deterioration of the uMngeni river catchment in South Africa to the multifaceted relationships between land use and land cover change and water quality parameters to be site specific.

Therefore, these findings help to understand the state of the environment in the upper Crocodile River catchment and aid in decision making on the implication on such findings on water resources which are considered be one of the most critical environmental problems in South Africa. The intensification of agricultural practices along the Crocodile River has had a negative impact on the receiving water through pollution as a result of the use of chemical fertilizers for cultivation profitable and more productive crop varieties (e.g. Fruits, grains and vegetables). The toxicity of water owing to the use of pesticides and other forms of chemical fertilizers draining into water bodies has resulted in the extinction of many marine organisms including serious effect on human health of those depending on the river as source for fish and domestic use [37, 84]. Furthermore, the decline in cropland from 2009 to 2018 may have a serious implication to food security and self-sufficiency for the province. This is further compounded by the increase in population growth urbanization, tourism, and other development activities are the principal drivers of LULC change in the Crocodile River catchment.

3.4 Prospect and implications for future studies

PTEs, even in trace amount in some cases, could pose a great risk to humans, exert harmful effects on the environment and other ecological receptors, as mentioned earlier. With this increased anthropogenic activity, considering the land use and landcover change, spanning from 1999 to 2018, with the concomitant rise in PTEs observed in the study area, as one of the African water bodies, the need for continuous environmental monitoring of the safety of the river water body has emerged, of great importance. Standard techniques for detection of the PTEs such as inductively coupled plasma optical emission spectrometry (ICP-OES) [108], Uv–Vis spectrometry [109], atomic absorption/emission spectroscopy [110], laser-induced breakdown spectroscopy (LIBS) [111] and even the inductively coupled plasma mass spectroscopy (ICP-MS) employed in this study, are not generally suitable for in situ, fast, easy and low cost operations [112]. Gross setbacks like tedious sample preparation and pre-concentration, professionalism needed in personnel operation, and high cost of procuring and maintaining equipment have surrounded the use of such techniques. Such growing mandatory demand for real-time on-site tracking of water quality for human health and the environmental monitoring requires a competitively sensitive and reliable technique which is affordable and exerts less pressure on the environment.

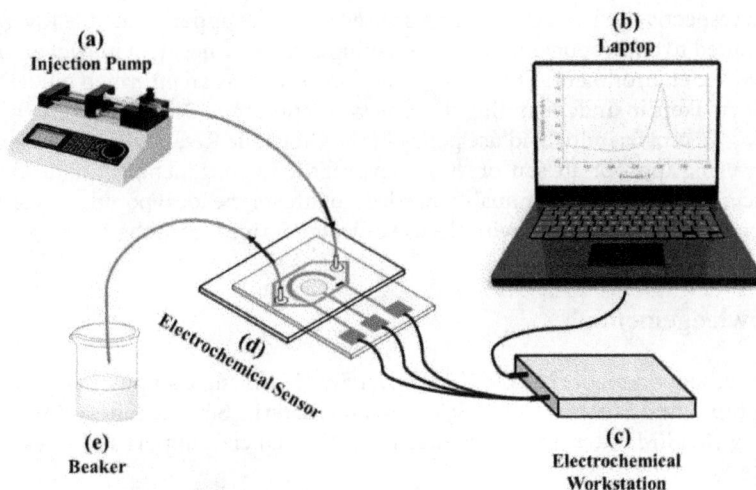

Figure 8.
Typical electrochemical set-up [114].

Hence, it is proposed that electrochemical monitoring technique could be a promising portable, low cost alternative with high selectivity and low detection limit [112]. Consequently, electrochemical sensors could be simply assembled into a compact system that is cheaper, simple to operate and possible for the desirable on-the-field application. These techniques leverage on the electro-catalytic oxidation of pre-concentrated deposited analyte on the surface of a prepared electrode. They have been engaged in extensive scope of applications such as environmental safety monitoring, control of food quality, medical diagnostics, and chemical threat detection. Some of the electrochemical methods most commonly in use nowadays include voltammetry, amperometry, impedemetry, potentiometry and conductometry [113]. Therefore, the safety assessment of African water bodies could profit immensely from the synergic integration of remote sensing and electrochemical technique in a way that is comparably affordable and efficient. **Figure 8** [114] illustrates a typical electrochemical setup.

4. Conclusion

The physicochemical parameters and PTEs contamination in the Crocodile River were analyzed to highlight the effect of the PTEs have on the river health. The results of this study revealed that the Crocodile River is contaminated with the following PTEs, (Al, Mn and Fe) as their contamination level were above the stipulated permissible guideline of DWAF of 0.005, 0.18 and 0.1 mg/L respectively. Non-point sources of metals in the river could possibly be attributed to anthropogenic activities such as agriculture, mining, resorts, and privately owned accommodation, commercial activities and the increasing population along the Crocodile River. A measure to curb metal pollution in the Crocodile River would be to avoid tannery discharge effluent into the river and farmland without prior treatment. Apart from the treatment of wastewater, effluent discharged into the Crocodile River. The different classes of land use and land cover revealed the following change patterns; bare land and built-up declined from 1999 to 2018, with a net change of −42 938 ha and − 2 663 ha respectively. Whereas, land cover category for grassland, cropland and water bodies exhibited an increase of 26 612, 17 578 and

1 411 ha respectively. The LULC changes observed in the upper Crocodile River can be attributed to anthropogenic activities having a range of negative impact on the river and the environment. This result, therefore, serves as an informed guideline for policymakers in understanding the effects of land use and land cover change in designing an eco-friendly land use policy in the Crocodile River. Electrochemical strategy using appropriate sensors has been proposed a congruent technique for periodic monitoring of water quality needed, to inform the local population of the human health risk associated with the use of water derived from the river.

Acknowledgements

The authors thank the North-West University (Mafikeng Campus), Department of Geography and Environmental Sciences, and Material Science Innovation and Modeling (MaSIM) Research Focus Area for their financial support and research facilities.

Conflict of interest

All authors declared no conflicts of interest.

Author details

Nde Samuel Che[1*], Sammy Bett[1], Enyioma Chimaijem Okpara[2,3], Peter Oluwadamilare Olagbaju[4], Omolola Esther Fayemi[2,3] and Manny Mathuthu[5]

1 Department of Geography and Environmental Sciences, North-West University, Mafikeng Campus, Mmabatho, South Africa

2 Department of Chemistry, North-West University, Mmabatho, South Africa

3 Material Science Innovation and Modelling (MaSIM) Research Focus Area, North-West University, Mmabatho, South Africa

4 Department of Physics, North-West University, Mmabatho, South Africa

5 Centre for Applied Radiation Science and Technology, North-West University, Mmabatho, South Africa

*Address all correspondence to: ndesamuelche@gmail.com

IntechOpen

References

[1] Rahmanian N, Ali SHB, Homayoonfard M, Ali N, Rehan M, Sadef Y, et al. Analysis of physico-chemical parameters to evaluate the drinking water quality in the State of Perak, Malaysia. Journal of Chemistry. 2015;2015.

[2] Ngwenya B, Thakadu O, Phaladze N, Bolaane B. Access to water and sanitation facilities in primary schools: A neglected educational crisis in Ngamiland district in Botswana. Physics and Chemistry of the Earth, Parts A/B/C. 2018;105:231-238.

[3] Moss T. The governance of land use in river basins: prospects for overcoming problems of institutional interplay with the EU Water Framework Directive. Land use policy. 2004;21(1):85-94.

[4] Peters NE, Meybeck M. Water quality degradation effects on freshwater availability: impacts of human activities. Water International. 2000;25(2):185-193.

[5] Luo Z, Zuo Q, Shao Q. A new framework for assessing river ecosystem health with consideration of human service demand. Science of the Total Environment. 2018;640:442-453.

[6] Shiferaw H, Bewket W, Alamirew T, Zeleke G, Teketay D, Bekele K, et al. Implications of land use/land cover dynamics and Prosopis invasion on ecosystem service values in Afar Region, Ethiopia. Science of the total environment. 2019;675:354-366.

[7] Zhang Y, Huang G, Lu H, He L. Planning of water resources management and pollution control for Heshui River watershed, China: a full credibility-constrained programming approach. Science of The Total Environment. 2015;524:280-289.

[8] Tekken V, Costa L, Kropp JP. Increasing pressure, declining water and climate change in north-eastern Morocco. Journal of Coastal Conservation. 2013;17(3):379-388.

[9] Cosgrove WJ, Loucks DP. Water management: Current and future challenges and research directions. Water Resources Research. 2015;51(6):4823-4839.

[10] Brack W, Dulio V, Ågerstrand M, Allan I, Altenburger R, Brinkmann M, et al. Towards the review of the European Union Water Framework management of chemical contamination in European surface water resources. Science of the Total Environment. 2017;576:720-737.

[11] Corcoran E. Sick water?: the central role of wastewater management in sustainable development: a rapid response assessment: UNEP/Earthprint; 2010.

[12] Rehman F, Rehman F. Water importance and its contamination through domestic sewage: Short review. Greener J Phys Sci. 2014;4(3):045-048.

[13] Dube T, Shoko C, Sibanda M, Baloyi MM, Molekoa M, Nkuna D, et al. Spatial modelling of groundwater quality across a land use and land cover gradient in Limpopo Province, South Africa. Physics and Chemistry of the Earth, Parts A/B/C. 2020;115:102820.

[14] Anderko L, Chalupka S. Climate Change and Health. The American journal of nursing. 2014;114(8):67-69.

[15] Abbasi S, Vinithan S. Water quality in and around an industrialized suburb of Pondicherry. Indian Journal of Environmental Health. 1999;41(4):253-263.

[16] El-Sheekh MM. Impact of water quality on ecosystems of the Nile River. The Nile River: Springer; 2016. p. 357-385.

[17] Hellmuth ME, Moorhead A, Thomas MC, Williams J. Climate risk management in Africa: Learning from practice. 2007.

[18] Bloch R. The Future of Water in African Cities: Why Waste Water? Integrating Urban Planning and Water Management in Sub-Saharan Africa, Background Report. 2012.

[19] Shrestha S, Bhatta B, Shrestha M, Shrestha PK. Integrated assessment of the climate and landuse change impact on hydrology and water quality in the Songkhram River Basin, Thailand. Science of The Total Environment. 2018;643:1610-1622.

[20] Giri S, Qiu Z. Understanding the relationship of land uses and water quality in Twenty First Century: A review. Journal of environmental management. 2016;173:41-48.

[21] Mintz E, Bartram J, Lochery P, Wegelin M. Not just a drop in the bucket: expanding access to point-of-use water treatment systems. American journal of public health. 2001;91(10):1565-1570.

[22] Zeinu KM, Hou H, Liu B, Yuan X, Huang L, Zhu X, et al. A novel hollow sphere bismuth oxide doped mesoporous carbon nanocomposite material derived from sustainable biomass for picomolar electrochemical detection of lead and cadmium. Journal of Materials Chemistry A. 2016;4(36):13967-13979.

[23] Yeboah-Assiamah E. Involvement of private actors in the provision of urban sanitation services; potential challenges and precautions. Management of Environmental Quality: An International Journal. 2015.

[24] Dos Santos S, Adams E, Neville G, Wada Y, De Sherbinin A, Bernhardt EM, et al. Urban growth and water access in sub-Saharan Africa: Progress, challenges, and emerging research directions. Science of the Total Environment. 2017;607:497-508.

[25] Organization WH. Health in 2015: from MDGs, millennium development goals to SDGs, sustainable development goals. 2015.

[26] Hopewell MR, Graham JP. Trends in access to water supply and sanitation in 31 major sub-Saharan African cities: an analysis of DHS data from 2000 to 2012. BMC public health. 2014;14(1):208.

[27] Hickling S. Status of sanitation and hygiene in Africa. Sanitation and Hygiene in Africa: Where do We Stand? 2013;36:11.

[28] Ding J, Jiang Y, Liu Q, Hou Z, Liao J, Fu L, et al. Influences of the land use pattern on water quality in low-order streams of the Dongjiang River basin, China: A multi-scale analysis. Science of The Total Environment. 2016;551-552:205-16.

[29] Peter A, Mujuru M, Dube T. An assessment of land cover changes in a protected nature reserve and possible implications on water resources, South Africa. Physics and Chemistry of the Earth, Parts A/B/C. 2018;107:86-91.

[30] Gyamfi C, Ndambuki JM, Salim RW. Hydrological responses to land use/cover changes in the Olifants Basin, South Africa. Water. 2016;8(12):588.

[31] Wijesiri B, Deilami K, Goonetilleke A. Evaluating the relationship between temporal changes in land use and resulting water quality. Environmental Pollution. 2018;234:480-486.

[32] Tong ST, Chen W. Modeling the relationship between land use and surface water quality. Journal of environmental management. 2002;66(4):377-393.

[33] Yang J, Ma S, Zhou J, Song Y, Li F. Heavy metal contamination in soils and vegetables and health risk assessment of inhabitants in Daye, China. Journal of International Medical Research. 2018;46(8):3374-3387.

[34] Camara M, Jamil NR, Abdullah AFB. Impact of land uses on water quality in Malaysia: a review. Ecological Processes. 2019;8(1):10.

[35] Hua AK. Land use land cover changes in detection of water quality: A study based on remote sensing and multivariate statistics. Journal of environmental and public health. 2017;2017.

[36] Pujol L, Evrard D, Groenen-Serrano K, Freyssinier M, Ruffien-Cizsak A, Gros P. Electrochemical sensors and devices for heavy metals assay in water: the French groups' contribution. Frontiers in chemistry. 2014;2:19.

[37] Nde SC, Mathuthu M. Assessment of Potentially Toxic Elements as Non-Point Sources of Contamination in the Upper Crocodile Catchment Area, North-West Province, South Africa. International journal of environmental research and public health. 2018; 15(4):576.

[38] Antoniadis V, Shaheen SM, Levizou E, Shahid M, Niazi NK, Vithanage M, et al. A critical prospective analysis of the potential toxicity of trace element regulation limits in soils worldwide: Are they protective concerning health risk assessment?-A review. Environment international. 2019;127:819-847.

[39] Olaniran AO, Balgobind A, Pillay B. Bioavailability of heavy metals in soil: impact on microbial biodegradation of organic compounds and possible improvement strategies. International journal of molecular sciences. 2013;14(5):10197-10228.

[40] Cipullo S, Prpich G, Campo P, Coulon F. Assessing bioavailability of complex chemical mixtures in contaminated soils: Progress made and research needs. Science of the Total Environment. 2018;615:708-723.

[41] Oyekunle ASAJA, Suliat O. Speciation Study of the Heavy Metals in Commercially. Environ Monit Assess. 2011;169:597-606.

[42] Devi SS, Sethu M, Priya PG. Effect of Artemia franciscana on the Removal of Nickel by Bioaccumulation. Biocontrol science. 2014;19(2):79-84.

[43] Tortora F, Innocenzi V, Prisciandaro M, Vegliò F, Di Celso GM. Heavy metal removal from liquid wastes by using micellar-enhanced ultrafiltration. Water, Air, & Soil Pollution. 2016;227(7):240.

[44] Borba C, Guirardello R, Silva E, Veit M, Tavares C. Removal of nickel (II) ions from aqueous solution by biosorption in a fixed bed column: experimental and theoretical breakthrough curves. Biochemical Engineering Journal. 2006;30(2): 184-191.

[45] Ihedioha J, Okoye C. Levels of some trace metals (Zn, Cr, and Ni) in the muscle and internal organs of cattle in Nigeria. Human and Ecological Risk Assessment: An International Journal. 2013;19(4):989-998.

[46] Oyaro N, Ogendi J, Murago EN, Gitonga E. The contents of Pb, Cu, Zn and Cd in meat in nairobi, Kenya. 2007.

[47] Al Moharbi SS, Devi MG, Sangeetha B, Jahan S. Studies on the removal of copper ions from industrial effluent by Azadirachta indica powder. Applied Water Science. 2020;10(1):23.

[48] Zhang X, Yang L, Li Y, Li H, Wang W, Ye B. Impacts of lead/zinc mining and smelting on the

environment and human health in China. Environmental monitoring and assessment. 2012;184(4):2261-2273.

[49] Naseem R, Tahir S. Removal of Pb (II) from aqueous/acidic solutions by using bentonite as an adsorbent. Water Research. 2001;35(16):3982-3986.

[50] Fu F, Wang Q. Removal of heavy metal ions from wastewaters: a review. Journal of environmental management. 2011;92(3):407-418.

[51] Chedrese PJ, Piasek M, Henson MC. Cadmium as an endocrine disruptor in the reproductive system. Immunology, Endocrine & Metabolic Agents in Medicinal Chemistry (Formerly Current Medicinal Chemistry-Immunology, Endocrine and Metabolic Agents). 2006;6(1):27-35.

[52] de Angelis C, Galdiero M, Pivonello C, Salzano C, Gianfrilli D, Piscitelli P, et al. The environment and male reproduction: The effect of cadmium exposure on reproductive function and its implication in fertility. Reproductive Toxicology. 2017;73:105-127.

[53] Żukowska J, Biziuk M. Methodological evaluation of method for dietary heavy metal intake. Journal of food science. 2008;73(2):R21-RR9.

[54] Gao X, Schulze DG. Chemical and mineralogical characterization of arsenic, lead, chromium, and cadmium in a metal-contaminated Histosol. Geoderma. 2010;156(3-4):278-286.

[55] Royer MD, Smith LA. Contaminants and remedial options at selected metals contaminated sites-a technical resource document. Citeseer; 1995.

[56] Sparks DL. Environmental soil chemistry: Elsevier; 2003.

[57] Yabe J, Ishizuka M, Umemura T. Current levels of heavy metal pollution in Africa. Journal of Veterinary Medical Science. 2010;72(10):1257-1263.

[58] Mohod CV, Dhote JJIJoIRiS, Engineering, Technology. Review of heavy metals in drinking water and their effect on human health. 2013;2(7):2992-2996.

[59] Olade M. Heavy Metal Pollution and the Need for Monitoring: Illustratedfor Developing Countries in West Africa. 1987.

[60] Jackson VA, Paulse A, Odendaal JP, Khan WJW, Air,, Pollution S. Identification of point sources of metal pollution in the Berg River, Western Cape, South Africa. 2013;224(3):1477.

[61] Okoro HK, Fatoki OS, Adekola FA, Ximba BJ, Snyman RG. A review of sequential extraction procedures for heavy metals speciation in soil and sediments. 2012.

[62] Edokpayi JN, Odiyo JO, Olasoji SOJIJNSR. Assessment of heavy metal contamination of Dzindi river, in Limpopo Province, South Africa. 2014;2(10):185-194.

[63] Lalah J, Ochieng E, Wandiga S. Sources of heavy metal input into Winam Gulf, Kenya. Bulletin of Environmental Contamination and Toxicology. 2008;81(3):277-284.

[64] Nthunya LN, Masheane ML, Malinga SP, Nxumalo EN, Mamba BB, Mhlanga SD. Determination of toxic metals in drinking water sources in the Chief Albert Luthuli Local Municipality in Mpumalanga, South Africa. Physics and Chemistry of the Earth, Parts A/B/C. 2017;100:94-100.

[65] Reza R, Singh GJIJoES, Technology. Heavy metal contamination and its indexing approach for river water. 2010;7(4):785-792.

[66] Caruso B, Cox T, Runkel RL, Velleux M, Bencala KE, Nordstrom DK, et al. Metals fate and transport modelling in streams and watersheds: state of the science and USEPA workshop review. 2008.

[67] Mohanty J, Misra S, Nayak B. Sequential leaching of trace elements in coal: A case study from Talcher coalfield, Orissa. JOURNAL-GEOLOGICAL SOCIETY OF INDIA. 2001;58(5):441-448.

[68] Cravotta III CA. Dissolved metals and associated constituents in abandoned coal-mine discharges, Pennsylvania, USA. Part 1: Constituent quantities and correlations. Applied Geochemistry. 2008;23(2):166-202.

[69] SHAHTAHERI S, Abdollahi M, Golbabaei F, RAHIMI FA, Ghamari F. Monitoring of mandelic acid as a biomarker of environmental and occupational exposures to styrene. 2008.

[70] Rim-Rukeh A, Ikhifa OG, Okokoyo A. Effects of agricultural activities on the water quality of Orogodo River, Agbor Nigeria. Journal of applied sciences research. 2006;2(5):256-259.

[71] Khadse G, Patni P, Kelkar P, Devotta S. Qualitative evaluation of Kanhan river and its tributaries flowing over central Indian plateau. Environmental monitoring and assessment. 2008;147(1-3):83-92.

[72] Juang D, Lee C, Hsueh S. Chlorinated volatile organic compounds found near the water surface of heavily polluted rivers. International Journal of Environmental Science & Technology. 2009;6(4):545-556.

[73] Venugopal T, Giridharan L, Jayaprakash M. Characterization and risk assessment studies of bed sediments of River Adyar-An application of speciation study. 2009.

[74] Sekabira K, Origa HO, Basamba T, Mutumba G, Kakudidi E. Assessment of heavy metal pollution in the urban stream sediments and its tributaries. International journal of environmental science & technology. 2010;7(3):435-446.

[75] Masindi V, Muedi KL. Environmental contamination by heavy metals. Heavy metals. 2018;10:115-132.

[76] Ali H, Khan E, Ilahi I. Environmental chemistry and ecotoxicology of hazardous heavy metals: environmental persistence, toxicity, and bioaccumulation. Journal of chemistry. 2019;2019.

[77] Kinge CW, Mbewe M. Bacterial contamination levels in river catchments of the North West Province, South Africa: Public health implications. African Journal of Microbiology Research. 2012;6(7):1370-1375.

[78] Rimayi C, Odusanya D, Weiss JM, de Boer J, Chimuka L. Contaminants of emerging concern in the Hartbeespoort Dam catchment and the uMngeni River estuary 2016 pollution incident, South Africa. Science of the Total Environment. 2018;627:1008-1017.

[79] Mbiza NX. Investigation of the effectiveness of techniques deployed in controlling cyanobacterial growth in Rietvlei Dam, Roodeplaat Dam and Hartbeespoort Dam in Crocodile (West) and Marico Water Management Area 2014.

[80] Benabdelkader A, Taleb A, Probst J-L, Belaidi N, Probst A. Anthropogenic contribution and influencing factors on metal features in fluvial sediments from a semi-arid Mediterranean river basin (Tafna River, Algeria): A multi-indices approach. Science of The Total Environment. 2018;626:899-914.

[81] Pavlović P, Marković M, Kostić O, Sakan S, Đorđević D, Perović V, et al.

Evaluation of potentially toxic element contamination in the riparian zone of the River Sava. Catena. 2019;174:399-412.

[82] Vareda JP, Valente AJ, Durães L. Assessment of heavy metal pollution from anthropogenic activities and remediation strategies: A review. Journal of environmental management. 2019;246:101-118.

[83] Chetty S, Pillay L. Assessing the influence of human activities on river health: a case for two South African rivers with differing pollutant sources. Environmental Monitoring and Assessment. 2019;191(3):168.

[84] Alam A, Bhat MS, Maheen M. Using Landsat satellite data for assessing the land use and land cover change in Kashmir valley. GeoJournal. 2019:1-15.

[85] Hussain M, Chen D, Cheng A, Wei H, Stanley D. Change detection from remotely sensed images: From pixel-based to object-based approaches. ISPRS Journal of photogrammetry and remote sensing. 2013;80:91-106.

[86] Dewan AM, Yamaguchi Y. Using remote sensing and GIS to detect and monitor land use and land cover change in Dhaka Metropolitan of Bangladesh during 1960-2005. Environmental monitoring and assessment. 2009;150(1-4):237.

[87] Witharana C, Civco DL. Optimizing multi-resolution segmentation scale using empirical methods: exploring the sensitivity of the supervised discrepancy measure Euclidean distance 2 (ED2). ISPRS Journal of Photogrammetry and Remote Sensing. 2014;87:108-121.

[88] Benz UC, Hofmann P, Willhauck G, Lingenfelder I, Heynen M. Multi-resolution, object-oriented fuzzy analysis of remote sensing data for GIS-ready information. ISPRS Journal

of photogrammetry and remote sensing. 2004;58(3-4):239-258.

[89] Marara T, Palamuleni L. A spatiotemporal analysis of water quality characteristics in the Klip river catchment, South Africa. Environmental Monitoring and Assessment. 2020;192(9):1-28.

[90] Paerl HW. Mitigating toxic planktonic cyanobacterial blooms in aquatic ecosystems facing increasing anthropogenic and climatic pressures. Toxins. 2018;10(2):76.

[91] du Preez GC, Wepener V, Fourie H, Daneel MS. Irrigation water quality and the threat it poses to crop production: evaluating the status of the Crocodile (West) and Marico catchments, South Africa. Environmental Monitoring and Assessment. 2018;190(3):127.

[92] Ogoyi DO, Mwita C, Nguu EK, Shiundu PM. Determination of heavy metal content in water, sediment and microalgae from Lake Victoria, East Africa. 2011.

[93] Schwartz JDM. The functional role of fish diversity in Lake Victoria, East Africa: Boston University; 2002.

[94] Cobbina S, Myilla M, Michael KJIJSTR. Small scale gold mining and heavy metal pollution: Assessment of drinking water sources in Datuku in the Talensi-Nabdam District. 2013;2(1).

[95] Witte F, Goldschmidt T, Wanink J, van Oijen M, Goudswaard K, Witte-Maas E, et al. The destruction of an endemic species flock: quantitative data on the decline of the haplochromine cichlids of Lake Victoria. Environmental biology of fishes. 1992;34(1):1-28.

[96] Mambo M, Jonathan OO, Nana AM, editors. HRP biosensor based on carbonized maize tassel-MWNTs

modified electrode for the detection of divalent trace metal ions. SENSORS, 2013 IEEE; 2013: IEEE.

[97] Lapworth D, Nkhuwa D, Okotto-Okotto J, Pedley S, Stuart M, Tijani M, et al. Urban groundwater quality in sub-Saharan Africa: current status and implications for water security and public health. Hydrogeology Journal. 2017;25(4):1093-1116.

[98] Orata F, Birgen F. Fish tissue bio-concentration and interspecies uptake of heavy metals from waste water lagoons. Journal of Pollution Effects & Control. 2016;4(2):157.

[99] Oluyemi E, Adekunle A, Adenuga A, Makinde W. Physico-chemical properties and heavy metal content of water sources in Ife North Local Government Area of Osun State, Nigeria. African Journal of Environmental Science and Technology. 2010;4(10):691-697.

[100] Edokpayi JN, Odiyo JO, Popoola OE, Msagati TA. Assessment of trace metals contamination of surface water and sediment: a case study of Mvudi River, South Africa. Sustainability. 2016;8(2):135.

[101] Okonkwo JO, Mothiba M. Physico-chemical characteristics and pollution levels of heavy metals in the rivers in Thohoyandou, South Africa. Journal of Hydrology. 2005;308(1-4):122-127.

[102] Somerset V, Van der Horst C, Silwana B, Walters C, Iwuoha E. Biomonitoring and Evaluation of Metal Concentrations in Sediment and Crab Samples from the North-West Province of South Africa. Water, Air, & Soil Pollution. 2015;226(3):43.

[103] Bouaroudj S, Menad A, Bounamous A, Ali-Khodja H, Gherib A, Weigel DE, et al. Assessment of water quality at the largest dam in Algeria (Beni Haroun Dam) and effects of irrigation on soil characteristics of agricultural lands. Chemosphere. 2019;219:76-88.

[104] Wongsasuluk P, Chotpantarat S, Siriwong W, Robson M. Heavy metal contamination and human health risk assessment in drinking water from shallow groundwater wells in an agricultural area in Ubon Ratchathani province, Thailand. Environmental geochemistry and health. 2014;36(1):169-182.

[105] Ahmad JU, Goni MA. Heavy metal contamination in water, soil, and vegetables of the industrial areas in Dhaka, Bangladesh. Environmental Monitoring and Assessment. 2010;166(1):347-357.

[106] Li S, Zhang Q. Risk assessment and seasonal variations of dissolved trace elements and heavy metals in the Upper Han River, China. Journal of Hazardous Materials. 2010;181(1-3):1051-1058.

[107] Namugize JN, Jewitt G, Graham M. Effects of land use and land cover changes on water quality in the uMngeni river catchment, South Africa. Physics and Chemistry of the Earth, Parts A/B/C. 2018;105:247-264.

[108] Schunk PFT, Kalil IC, Pimentel-Schmitt EF, Lenz D, de Andrade TU, Ribeiro JS, et al. ICP-OES and micronucleus test to evaluate heavy metal contamination in commercially available Brazilian herbal teas. Biological trace element research. 2016;172(1):258-265.

[109] Mehder A, Habibullah Y, Gondal M, Baig U. Qualitative and quantitative spectro-chemical analysis of dates using UV-pulsed laser induced breakdown spectroscopy and inductively coupled plasma mass spectrometry. Talanta. 2016;155:124-132.

[110] Yu J, Yang S, Sun D, Lu Q, Zheng J, Zhang X, et al. Simultaneously determination of multi metal elements in water samples by liquid cathode glow discharge-atomic emission spectrometry. Microchemical Journal. 2016;128:325-330.

[111] dos Santos Augusto A, Batista ÉF, Pereira-Filho ER. Direct chemical inspection of eye shadow and lipstick solid samples using laser-induced breakdown spectroscopy (LIBS) and chemometrics: proposition of classification models. Analytical Methods. 2016;8(29):5851-5860.

[112] Hou H, Zeinu KM, Gao S, Liu B, Yang J, Hu J. Recent advances and perspective on design and synthesis of electrode materials for electrochemical sensing of heavy metals. Energy & Environmental Materials. 2018;1(3):113-131.

[113] Ahammad A, Lee J-J, Rahman M. Electrochemical sensors based on carbon nanotubes. sensors. 2009;9(4):2289-2319.

[114] Huang A, Li H, Xu D. An on-chip electrochemical sensor by integrating ITO three-electrode with low-volume cell for on-line determination of trace Hg (II). Journal of Electroanalytical Chemistry. 2019;848:113189.

Chapter 3

Wave-Forced Dynamics at Microtidal River Mouths

Maurizio Brocchini, Matteo Postacchini, Lorenzo Melito,
Eleonora Perugini, Andrew J. Manning, Joseph P. Smith
and Joseph Calantoni

Abstract

Microtidal river mouths are dynamic environments that evolve as a consequence of many forcing actions. Under the hydrodynamic viewpoint, river currents, sea waves and tides strongly interact, and their interplay determines specific sediment transport and morphological patterns. Beyond literature evidence, information comes from field observations made at the Misa River study site, a microtidal river along the Adriatic Sea (Italy), object of a long-going monitoring. The river runs for 48 km in a watershed of 383 km^2, providing a discharge of about 400 m^3/s for return periods of 100 years. The overall hydrodynamics, sediment transport and morphological evolution at the estuary are analyzed with particular attention to specific issues like: the generation of vortical flows at the river mouth, the influence of various wave modes (infragravity to tidal) propagating upriver, the role of sediment flocculation, the generation and evolution of bed features (river-mouth bars and longitudinal nearshore bars). Numerical simulations are also used to clarify specific mechanisms of interest.

Keywords: estuarine dynamics, river-sea interaction, river current, tide, infragravity waves, sandbars, microtidal

1. Introduction

Estuaries are dynamic and complex environments. Estuarine hydrodynamics are the product of nonlinear interactions between freshwater flow, tidal, wave, and wind forcing, and bathymetric and topographic changes [1, 2]. Such processes directly affect sediment transport and the morphodynamics of the estuary, which may lead to the formation of complex morphological patterns like river mouth bars [3] and submerged sandbars [4].

Although the influence of classical riverine and marine hydrodynamic forcing mechanisms, such as gravity waves and tides, are typically accounted for to describe the estuarine dynamics, an increasing number of studies is focusing on infragravity (IG hereafter) waves, which are seen to play a non-negligible role in estuary evolution [5]. IG waves are a specific type of low-frequency waves with periods between 20 and 30 seconds to 5 minutes, larger than those of sea/swell waves [6, 7]. Typically, IG waves are generated either (i) as long waves bound to short wave groups, or (ii) by a temporal variation of the breakpoint, or (iii) from swell–swell interactions [8–10].

IG waves are regarded as an important trigger of sandbar generation in the coastal area, especially in the short term, although their role has not been properly understood so far [11, 12]. IG waves are also thought to be of some importance for sediment transport and nearshore morphodynamics, though not a primary forcing [13]. For instance, during storms, breakers mobilize a large quantity of sediment, whose transport is modulated by IG waves [14].

In general, subtidal bars generating and evolving along sandy coasts are typical of wave-dominated environments as a whole and are the result of complex hydro-dynamic and sediment transport patterns taking place within the surf zone [11, 15]. Sandbars exist both close to and far from river mouths, due to their strong link to the marine forcing, although their behavior is largely affected by river current and sediment transport, as well as by artificial structures existing at the mouth [4, 16, 17]. Since sandbars promote wave breaking and energy dissipation, they represent a natural solution for beach protection and are fundamental for coastal stability in both short term (storm scale) and long term (scale of years/decades) timescales [18, 19]. Several studies focused on the definition of evolutionary patterns for sand bars. It is commonly acknowledged that waves break over the bars during sea storms and generate undertow profiles, which lead to sediment being advected seaward and sandbars migrating offshore. Conversely, although with some notable exceptions [20], onshore bar motion is typically observed under non-breaking waves and relatively mild wave climate, like that occurring in summertime [12, 21–23].

Evidence shows that IG waves can easily enter a river mouth and propagate upriver for long distances, even during fast river flow conditions. The presence of IG waves in the estuarine regions has been put in relation to edge waves entering the estuary and producing resonance, while IG wave modulation by tides has been observed along the Pescadero River (North California), Ría de Santiuste (Spain), the Albufeira Lagoon (Portugal) and the Misa River (Italy) [1, 24–26].

River mouth bars are morphological features that generate and evolve due to the direct effect of riverine currents and marine actions (waves, tides). Typically, the sediment transported via river flow deposits out of the mouth, at a distance of about twice the river width, due to flow expansion [27–29]. Nevertheless, the location where the bar is formed depends on the amount of sediment transport induced by the river discharge and the net residuals of marine actions. Therefore, in a wave-dominated environment, sediment deposits can also be present inside the river mouth and lead to mouth bars. Such bars are extremely relevant for the overall estuarine dynamics, potentially obstructing the river cross-section to the point of contributing to flooding or river overflows. Downriver migration of the bar occurs when the river discharge dominates the estuarine dynamics and the increased flow velocity on the bar crest is able to erode the sediment on the top, which then deposits seaward because of flow deceleration [30]. On the other hand, weak river flows associated with comparatively more intense sea actions induce sediment accumulation and upriver migration of the bar.

Due to higher concentration values with respect to both upriver and river-mouth regions, a "sediment trap" can be generated in the upper estuary due to both suspended matter and residual circulation. A clay content of >5–10% can make fine-grained sediments behave in a cohesive manner [31, 32]. The sediment grains, and often "sticky" natural organic matter that is present, cause sediment particles to cease acting independently. The behavior is typically characterized by flocculation (i.e., individual sediment particles interact and "bond"), a process whereby cohesive and fine-grained mixed sediment particles have the potential to aggregate into flocs [33, 34], which contribute to the formation of near-bed suspension layers in estuaries and may alter water column turbulent mixing, rheology, and sedimentary particle residence (retention) times.

Experiments have shown that when the fine fraction and the larger non-cohesive sediments coexist as a single mixture in an estuary [31], there exists the potential for the two fractions to combine and exhibit some degree of interactive flocculation [35, 36]. Further, cohesive sediments that are mixed into a predominately cohesionless sandy region can create a 'cage-like' structure that can fully encompass the sand grains, thereby trapping the sand within a clay floc envelope [37]. The degree of cohesion between the various sediment fractions tends to increase with the content of fine clay minerals within the sediment, especially for clay contents larger than 5–10% [37–40], with the biological activity playing an important role in the flocculation of mixed suspensions [41]. In terms of properties, the floc size ranges from microns to centimeters and their effective density generally decreases with size [42–44], while their settling velocity follows a size-dependent Stokes law settling relationship [45, 46]. Due to their fast settling velocities [47, 48], macroflocs tend to have the most influence on the mass settling flux [49].

To investigate the interplay between the characteristic processes of microtidal estuarine environments and their effects in the nearby river reach and coastal area, field observations have been performed along the Misa River (MR hereafter) and its estuary, with dedicated field campaigns [50, 51] and long-term measurements collected using both on-site instrumentation and remote sensors [4, 26, 52]. The present work illustrates a comprehensive overview of recent observations and analyses performed to achieve a better understanding of estuarine dynamics at a microtidal river mouth, from the hydrological, morphodynamic, and sedimentological viewpoints.

The chapter is divided as follows. Section 2 describes the field campaigns and the instrumentation deployed along the MR and at the estuary, while the main results are reported in Section 3. An overall discussion, followed by some concluding remarks, is presented in Section 4.

2. Materials and methods

Both short-term and long-term monitoring was carried out at the MR estuary. Specifically, two field campaigns were carried out in September 2013 and January 2014 in the most downstream part of the MR and within the nearby coastal area [50, 51]. Additionally, long-term measurements are being collected since 2015, with a set of instrumentation installed in river and sea, as well as using remote sensors. Numerical simulations are also used to support both experimental findings and speculations, and to better understand the complex dynamics at the MR estuary.

2.1 Study site

The MR runs from the Apennine Mountains (central Italy) to Senigallia (Marche Region), where it flows into the Middle Adriatic Sea after about 48 km. The MR is characterized by a 383 km^2 watershed and its flow rate is around 400, 450, and 600 m^3/s for return periods of 100, 200, and 500 years, respectively. While a low-flow regime is promoted by a relatively small amount of precipitation in the MR watershed during the summer, significantly higher flow regime characterizes the wintertime [50, 51].

An illustration of the coastal and offshore area is provided in **Figure 1a**, with a close-up view of the nearshore and estuarine regions shown in **Figure 1b**. The most downstream portion of the MR is also shown in **Figure 1c**. The final reach features a heavily engineered river mouth, characterized by cement walls in place of classical riverbanks, which allows one to easily collect and analyze the hydro-morphodynamic data of the microtidal environment. The beach located north of the estuary and harbor is protected by emerged rubble-mound breakwaters, while the southern natural beach is protected by multiple alignments of longshore sandbars (**Figure 1b**).

From a hydrological viewpoint, the MR estuary is classified as a salt-wedge estuary, where the water column is split into two parts: the outgoing river current flows in the upper portion, while the entering seawater flows in the lower portion [53]. The salt-wedge behavior, i.e. the simultaneous existence of saltwater and freshwater within the water column extends some kilometers seaward, as confirmed by salinity measurements collected up to 2 km from the estuary [50].

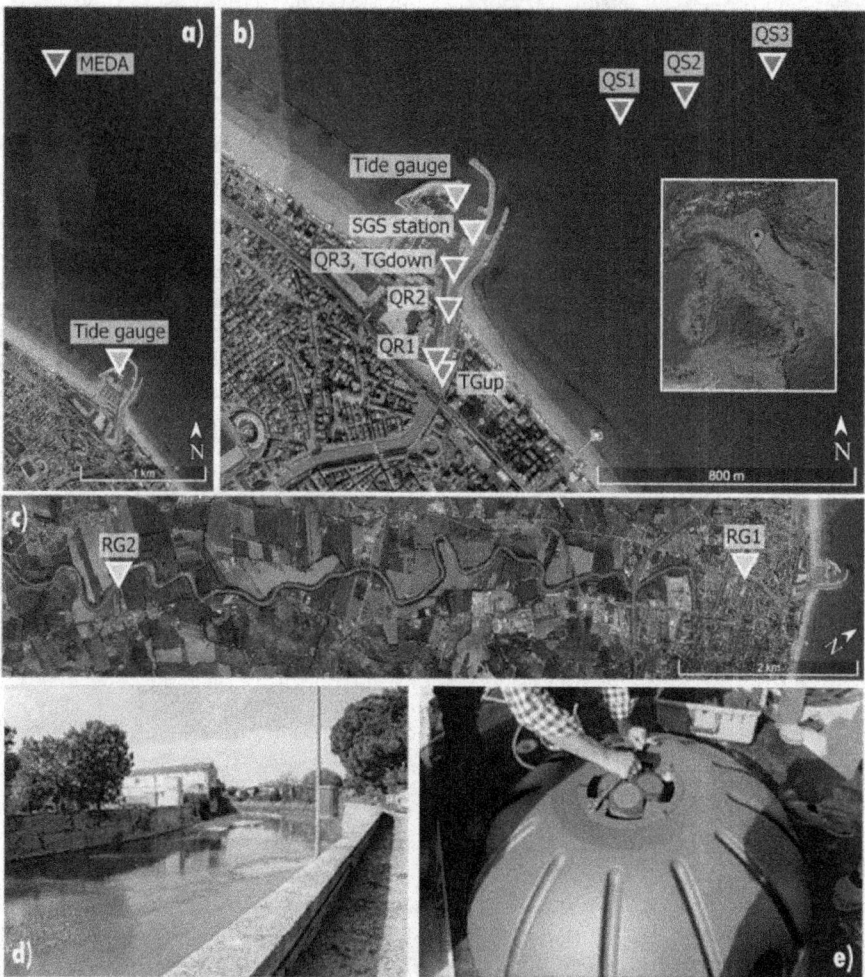

Figure 1.
Instrumentation employed for the MR monitoring: a) locations of the ADCP and tide gauge in the coastal region; b) locations of sensors displaced within the final reach of the river and the offshore area; c) locations of the two river gauges. Photos illustrating: d) the river portion near the RG1 instrument; e) the ADCP deployed at the MEDA station.

Soil samples were collected within the lower MR reach and confirmed the presence of clay, characterized by high percentages of montmorillonite minerals. Such fine-grained clay sediments promote formation of flocs, i.e. aggregations of individual clay and other fine particles, organic matter, micro-algae and bacteria, and other reactive constituents. Based on both salinity measurements and the sampled clayey material at the riverbed, it was speculated the existence of an upriver flocculation zone (between 2 km and 700 m from the mouth), with salinity S < 10 psu, and a downstream deposition zone (between 700 m from the mouth and the mouth itself), with S > 10 psu [50, 51].

The MR is characterized by large sediment transport rates despite the moderate flow rate, similar to many rivers originating within the Apennine Mountains. Such intense sediment outflow supplies a large amount of material to the coastal areas around the estuary, this being true especially for the natural beach located south of the estuary, which presents an array of longshore sandbars with a long-term morphological evolution [4, 16].

2.2 Short-term and long-term measurements

Within the framework of two international projects funded by the Office of Naval Research Global (UK), named EsCoSed ("EStuarine COhesive SEDiments") and MORSE ("Modeling and Observation of River-Sea Exchanges at a microtidal estuary"), a series of activities have been planned in the last decade within the microtidal estuary of the MR.

2.2.1 The EsCoSed project

During the EsCoSed project, two field campaigns were carried out in September 2013 and January 2014 with the purpose of investigating, respectively, the summertime regime, mainly characterized by low-flow conditions, and the wintertime regime, where low-flow conditions alternate to high-flow conditions [50, 51]. Observations of meteorology, hydrodynamics and morphodynamics were performed using instrumentation deployed for some days during both investigated periods. Furthermore, in the wintertime experiment, water and sediment were sampled within the river and estuary, as well as in the sediment plume generated during high-flow/stormy conditions.

The hydro-morphodynamic parameters were recorded using small quadpods (an overall height of ~1 m and a roughly square base of ~1 m^2) both in the final MR reach and in the nearby sea, up to 7-m depth. The quadpods were specifically devised to accommodate a set of instruments for the measurement of water velocity profiles along the water column (six velocity profilers), seabed variations (two pencil beams), the turbidity along the lower portion of the water column (two CT probes), and wave characteristics (one Sentinel 1200KHz Acoustic Doppler Current Profiler - ADCP, from Teledyne RDI®, deployed at a depth of ~7 m). The quadpod locations in the 2013–2014 experiments are shown in **Figure 1b** as red and blue triangles, respectively indicating quadpods deployed in the river (QR1, QR2, QR3) and sea (QS1, QS2, QS3). Additionally, information on both water surface level at further locations and surface current were collected using, respectively, two tide gauges (TGup, TGdown) and Lagrangian drifters launched into the final reach of the MR.

Finally, a video-monitoring station named "Sena Gallica Speculator" (SGS) was installed at the Senigallia harbor in 2015 and is currently operating. The station is composed of four cameras located on top of a tower and is oriented to encompass the mouth of the MR and a coastal area located between the MR estuary and a pier, called "Rotonda a Mare" and located 500 m south of the estuary [4].

2.2.2 The MORSE project

The MORSE project aimed at providing a long-term monitoring of the MR estuary and adjacent portions or river and sea, through the deployment of onsite instrumentation in 2018 [52, 54].

First, a Workhorse Sentinel 300 kHz ADCP (Teledyne RDI®) for the measurements of offshore wave conditions (**Figure 1e**) was installed within the MEDA station, a meteo-marine station located ~1.5 nm north of Senigallia and 1.5 nm from the coast (purple triangle in **Figure 1a**). The station is property of the Italian National Research Council (CNR) and is additionally equipped with several sensors for the monitoring of both atmosphere and sea water.

A tide gauge was deployed within a protected area at the entrance of the Senigallia harbor to record tidal excursions and storm surges (green triangle in **Figure 1a**).

Finally, a stream gauge was installed within the MR for flow-rate measurements, which is fundamental for the understanding of river-sea interactions. The chosen site is located about 1.2 km upriver of the mouth, near the bridge known as "Ponte Garibaldi", and very close to a pre-existent hydrometer, property of the Civil Protection (Marche Region), installed for the measurement of the river stages. Both the hydrometer and the stream gauge, the location of which is collectively indicated here as RG1 (**Figure 1c**), are employed to observe upriver wave propagation from the MR mouth. The RG1 hydrometer complements another hydrometer located at "Bettolelle" (here referred as RG2), about 10 km from the estuary (**Figure 1c**).

The data collected by some of the above-mentioned instruments are represented in **Figure 2**. These can be profitably used as initial or boundary conditions for the initialization and validation of the numerical modeling of the MR estuary, as illustrated in the following.

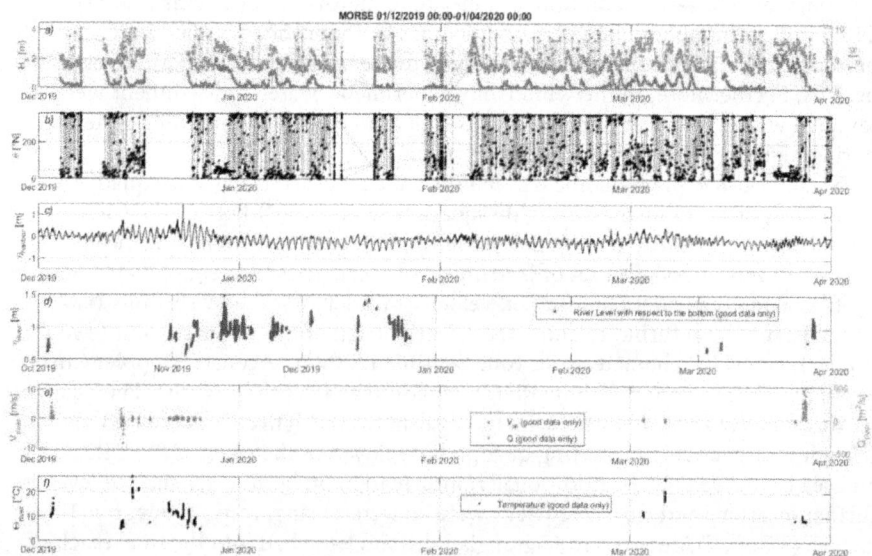

Figure 2.
Overall view of the data collected from December 2019 to April 2020. (a) Significant wave height and (b) peak period measured by the offshore ADCP at the MEDA station. (c) Water-surface level at Senigallia harbor. (d) Water-surface level, (e) mean speed and discharge, (f) temperature at the RG1 stream gauge.

2.3 Numerical modeling

The large amount of observed data allowed us to set up a numerical model of the MR estuarine area using the Delft3D software suite [55, 56]. A two-dimensional, depth-averaged model was used to reproduce the hydro-morphodynamics occurring along the final stretch of the MR estuarine channel [54]. Specifically, a coupled WAVE-FLOW simulation was performed to investigate the deformation and displacement of the river mouth bar under the actions of river current and waves. The WAVE model was forced using time series of wave height, period and direction recorded at the offshore MEDA station. The time series of the water level recorded at the tide gauge was instead used for a first validation of the model.

3. Results

Observations and results coming from both short-term and long-term recordings are reported in the following. The local dynamics observed during the 2013–2014 experiments are first described, with focus on the main seasonal differences in the hydrodynamics and the interplay between river and marine forcing actions in winter, as well as the saltwater-freshwater interaction across the water column (Section 3.1). Insight from a flocculation model supports the observed local dynamics (Section 3.2). The wave entrance in the MR estuary and the upriver propagation of long-wave modes are detailed in Section 3.3. The morphological processes occurring within the lower river reach (Section 3.4) and in the nearshore area south of the MR estuary (Section 3.5) are finally described, with special focus on the bar evolution.

3.1 Local dynamics

The estuarine area of the MR is subjected to sea storms mainly coming from two directions, as typically observed in coastal regions of the Middle Adriatic Sea. Short, steeper waves generated by WNW, N or NNE (Bora) winds typically enter the MR mouth, since their incoming direction is almost perfectly aligned to the estuary orientation. Also relevant in the MR environment are the ESE-approaching waves, induced by Sirocco winds. Such waves are significantly angled and cannot easily enter the MR. Consequently, Sirocco waves are reflected by the river walls and strongly affect the morphology around the estuary, thus impacting on the evolution of nearby sandbars [4].

Significant differences exist between summertime and wintertime conditions in terms of wind and rainfall, these directly affecting the wave action and the river current interplaying at the MR estuary. Specifically, the wind blowing during the investigated periods promoted the generation of waves of different heights, which mainly depended on the wind direction rather than on its velocity. The wind direction was frequently changing during the summertime experiment, whereas two intense storms were observed in wintertime and were characterized by almost constant wind directions.

During mild/quiescent conditions in both summer and winter (i.e. time intervals before and after storms), an upriver flow propagation was observed very close to the MR estuary (at QR3, i.e. 290 m upriver), with farther inland locations (QR2, i.e. around 400 m upriver) presenting a significant tidal modulation of the water column. An enhanced salt-wedge behavior was also observed during the wintertime experiment [57]. However, some differences arise between quiescent conditions occurring in summer and winter. Specifically, the marine action in summertime was comparable to the river forcing, as confirmed by the observed tracks of surface drifters deployed within the final reach of the MR. Recorded surface speeds in summer are generally smaller than those recorded in winter, with the surface flow being slowed down and sometimes reversed due to upriver-propagating waves and tide. Conversely, drifter deployed before and after the two winter storms showed an increasing surface velocity while moving downstream. Such behavior is further supported by recorded velocity profiles across the water column at QR2 and QR3 locations, which followed a marked salt-wedge pattern, additionally modulated by the tidal motion [51].

Hydraulic data recorded during the January 2014 experiment are illustrated in **Figure 3**. The stage at RG2 and the mean precipitation within the MR watershed are shown in **Figure 3a**. The two floods are here highlighted by the stage peaks occurred in the mornings of 25 and 28 January, almost simultaneous to high-tide conditions recorded at Ancona harbor (**Figure 3b**) and to the sea storms recorded at QS3 (significant height and peak periods are illustrated in **Figure 3c**). The instantaneous velocity magnitude recorded along the water column at QR2 are plotted in **Figure 3d**.

The increased river outflow during the January 2014 storms generated a large river plume with a considerable amount of sediment extending up to around 1.3 km offshore of the mouth. The occurrence of strong river outflow was coupled with a negligible modulation provided by tidal oscillation on the velocity distribution along the water column, especially at a relatively far distance from the mouth (e.g., at QR2). Additionally, during the winter storms a large sediment deposition was observed closer to the mouth (i.e., near the bend, at QR3), due to the convergence of hydrodynamic fluxes and suspended sediments from both river and sea. This also suggests the existence of a turbidity maximum zone (TMZ), typically observed in macro- and meso-tidal estuaries, but rarely in microtidal environments [58–60]. Although TMZs were observed during both winter storms, their vertical structure was different and depended on the energetic nature of the storms. The storm occurred during 28–29 January 2014 was much less intense than that occurred

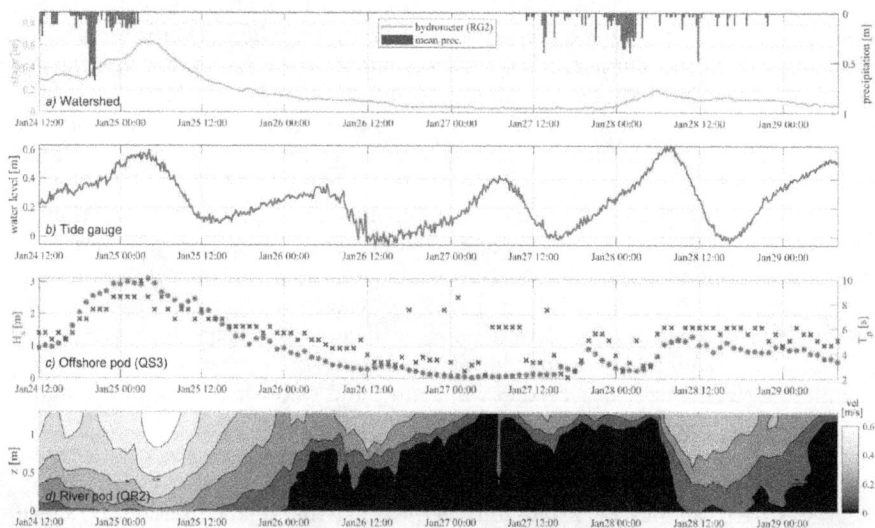

Figure 3.
Observed environmental conditions during the January 2014 experiment: (a) precipitation in the watershed and stage at RG2, (b) water surface level recorded by tide gauge (Ancona), (c) offshore wave characteristics recorded at QS3, (d) velocity distribution along the water column.

during 24–25 January 2014, especially in terms of incoming wave height and outgoing river flow. A relatively larger turbidity was observed during the smaller storm [57].

While the summer period is supposed to be characterized by a flocculation region at distances larger than 600 m upstream of the MR mouth and by large sediment deposition throughout the final river reach, wintertime stormy conditions enhanced the sediment transport and provided different morphological patterns due to the strong interplay between incoming and outgoing fluxes: 1) a relevant erosion upstream of QR3, where the river discharge dominates on the marine forcing; 2) an erosion/deposition pattern downstream of QR3 and at the mouth; 3) a modification of the rheological properties of the soil both at the MR mouth and within the river-plume area [50, 51].

3.2 Floc dynamics

To investigate the potential depositional effects within the Misa system, a rigorously proven flocculation model (FM) suite was used [47, 48, 61, 62]. The FM is built on a series of empirical-derived algorithms that can quantitatively assess the relative settling and mass flux dynamics for both pure mud flocs and floc populations derived from a range of mud-sand mixtures. It is based on flocculation concept of macroflocs (D > 160 μm) and microflocs (D < 160 μm), as outlined in the classic 'order of aggregation' [63, 64]. FM outputs include macrofloc/microfloc settling velocity, ratio of floc mass between the two size fractions and the total mass settling flux (MSF).

The MR estuary has been subdivided into three regions with different representative sediment compositions, two located 600 m and 100 m upriver of the estuary, one located 400 m off the estuary. A nominal representative mud:sand (M:S) composition of 100 M:0S is chosen at the inland site, 50 M:50S equal mud/sand mixture at the mid-zone, and pure sand (0M:100S) in the seaward region, based on the collected field data [50, 51]. The level of flocculation primarily depends upon the combined effects of Suspended Sediment Concentration (SSC) and turbulent mixing. Hence, the FM indicated a fast macrofloc settling (3.5 mm/s at shear stress of 0.35 Pa) and a relatively quick microfloc settling at the pure cohesive inland generated during small storm conditions (28–29 January 2014). Further, both the lower river and seaward zones were much less turbid. In the lower river, the less cohesive equally mixed sediment composition produced slower macroflocs and quicker microflocs than within the inland region, with an equal apportioning of floc mass between macro- and microfloc fractions. The seaward (pure-inert/non-flocculating) sediment settled fast at ∼6.8 mm/s.

Throughout the Misa system, the MSF was only ∼40% of that occurring during small storm conditions. Quiescent conditions (27 January 2014) saw a further 20–25% slowing in the floc settling velocities, with SSC being only 30–40% of that found during the small storm at each site. Specifically, the FM indicated that quiescent conditions favored smaller microfloc fraction dynamics, a much tighter spatial grouping and significantly smaller settling flux values.

3.3 Upriver propagation of long waves

IG waves are long-wave modes that are seen to easily propagate upriver in meso- and macro-tidal estuaries [1, 24, 25], while their impact on microtidal estuaries has been rarely investigated. Upriver propagation of IG waves were observed in the microtidal environment of the MR during the January 2014 experiment (**Figure 4a–c**) [26]. The dominant generation mechanism of IG waves in this case is

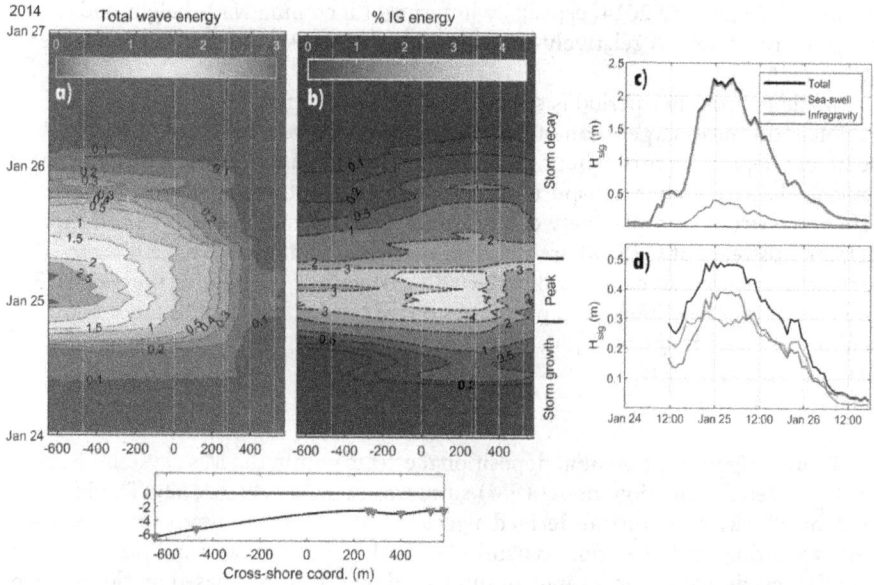

Figure 4.
Temporal and spatial evolution of wave energy during 24–26 January 2014: (a) total energy, (b) percentage of sea-swell contribution, (c) percentage of IG contribution. Evolution of band-specific significant wave height (d) in the offshore area, at QS2, and (e) in the river, at QR2. The bottom panel shows a schematic of the bed profile at the estuary and further offshore. Purple triangles represent the sensor location (x = 0 m at the river mouth, x > 0 moving upriver).

probably the bound wave mechanism [10], as suggested by a normalized slope parameter calculated off the mouth

$$\beta_b = \frac{h_x}{\omega}\sqrt{\frac{g}{h}} = 0.010 - 0.014 \qquad (1)$$

where $\omega = 0.63$ Hz is the wave angular frequency during the storm climax, g is the gravity acceleration, $h_x = 1/200$ is the mean seabed slope, while $h = (3 - 6)$ m is the depth range off the MR mouth.

It was observed that, while the IG contribution in the open sea was significantly smaller than that provided by sea-swell waves, it gained a more relevant role within the final reach of the MR. This is exemplified in **Figure 4c** and **d**, which shows the time evolution of band-specific significant wave heights during the storm event occurred on 25–26 January 2014, at two locations representative of offshore (QS2) and riverine (QR2) areas. While the total wave height in the sea was mainly due to sea-swell throughout the storm (**Figure 4c**), the IG contribution was seen to be much more important in the MR, especially during the storm climax (**Figure 4d**). This is due to the intense wave breaking affecting the shortest wave components at the mouth, and further enhanced by strong opposing currents. Inspection of wave energy levels revealed a large energy decay across a wide region off the MR mouth (between −400 m and 0 m, at (2–5) m depths, as shown in **Figure 4a**). This decay especially involved the short-wave components, while the normalized IG component seemed to peak just past the MR mouth and values larger than the reference offshore value characterize the region between 0 and 200 m upriver (**Figure 4b**). Such behavior is connected to the large damping and deviation of storm-driven sea currents promoted by the river discharge [51], not enough to block IG components.

Hence, the MR estuarine area removes higher-frequency waves and retains low-frequency energy, thus operating as a low-pass filter. This process has already been observed in field experiments carried out in other environments, like microtidal estuaries or energetic oceanic inlets [24, 25, 65].

Although IG waves were observed to propagate upriver for hundreds of meters and tide propagate upriver for kilometers, the interplay between such low-frequency modes is of importance, with the tide controlling the upriver propagation of IG components [65]. In addition, despite the low tidal range at the MR estuary (less than 0.6 m during the January 2014 experiment), tidal oscillations reach distances larger than 1.2 km within the MR (e.g., at RG1), although tidal currents are low and their effect on the river current is negligible. While low-flow river conditions imposes very small resistance to the tide upriver propagation, the tidal forcing was easily suppressed by river floods and high-flow conditions [26, 51]. Furthermore, the tidal effect is important just off the estuary, in connection to a persistence of wave-induced setup during the storm decay.

On the other hand, no tidal influence was observed at RG2, as confirmed by more recent observations performed in 2019 (**Figure 5a–c**) [52]. These observations also confirmed the presence of tide at RG1, as well as the upriver propagation of additional low-frequency modes. Specifically, the stream-gauge recordings of some flood events occurred in 2019 allowed for the observation of modes not detected by the hydrometer, due to the difference in the sampling rate of the two instruments (2′ for the stream gauge versus 30′ for the hydrometer). The spectral content of two storm events recorded by the stream gauge and lasting a bit more than one day each (12–13 and 15–16 November 2019) has been analyzed, in conjunction with the spectral content of the signals recorded by both hydrometers (RG1 and RG2) and the tide gauge (**Figure 5d**). A long-time range has been used for hydrometer and tide signals, i.e. between 10 September and 17 December 2019. The hydrometers at RG1 and RG2 showed a similar spectral pattern, especially for frequencies $f > 0.5 \cdot 10^{-4}$ Hz, while significant differences occurred for $f < 0.5 \cdot 10^{-4}$ Hz, due to diurnal ($\sim$25.6 h) and semi-diurnal ($\sim$12.8 h) tidal constituents observed at RG1, but not at RG2.

Furthermore, the analysis of the stream-gauge signals during the recorded events at RG1 does not show relevant peaks referring to tidal constituents, due to

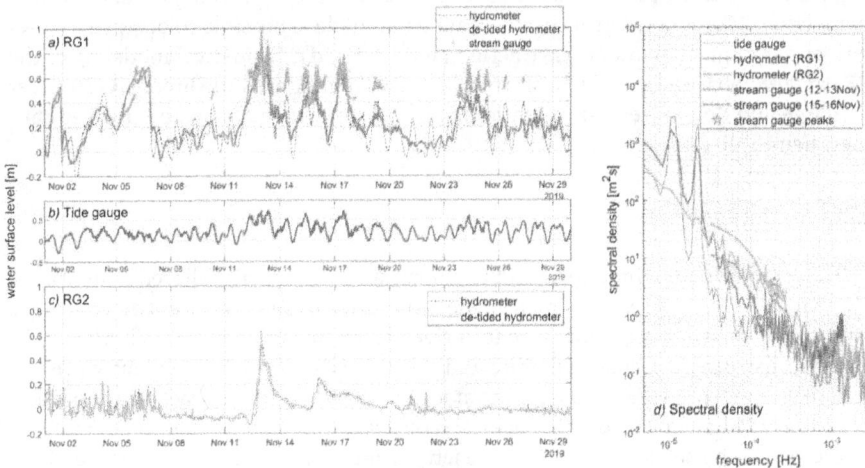

Figure 5.
Time series recorded in 2019 by: (a) hydrometer and stream gauge at RG1, (b) tide gauge at Senigallia harbor, (c) hydrometer at RG2. (d) Spectral density of recorded signals.

the reduced event duration. However, local peaks exist at ~1.42 hours during both events, which can be probably ascribed to other long-wave modes generating in the Adriatic Sea. Enclosed and semi-enclosed basins, like the Adriatic Sea, show low-frequency oscillations like the seiche motion, which can be described by the natural period at a specific mode n:

$$T_{seiche,n} = \frac{2L}{n\sqrt{gh}} \tag{2}$$

Selecting the mode-2 oscillation ($n = 2$), with $L \sim 130$ km being the transversal basin length and $h = (50–70)$ m being a representative water depth along the transversal direction, Eq. (2) yields $T_{seiche,n} = (1.38 - 1.63)$ hours. Hence, the peak period observed in **Figure 5d** falls in the calculated range and can be thus motivated by the transversal oscillation of the Adriatic basin, only detectable using high-frequency recordings during flood conditions. This might suggest that seiche modes are not particularly relevant in the long term, but they cannot be neglected during short-term events (order of hours to days).

3.4 River mouth bar

The dynamics of river mouth bars is strongly correlated with the mutual inter-action of the river discharge and wave actions. The long-term monitoring of the MR estuary allowed us to correlate the behavior of the emerged mouth bar with the river and sea forcing. Hourly videos recorded by the SGS video-monitoring station were post-processed to create a 10-minute time-exposure image (timex). For each hour, the four timex images were stabilized, geo-rectified and merged to obtain the plan-view images. The ortho-rectified timex images, from 2017 to 2019, were ana-lyzed using a semi-automatic procedure[1] to detect the presence of the emerged bar and its geometric features. The evolution of the area and center of mass of the bar was correlated with the sea state (wave height and direction) time-series recorded by the offshore ADCP, the tidal levels, as well as flow rate and water level recorded along the MR.

The results showed a net downriver migration due to relevant flood events (**Figure 6e–h**) and a persistence of sediment accumulation during periods of weak river action. Moreover, the accurate monitoring made it possible to observe a slow upriver bar migration under wave action (**Figure 6a–d**). Numerical modeling of the MR estuary, performed by Delft3D software, also highlighted the link between river discharge and downriver bar migration, and between wave action and upriver bar migration [54].

3.5 Coastal impact

The evolution of submerged sand bars may well be crucially altered (intention-ally or not) by the presence of man-made structures, like breakwaters or jetties, the latter of which are commonly present in riverine environments [66, 67]. Melito et al. [4] discussed sandbar dynamics and their correlations with incident wave climate and morphological constraints at a portion of sheltered beach south of the MR estuary (**Figures 1b** and **7**). The beach object of the study is delimited, at its northern side, by the southern concrete jetty delimiting the final reach of the MR. The jetty provides a partial sheltering from wave attack coming from northern

---

[1] https://github.com/Coastal-Imaging-Research-Network/River-Bar-Toolbox

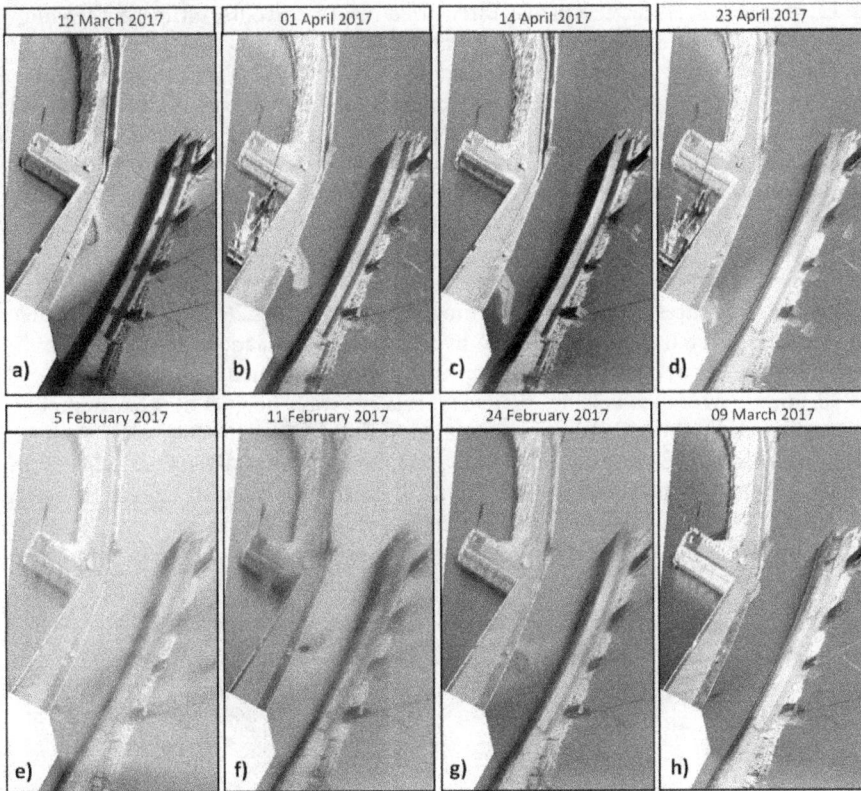

Figure 6.
Ortho-rectified timex images recorded from the SGS video-monitoring station. (a, b, c, d) is an example of observed upriver migration. (e, f) and (g, h) are two examples of observed downriver migration.

quadrants (mainly originated by Bora storms), but leaves the coastline exposed to incident waves from eastern directions.

The submerged beach is characterized by an array of three shallow bars, whose displacement in response to seasonal climate and storm events is monitored since 2015 with the aid of remote sensing products from the SGS station. The bars at the sheltered beach show a response dominated by seasonal oscillations in wave climate, presenting occasionally consistent onshore displacements in milder climates typical of summer months, and offshore migration in winter months. The bar array is generally poorly responsive to single storms; a circumstance shared with other portions of unprotected coastline far from the influence of the river jetty. The overall behavior of the bar structure can be therefore assimilated to a pattern of bars oscillating around a more or less well-defined point of equilibrium (OPE pattern) [68].

Two short storm events from SSE occurred in February 2016 and October 2018, however, imposed a remarkable change in this established pattern by generating storm-scale displacements in the order of 30–50 m to all submerged bars. Migrations of such entity and with such short response times are not repeated anytime during the investigation period, spanning from 2015 to 2019, even during much more intense wave attacks. This exceptional event is likely connected to the peculiar interaction between incoming waves from eastern directions and the presence of the concrete jetty. Storm waves approach almost normally to the river jetty, generating enhanced reflection and intensified return currents, ultimately leading to enhanced offshore displacements of bars.

The eventuality of different beach response to storm waves with different incidence was explored with a campaign of numerical simulations run with FUNWAVE [69] and focused on the interaction of the man-made landmark with the dominant wave field at the MR estuary [4]. Two simulations were devised to represent wave attacks from a typical Bora storm (NNE direction) and a typical Sirocco storm (ESE). The two prototypical storms used as wave input are rather different in terms of wave period: while the Bora storm is characterized by steeper waves ($H_s = 2.28$ m, $T_p = 7.6$ s) due to the more intense wind and limited fetch from northern quadrants, the Sirocco-driven storm features longer, swell-dominated waves ($H_s = 2.26$ m, $T_p = 9.3$ s) thanks to an increased fetch distance. The two hydrodynamic simulations are run over a bathymetry generated using a DTM featuring two well-defined bars.

30-minute-averaged velocity fields and relative wave heights in the nearshore area delimited by the MR jetty are presented in **Figure 7** for the Bora (NNE) storm and the Sirocco (ESE) storm. The FUNWAVE model predicts longshore currents on top of the bar in both cases; however, while the current is directed towards the jetty during the Bora storm (**Figure 7c**), Sirocco waves force a stronger current, directed away from the jetty, and escaping the nearshore giving birth to a marked circulation

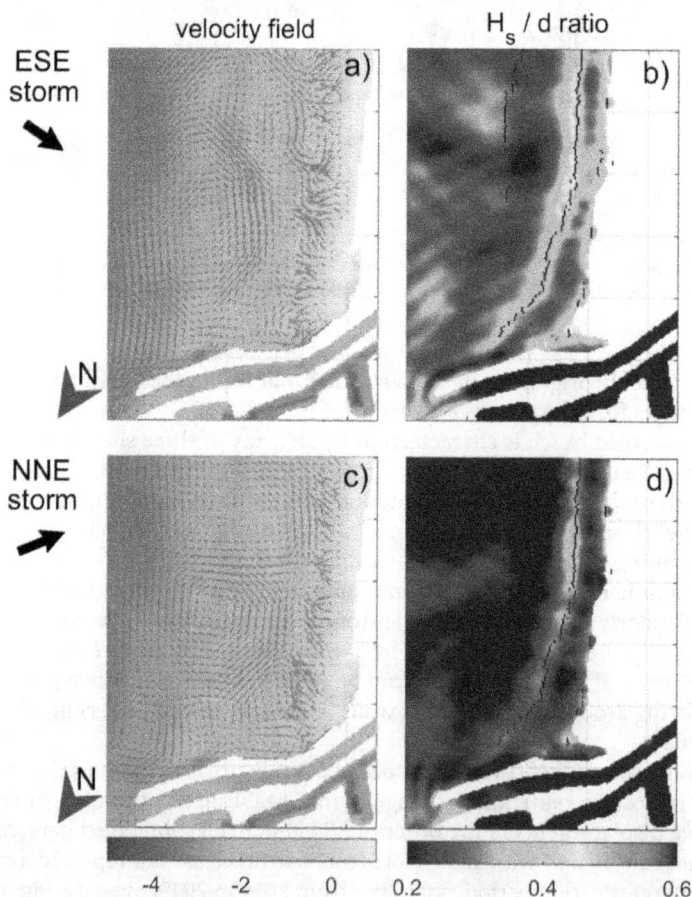

Figure 7.
Numerical simulations of an ESE storm (a, b) and NNE storm (c, d). 30-minute-averaged velocity fields (a, c) and relative wave heights (b, d) in the nearshore area south of the MR jetty.

cell (**Figure 7a**). Stronger currents, along with higher relative wave heights over the bars for the Sirocco (ESE) event (**Figure 7b**) in comparison to those modeled during the Bora (NNE) event (**Figure 7d**), can thus be linked to a greater potential for sediment stirring and motion by Sirocco waves and, ultimately, bar migration during ESE storms.

4. Discussion and concluding remarks

The overall dynamics of the salt-wedge estuary of the MR have been observed exploiting short-term measurements in both river and sea during September 2013 and January 2014, as well as long-term monitoring at different and farther locations. The short-term experiments revealed that the marine forcing is able at propagating upriver for long distances, during sea storms but also during relatively mild/quiescent conditions.

In detail, the large waves observed during both big and small storms propagated upriver for some hundreds of meters, also suggesting the existence of a TMZ developing at different locations within the MR, similar to what observed in meso- and macro-tidal estuaries [57, 59]. Further, results of the FM in the final reach of the MR suggest a fast macrofloc settling and high flocculation occurring about 600 m upriver of the mouth during the small storm, and an efficient flocculation throughout the range of turbulent stimulation, while both lower-river and seaward zones were characterized by a much smaller MSF compared to that estimated inland [57]. The impact of waves in the estuarine area is also suggested by the relevant seabed variations obtained comparing the bathymetric surveys of September 2013 and January 2014, which show significant erosion/deposition patterns just off the jetty (bed variations of ±1.5 m), as well as important changes in the final river reach (erosion up to 1 m) [51].

However, the wave forcing is not the only action that pushes upriver marine waters along the MR. Evidence of the upriver propagation of marine fluxes was given in the summertime experiment (September 2013), when the salinity level in the MR, at about 1.8 km from the mouth, was larger than zero and suggested the existence of a flocculation zone at about (1.8–0.6) km from the mouth. Further, a floc deposition was supposed to occur in the final 600 m of the MR, as also demonstrated by the seabed increase and sediment deposition occurred in the 2013 summer, between May and September [50, 51]. In addition, more recent observations confirmed that the marine forcing can propagate upriver for kilometers, although not as IG waves, which mainly affect the final reach of the MR like the sea-swell waves do, as observed during the January 2014 storms [26]. Differently, very low-frequency waves like tides or seiches ($T \sim 1.4$ hours) were recorded 1.2 km from the estuary and strongly affected the local hydrodynamics, being the water surface level generally more altered by tidal excursion than by river forcing during low-flow/quiescent conditions [52].

In terms of sediment transport and bed morphology, river mouth bars and coastal sandbars are directly linked to the forcing actions existing in the investigated microtidal area. The interplay among such actions is summarized by the evolution of a river mouth bar. While significant flood events promote a net downriver migration, a slow upriver bar migration is observed under wave action.

Linked to the main processes occurring within river and estuary are also the dynamics characterizing the coastal area just south of the MR jetty, where submerged sandbars evolve while protecting the natural beach. While the sediment supply from the MR significantly affects the sediment transport in such area, the main incoming direction of sea storms is thought to largely affect the SSC just off the estuary. Numerical hydrodynamic simulations suggested that during NNE storms a recirculation cell is generated south of the jetty and over the sandbars, while the velocity field at the estuary is relatively small [4]. Hence, the river-induced plume propagating far offshore from the estuary seems to be not significantly affected by NNE waves, whereas a different velocity field seems to generate during ESE storms, providing a seaward-directed fast flow in correspondence of the jetty. Hence, on the one side, the MR plume seems to be enhanced by the wave-induced cross-shore current. On the other side, the sandbar array is supposed to evolve and migrate, due to a larger relative height over the bars themselves and possibly related sediment mobilization.

Many of the above-described aspects, especially the main processes related to the evolution of river mouth bars and river plume, as well as the accurate spatial analysis of the sea-forcing propagation within the MR during mild and stormy conditions, are currently under investigation and will be detailed in future contributions.

Acknowledgements

The financial support from the MORSE Project (Office of Naval Research Global - UK, Research Grant Number N62909-17-1-2148) and the FUNBREAK Project (MIUR PRIN 2017 - Italy, Grant Number 20172B7MY9) is gratefully acknowledged. AJM's contribution towards this research was partly supported by the US National Science Foundation under grants OCE-1736668 and OCE-1924532, and HR Wallingford company research FineScale project (ACK3013_62). The authors would like to thank all colleagues who made significant contributions during the planning and execution of the field experiments, as well as for the following activities, including Edward F. Braithwaite III, Sara Corvaro, Giovanna Darvini, Michael Fuller, Kevin Lois, Carlo Lorenzoni, Alessandro Mancinelli, Pierluigi Penna, Allen Reed, Aniello Russo, Alex Sheremet, Luciano Soldini, Tracy Staples, Gianluca Zitti. The following authorities and companies are also acknowledged: the Municipality of Senigallia, the Capitaneria di Porto of Senigallia and of Ancona, MARIDIPART La Spezia and MARIFARI Venezia, GESTIPORT (Senigallia), Club Nautico (Senigallia), NOTA srl (Senigallia), Carmar Sub (Ancona), Sena Gallica (Senigallia), METIS S.R.L. (Senigallia). The data used in this paper are available at this link.

Conflict of interest

The authors declare no conflict of interest.

Author details

Maurizio Brocchini[1,2], Matteo Postacchini[1*], Lorenzo Melito[1], Eleonora Perugini[1], Andrew J. Manning[2,3,4,5,6,9], Joseph P. Smith[7] and Joseph Calantoni[8]

1 Università Politecnica delle Marche, Ancona, Italy

2 University of Florida, Gainesville, FL, USA

3 HR Wallingford, Coasts and Oceans Group, Wallingford, UK

4 Environment and Energy Institute, University of Hull, Hull, UK

5 Stanford University, Stanford, California, USA

6 University of Delaware, Delaware, USA

7 Oceanography Department, U.S. Naval Academy, Annapolis, MD, USA

8 Ocean Sciences Division, U.S. Naval Research Laboratory, Stennis Space Center, MS, USA

9 University of Plymouth, Plymouth, UK

*Address all correspondence to: m.postacchini@staff.univpm.it

IntechOpen

References

[1] G. Dodet, X. Bertin, N. Bruneau, A.B. Fortunato, A. Nahon, A. Roland, Wave-current interactions in a wave-dominated tidal inlet, J. Geophys. Res. Ocean. (2013). https://doi.org/10.1002/jgrc.20146.

[2] M. Olabarrieta, W.R. Geyer, N. Kumar, The role of morphology and wave-current interaction at tidal inlets: An idealized modeling analysis, J. Geophys. Res. Ocean. (2014). https://doi.org/10.1002/2014JC010191.

[3] L. Melito, M. Postacchini, A. Sheremet, J. Calantoni, G. Zitti, G. Darvini, M. Brocchini, Wave-Current Interactions and Infragravity Wave Propagation at a Microtidal Inlet, Proceedings. (2018). https://doi.org/10.3390/proceedings2110628.

[4] L. Melito, L. Parlagreco, E. Perugini, M. Postacchini, S. Devoti, L. Soldini, G. Zitti, L. Liberti, M. Brocchini, Sandbar dynamics in microtidal environments: Migration patterns in unprotected and bounded beaches, Coast. Eng. (2020). https://doi.org/10.1016/j.coastaleng.2020.103768.

[5] M. Brocchini, Wave-forced dynamics in the nearshore river mouths, and swash zones, Earth Surf. Process. Landforms. (2020). https://doi.org/10.1002/esp.4699.

[6] W.H. Munk, Origin and Generation of Waves, Coast. Eng. Proc. 1 (1950) 1. https://doi.org/10.9753/icce.v1.1.

[7] R. Davidson-Arnott, An Introduction to Coastal Processes and Geomorphology, 2009. https://doi.org/10.1017/cbo9780511841507.

[8] M. Tucker, Surf beats: sea waves of 1 to 5 min. period, Proc. R. Soc. London. Ser. A. Math. Phys. Sci. 202 (1950) 565–573. https://doi.org/10.1098/rspa.1950.0120.

[9] G. Symonds, D.A. Huntley, A.J. Bowen, Two-dimensional surf beat: long wave generation by a time-varying breakpoint., J. Geophys. Res. (1982). https://doi.org/10.1029/JC087iC01p00492.

[10] X. Bertin, A. de Bakker, A. van Dongeren, G. Coco, G. André, F. Ardhuin, P. Bonneton, F. Bouchette, B. Castelle, W.C. Crawford, M. Davidson, M. Deen, G. Dodet, T. Guérin, K. Inch, F. Leckler, R. McCall, H. Muller, M. Olabarrieta, D. Roelvink, G. Ruessink, D. Sous, É. Stutzmann, M. Tissier, Infragravity waves: From driving mechanisms to impacts, Earth-Science Rev. (2018). https://doi.org/10.1016/j.earscirev.2018.01.002.

[11] K.M. Wijnberg, A. Kroon, Barred beaches, Geomorphology. (2002). https://doi.org/10.1016/S0169-555X(02)00177-0.

[12] B.G. Ruessink, J.H.J. Terwindt, The behaviour of nearshore bars on the time scale of years: A conceptual model, Mar. Geol. (2000). https://doi.org/10.1016/S0025-3227(99)00094-8.

[13] T.E. Baldock, P. Manoonvoravong, K.S. Pham, Sediment transport and beach morphodynamics induced by free long waves, bound long waves and wave groups, Coast. Eng. (2010). https://doi.org/10.1016/j.coastaleng.2010.05.006.

[14] T. Aagaard, B. Greenwood, Suspended sediment transport and the role of infragravity waves in a barred surf zone, Mar. Geol. (1994). https://doi.org/10.1016/0025-3227(94)90111-2.

[15] J.A. Roelvink, M.F.J. Stive, Bar-generating cross-shore flow mechanisms on a beach, J. Geophys. Res. (1989). https://doi.org/10.1029/JC094iC04p04785.

[16] M. Postacchini, L. Soldini, C. Lorenzoni, A. Mancinelli, Medium-Term

dynamics of a middle Adriatic barred beach, Ocean Sci. 13 (2017). https://doi.org/10.5194/os-13-719-2017.

[17] L. Parlagreco, L. Melito, S. Devoti, E. Perugini, L. Soldini, G. Zitti, M. Brocchini, Monitoring for coastal resilience: Preliminary data from five italian sandy beaches, Sensors (Switzerland). (2019). https://doi.org/10.3390/s19081854.

[18] T.C. Lippmann, R.A. Holman, Quantification of sand bar morphology: a video technique based on wave dissipation, J. Geophys. Res. (1989). https://doi.org/10.1029/JC094iC01p00995.

[19] M.S. Phillips, M.D. Harley, I.L. Turner, K.D. Splinter, R.J. Cox, Shoreline recovery on wave-dominated sandy coastlines: the role of sandbar morphodynamics and nearshore wave parameters, Mar. Geol. (2017). https://doi.org/10.1016/j.margeo.2017.01.005.

[20] T. Aagaard, J. Nielsen, B. Greenwood, Suspended sediment transport and nearshore bar formation on a shallow intermediate-state beach, Mar. Geol. (1998). https://doi.org/10.1016/S0025-3227(98)00012-7.

[21] T. Sunamura, I. Takeda, Landward Migration of Inner Bars, Dev. Sedimentol. (1984). https://doi.org/10.1016/S0070-4571(08)70141-9.

[22] F. Hoefel, S. Elgar, Wave-induced sediment transport and sandbar migration, Science (80-.). (2003). https://doi.org/10.1126/science.1081448.

[23] E.L. Gallagher, S. Elgar, R.T. Guza, Observations of sand bar evolution on a natural beach, J. Geophys. Res. Ocean. (1998). https://doi.org/10.1029/97jc02765.

[24] R.J. Uncles, J.A. Stephens, C. Harris, Infragravity currents in a small ría:

Estuary-amplified coastal edge waves?, Estuar. Coast. Shelf Sci. (2014). https://doi.org/10.1016/j.ecss.2014.04.019.

[25] M.E. Williams, M.T. Stacey, Tidally discontinuous ocean forcing in bar-built estuaries: The interaction of tides, infragravity motions, and frictional control, J. Geophys. Res. Ocean. (2016). https://doi.org/10.1002/2015JC011166.

[26] L. Melito, M. Postacchini, A. Sheremet, J. Calantoni, G. Zitti, G. Darvini, P. Penna, M. Brocchini, Hydrodynamics at a microtidal inlet: Analysis of propagation of the main wave components, Estuar. Coast. Shelf Sci. 235 (2020). https://doi.org/10.1016/j.ecss.2020.106603.

[27] V.N. Mikhailov, Hydrology and formation of river-mouth bars, in: Probl. Humid Trop. Zo. Deltas, 1966: pp. 59–64. ftp://ftp.ems.psu.edu/data/pub/geosc/pub/dedmonds/Mikhailov197.pdf.

[28] L.D. Wright, Morphodynamics of a Wave-Dominated River Mouth, in: Coast. Eng. 1976, American Society of Civil Engineers, New York, NY, 1976: pp. 1721–1737. https://doi.org/https://doi.org/10.9753/icce.v15.99.

[29] S. Fagherazzi, D.A. Edmonds, W. Nardin, N. Leonardi, A. Canestrelli, F. Falcini, D.J. Jerolmack, G. Mariotti, J.C. Rowland, R.L. Slingerland, Dynamics of river mouth deposits, Rev. Geophys. (2015). https://doi.org/10.1002/2014RG000451.

[30] D.A. Edmonds, R.L. Slingerland, Mechanics of river mouth bar formation: Implications for the morphodynamics of delta distributary networks, J. Geophys. Res. Earth Surf. (2007). https://doi.org/10.1029/2006JF000574.

[31] H. Mitchener, H. Torfs, Erosion of mud/sand mixtures, Coast. Eng. (1996). https://doi.org/10.1016/S0378-3839(96)00002-6.

[32] T.S. Mostafa, J. Imran, M.H. Chaudhry, I.B. Kahn, Erosion resistance of cohesive soils, J. Hydraul. Res. (2008). https://doi.org/10.3826/jhr.2008.2794.

[33] J.C. Winterwerp, W.G.M. van Kesteren, Introduction to the Physics of Cohesive Sediment Dynamics in the Marine Environment, 2004.

[34] A.J. Mehta, An introduction to hydraulics of fine sediment transport, World Scientific Publishing Company, 2013.

[35] A.J. Manning, J. V. Baugh, J.R. Spearman, R.J.S. Whitehouse, Flocculation settling Characteristics of mud: Sand mixtures, in: Ocean Dyn., 2010. https://doi.org/10.1007/s10236-009-0251-0.

[36] Manning, A.J., J.R. Spearman, R.J.S. Whitehouse, E.L. Pidduck, J.V. Baugh, K.L. Spencer, Laboratory Assessments of the Flocculation Dynamics of Mixed Mud:Sand Suspensions, in: A.J. Manning (Ed.), Sediment Transp. Process. Their Model. Appl., InTech (Rijeka, Croatia), 2013: pp. 119–164. https://doi.org/org/10.5772/3401.

[37] R. Whitehouse, R. Soulsby, W. Roberts, H. Mitchener, Dynamics of estuarine muds, 2000. https://doi.org/10.1680/doem.28647.

[38] K.R. Dyer, Coastal and estuarine sediment dynamics., (1986). https://doi.org/10.1016/0378-3839(88)90018-x.

[39] A.J. Raudkivi, Loose boundary hydraulics, 1998. https://doi.org/10.1139/l91-110.

[40] M. Van Ledden, Sand-mud segregation in estuaries and tidal basins, Commun. Hydraul. Geotech. Eng. (2003).

[41] D.M. Paterson, S.E. Hagerthey, Microphytobenthos in contrasting coastal ecosystems: Biology and dynamics, in: K. Reise (Ed.), Ecol. Comp. Sediment. Shores, Berlin: Springer, 2001: pp. 105–125.

[42] N. Tambo, Y. Watanabe, Physical characteristics of flocs-I. The floc density function and aluminium floc, Water Res. (1979). https://doi.org/10.1016/0043-1354(79)90033-2.

[43] R.C. Klimpel, R. Hogg, Effects of flocculation conditions on agglomerate structure, J. Colloid Interface Sci. (1986). https://doi.org/10.1016/0021-9797(86)90212-2.

[44] I.G. Droppo, D.E. Walling, E.D. Ongley, The influence of floc size, density and porosity on sediment and contaminant transport, in: IAHS-AISH Publ., 2000.

[45] G.G. Stokes, On the effect of the Internal friction of fluids on the motion of pendulums - Section III, Trans. Cambridge Philos. Soc. (1850).

[46] K.R. Dyer, A.J. Manning, Observation of the size, settling velocity and effective density of flocs, and their fractal dimensions, in: J. Sea Res., 1999. https://doi.org/10.1016/S1385-1101(98)00036-7.

[47] A.J. Manning, K.R. Dyer, Mass settling flux of fine sediments in Northern European estuaries: Measurements and predictions, Mar. Geol. (2007). https://doi.org/10.1016/j.margeo.2007.07.005.

[48] R.L. Soulsby, A.J. Manning, J. Spearman, R.J.S. Whitehouse, Settling velocity and mass settling flux of flocculated estuarine sediments, Mar. Geol. (2013). https://doi.org/10.1016/j.margeo.2013.04.006.

[49] A.J. Mehta, J.W. Lott, Sorting of fine sediment during deposition, in:

Coast. Sediments, ASCE, 1987: pp. 348–362.

[50] M. Brocchini, J. Calantoni, A.H. Reed, M. Postacchini, C. Lorenzoni, A. Russo, A. Mancinelli, S. Corvaro, G. Moriconi, L. Soldini, Summertime conditions of a muddy estuarine environment: The EsCoSed project contribution, Water Sci. Technol. 71 (2015). https://doi.org/10.2166/wst.2015.116.

[51] M. Brocchini, M. Postacchini, C. Lorenzoni, A. Russo, S. Corvaro, A. Mancinelli, L. Soldini, J. Calantoni, A.H. Reed, E.F. Braithwaite, A. Sheremet, T. Staples, J. Smith, Comparison between the wintertime and summertime dynamics of the Misa River estuary, Mar. Geol. 385 (2017). https://doi.org/10.1016/j.margeo.2016.12.005.

[52] M. Postacchini, L. Melito, A. Sheremet, J. Calantoni, G. Darvini, S. Corvaro, F. Memmola, P. Penna, M. Brocchini, Upstream Propagating Long-Wave Modes at a Microtidal River Mouth, Environ. Sci. Proc. (2020). https://doi.org/10.3390/environsciproc2020002015.

[53] A. Valle-Levinson, Contemporary issues in estuarine physics, 2010. https://doi.org/10.1017/CBO9780511676567.

[54] A. Baldoni, E. Perugini, L. Soldini, J. Calantoni, M. Brocchini, Long-term evolution of an inner bar at the mouth of a microtidal river, Estuar. Coast. Shelf Sci. (under review).

[55] Deltares, Delft3D-FLOW. Simulation of multi-dimensional hydrodynamic flows and transport phenomena, including sediments. User Manual. Hydro-Morphodynamics. Version: 3.15.34158, 2014.

[56] Deltares, Delft3D-Wave, Simulation of short-crested waves with SWAN,

User Manual, Version 3.05.58426, Deltares. (2019) 200.

[57] M. Postacchini, A.J. Manning, J. Calantoni, J.P. Smith, M. Brocchini, A Storm Driven Turbidity Maximum in a Microtidal Estuary, J. Geophys. Res. Ocean. (under review).

[58] R.J. Uncles, R.C.A. Elliott, S.A. Weston, Observed fluxes of water, salt and suspended sediment in a partly mixed estuary, Estuar. Coast. Shelf Sci. (1985). https://doi.org/10.1016/0272-7714(85)90035-6.

[59] R.J. Uncles, J.A. Stephens, Nature of the turbidity maximum in the Tamar Estuary, U.K., Estuar. Coast. Shelf Sci. (1993). https://doi.org/10.1006/ecss.1993.1025.

[60] A.J. Manning, W.J. Langston, P.J.C. Jonas, A review of sediment dynamics in the Severn Estuary: Influence of flocculation, Mar. Pollut. Bull. (2010). https://doi.org/10.1016/j.marpolbul.2009.12.012.

[61] J. Spearman, A.J. Manning, On the significance of mud transport algorithms for the modelling of intertidal flats, in: Proc. Mar. Sci., 2008. https://doi.org/10.1016/S1568-2692(08)80030-7.

[62] A.J. Manning, J. V. Baugh, J.R. Spearman, E.L. Pidduck, R.J.S. Whitehouse, The settling dynamics of flocculating mud-sand mixtures: Part 1-Empirical algorithm development, in: Ocean Dyn., 2011. https://doi.org/10.1007/s10236-011-0394-7.

[63] R.B. Krone, A study of rheologic properties of estuarial sediments, 1963.

[64] D. Eisma, Flocculation and de-flocculation of suspended matter in estuaries, Netherlands J. Sea Res. (1986). https://doi.org/10.1016/0077-7579(86)90041-4.

[65] X. Bertin, M. Olabarrieta, Relevance of infragravity waves in a wave-dominated inlet, J. Geophys. Res. Ocean. (2016). https://doi.org/10.1002/2015JC011444.

[66] M. Sedrati, P. Ciavola, J. Reyns, C. Armaroli, V. Sipka, Morphodynamics of a microtidal protected beach during low wave-energy conditions, in: J. Coast. Res., 2009.

[67] C. Bouvier, B. Castelle, Y. Balouin, Modeling the impact of the implementation of a submerged structure on surf zone sandbar dynamics, J. Mar. Sci. Eng. (2019). https://doi.org/10.3390/jmse7040117.

[68] R. Certain, J.P. Barusseau, Conceptual modelling of sand bars morphodynamics for a microtidal beach (Sète, France), Bull. La Soc. Geol. Fr. (2005). https://doi.org/10.2113/176.4.343.

[69] F. Shi, J.T. Kirby, J.C. Harris, J.D. Geiman, S.T. Grilli, A high-order adaptive time-stepping TVD solver for Boussinesq modeling of breaking waves and coastal inundation, Ocean Model. (2012). https://doi.org/10.1016/j.ocemod.2011.12.004.

Chapter 4

Environmental Evaluation of Surfactant: Case Study in Sediment of Tigris River, Iraq

Rana R. Al-Ani, Fikrat M. Hassan
and Abdul Hameed M. Jawad Al-Obaidy

Abstract

Many chemical pollutants take their way into different environment ecosystems. One of these pollutants is detergent, which these compounds used widely world-wide. There is less attention to their impact on the Iraqi environment, especially on an aquatic system; most of these compounds discharged into the river directly by non-urban communities, in addition to household uses that it had spent through-out the domestic drainage systems. Tigris river is the primary source of water in Baghdad City, Iraq, and passes throughout Baghdad city north to south of the city. This chapter deal with the qualitative and quantitative of these compounds in the sediment as it's considered the sink of most pollutant compounds. The four sampling sites were chosen along the river for 13 months, starting from Feb 2017 to Feb 2018 and represent as dry and wet seasons. Physicochemical parameters had measured during this study. For the sediment sample, two methods used Photolab and HPLC. The two types of surfactants were extracted from the sediment as follows anionic and nonionic surfactants which they had found at all the study sites, especially in some locations in the midstream. For temporal variation, the dry seasons noticed a high concentration for nonionic surfactant (56.19 and 467.3 µg/g) by Photolab and HPLC, respectively, and for anionic surfactant (135.74 µg/g) by HPLC. In contrast, by Photolab, only anionic surfactant was recorded a high concentration in wet seasons (72.05 µg/g). The lowest frequency of anionic and nonionic was recorded in wet seasons by Photolab and HPLC, respectively (41.83 µg/g and not detectable) unless for NS by HPLC in the dry season (10.80 µg/g). For spatial variation which according to the cluster diagram, the highest concentration for anionic and nonionic surfactants by Photolab had recorded (57.88 and 34.32 µg/g, respectively) at site1, while for HPLC anionic and nonionic surfactants was recorded highest values (48.37 and 235.79, respectively) at site 4. From this study concluded that sites 1 and 4 are the most pollutant than other sites because the activity of discharge of pollution.

Keywords: cluster analysis, lotic system, physicochemical facters, surfactants, water pollution

1. Introduction

Many of the detergents that had used worldwide comprised of anionic surfactant about 50–60% and nonionic surfactant 40% [1]. Surface active agents abbreviated to surfactants, which is one of the significant components of detergents that consisted of one or more hydrocarbon chains (organic compounds) and hydrophobic or hydrophilic characteristics [2]. In addition to the widespread uses of surfactants in washing purposes, besides used in the composition of emulsifiers, pesticide formulations, fibers, wetting agents, cosmetics, and treatment of textiles [3].

According to the charge of the hydrophilic part, the surfactants classify into anionic, cationic, nonionic, and amphoteric, and for this reason, it's applied in various domestic and industrial purposes. Thence, they could be passed into all ecosystem compartments (soil, water, and sediment) in multiple ways, after that subjected to different physicochemical processes in an environment like sorption, degradation, and transformation freely [4–6]. These compounds have a high propensity to adsorb in sediments, which represents an extreme concentration [7, 8]. Often, the solid wastes had thrown into the river from sewage treatment plants. These compounds in the environment are different in their fate, behavior, actions, and interaction with other components [9]. The environmental danger of surfactants is bioaccumulation, which has a detrimental effect on aquatic organisms, such as toxicity and endocrine homeostasis. It also improves the solubility of organic compounds in water, which can contribute to movement and aggregation in various divisions of the environmental [8].

The contamination of sediments is a major environmental problem worldwide. Weak ecological management in the past has contributed to natural bodies and erratic incidents, resulting in deposits being swept away by other pollutants [10, 11]. At low concentrations of surfactants in the environment are considered as safe as organic pollutants, while the toxicities at a high level had taken of great interest [12].

One of the main justifications for this study is that surfactants are very toxic and hazardous substances for aquatic organisms, and their everyday uses in domestic and industrial fields encourage their quantitative and qualitative examination in the Tigris River sediment. There is also a vast knowledge discrepancy that needs to explore concerning a surfactant product on the Tigris River. However, the quantitative and qualitative distribution of these surfactant compounds in river sediment had investigated in this study. Besides, this study also offers quantitative details on the effect of such surfactant classes on some of the river water's physicochemical properties and correlates this evidence with known standards.

2. Detergent components

Figure 1 illustrates the detergent ingredients, which consist of three groups [13].

A detergent is a surfactant that has cleaning characteristics in a dilute solution. Almost the alkylbenzene-sulfonates are usually substances of these compounds, and according to Authors [1, 13] which they mentioned that surfactant has a less ability to link with hard water compenent ions such as calcium in contrast with soap in hard water because its polar carboxyl. The word detergent in most domestic settings specifically refers traditionally the detergent known as agent of cleaning in restaurants and laundries, also as different home uses.

Detergents are widely present as powders or concentrated solutions. Detergents like soaps work because they are amphiphilic, partly hydrophilic (polar), and somewhat hydrophobic (nonpolar). These properties facilities the mixture of hydrophobic compounds (such as oil and grease) with water. One of the essential

Detergent Molecule

Surfactants	Builders	Additives
Anionic	Phosphates	Brighteners
Cationic	Silicates	Enzymes & perfumes
Nonionic	Carbonates	Anti-redeposition agents
Amphoteric	Sundry inorganic	Bacteriostatic
		Chelating agents
		Hydrotropes
		Amines and solvents

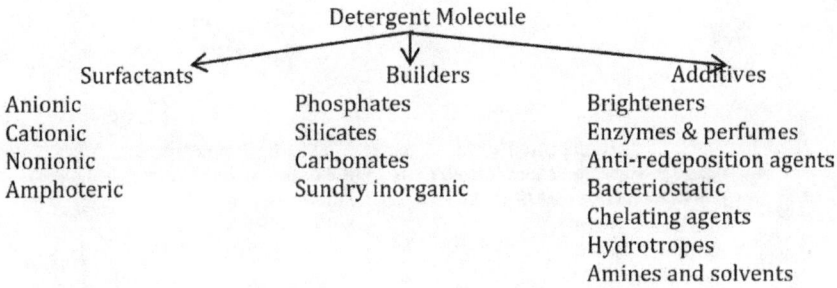

Figure 1.
Typical detergents group.

features of detergents is degraded in aquatic systems by microorganisms in the presence of O^2 into more toxic and harmless products (CO_2, H_2O, carbonates, and bicarbonates) [14].

3. Surfactants

3.1 Surfactants structure

Surfactants are a component that mainly responsible for the cleaning action of detergents [14].

Surfactants have a unique structure of molecules; one soluble part in polar media (hydrophilic), which is known as the head, while the other part nonpolar media (hydrophobic) is called the tail (**Figure 2**) [13]. They classified according to the head group into anionic, cationic, amphoteric, and nonionic compounds [2]. Anionic and cationic compounds have permanently, negative and positive charges, respectively, that are associated with the C-C chain (**Figure 2**). Anionic compounds have no charge. Instead, they have several atoms that are weakly electropositive and electronegative. That is because of the attraction of electrons to oxygen atoms [15].

The presence of polar and nonpolar groups in a surfactant molecule provides them with specific properties against all media; surfactant tends to absorb various surfaces. At a low concentration in water, the surfactant molecules are found as monomers [8], while at high levels, they exist as micelles (**Figure 3**) [16]. Such micelles are adsorbed at boundary phases in order to eliminate hydrophobic sections from water and the system's free energy [17, 18].

Because of a unique structure of surfactants, they found to use in different anthropogenic activities, including household or industrial products that improve the efficiency of the following processes:

1. Wetting/waterproofing,

2. Foaming,

3. Emulsification,

4. Dispersion or flocculation of vital objects in liquid forms,

5. Dissolving soluble reagents (non–/ in small quantities) in solvents,

6. And the viscosity of solution levels rises or decreases [19].

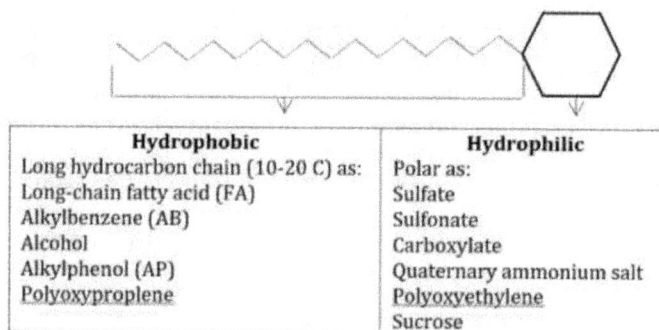

Hydrophobic	Hydrophilic
Long hydrocarbon chain (10-20 C) as:	Polar as:
Long-chain fatty acid (FA)	Sulfate
Alkylbenzene (AB)	Sulfonate
Alcohol	Carboxylate
Alkylphenol (AP)	Quaternary ammonium salt
Polyoxyproplene	Polyoxyethylene
	Sucrose

Figure 2.
Surfactant structure [13].

Figure 3.
A detergent and soap-micelle in water [16].

3.2 Surfactant applications

Areas of use of surfactants are shown in **Table 1** [6].

3.3 Ecotoxicity of surfactants

Surfactants show a significant impact on biological activities and function, especially AS when bound to proteins, enzymes, and DNA [20]. Quaternary ammonium compounds (QACs) (a type of cationic surfactant) can associate bacterial internal membranes [21]. One of the mixtures of surfactants that can bind to the components of the bacterial membrane is NS. It is found as anti-microbial compounds, as it increases the permeability properties that ultimately cause cell death [22].

Because of the high use of surfactants during everyday life needs to increase the study, the ecotoxicity of these compounds on aquatic life. A significant problem is the concentration of these surfactants in the sewage sludge, which is at high levels. Holt [23] noticed that despite the high concentrations of these surfactants found to degraded in wastewater treatment plants (WWTP), some of them remain in surface water, soil, or sediments [23]. The accumulation of these surfactants at high concentrations inhibits or prevents the sludge of microorganisms from the decomposition of pollutants in WWTP. Different types of surfactants exist in various environmental sections such as surface waters, sewage effluents, etc. Significant risks to the water surface ecosystem as a result of the extensive use and disposal of these surfactants [24]. Therefore, Croatian has identified specific criteria for their presence in the water body in **Table 2** [25]. Numerous studies have examined the toxicity of surfactants on bacteria, algae, invertebrates, and fish in the aquatic environment [26].

Cationic	Anionic	Nonionic	Amphoteric
i. Disinfectants & antiseptic agents.	i. Household detergents & surface cleaners.	i. Household & industrial detergents.	i. Shampoos.
ii. Ingredient of cosmetics, medicine, laundry detergents.	ii. Shampoos.	ii. Emulsifiers, wetting & dispersing agents.	
	iii. Hand dishwashing liquids.	iii. Cleaning products.	
iii. Fabric softeners.	iv. Laundry detergents.	iv. Cosmetics.	
iv. Antistatic agents.	v. Personal care products.	v. Paints.	
v. Corrosion inhibitors.	vi. Optical brighteners.	vi. Preservative coatings.	
vi. Flotation agents.	vii. Dyes.	vii. Ingredient of petroleum products.	
	viii. Dispersant, wetting, & suspending agents.	viii. Ingredient of pesticides.	
	ix. Ingredients of pesticides & pharmaceutical products.	ix. In textile, pulp, & paper industry.	

Table 1.
The areas of surfactants application [6].

Class of surfactants	MAC/mgL^{-1}	
	Surface waters	Sewage system
Anionic surfactants	1.0	10.0
Nonionic surfactants	1.0	10.0
Cationic surfactants	0.2	2.0

Table 2.
The maximum allowed concentrations (MAC) of surfactants in wastewater effluent, which can release in a natural aquatic recipient and sewage system in Croatia [25].

3.4 Emission of surfactants into the environment

Because of the particular structure of surfactant molecules, its use in different fields for human life activities. After the use of surfactants or their decomposition products, they will dispose of the WWTP. In case of the absence of WWTP, it will drop into surface water directly and impact aquatic ecosystems such as in the rural area. At the same time, sorption and biodegradation in the WWTP had observed to decompose all surfactants [8, 27].

After treatment processes in WWTP (second stage) for chemical compounds of surfactants, these compounds degraded under optimal conditions about 90–95% of initial surfactant concentration presented in inflowing streams can remove which depending on the efficiency of WWTP [28]. A large part of pollutants extracted as sewage sludge with a percentage ranging from 15% to more than 90%, while it notices that alkylphenol ethoxylates (APE) turned into more toxic when it decomposition [29, 30].

After all processes in WWTP effluents and sewage sludge, different types of surfactants and their degradation (several μg/L or g/Kg) can be existed [28, 31]. As a result of the toxic effects of surfactants through their concentration in different environmental departments, literary studies have increased significantly.

3.5 Fate of surfactants

In the water ecosystem, the surfactants are undergoing sorption and aerobic/anaerobic degradation processes. As a result of these processes, they lead to the elimination of pollutants and their transport to water systems.

Absorption and adsorption are considered a single process. While the sorption process prevents chemical compounds from degradation and hence their bioavailability can be decreased. Sorption processes are affected by some of the environmental parameters such as temperature, pH, salinity, carbon, or clay content of the particulate phase [8, 32]. Many researchers had observed that there is a relationship between higher salinity of water samples and higher sorption percentages for linear alkylbenzene sulfonate (LAS) on suspended solids such as calcium and magnesium [33]. Some surfactants have turned into more toxic decomposition products (e.g., for alkylphenol ethoxylates (APE) products). The researcher recorded a high concentration of the polar compounds in the dissolved form [33, 34], such as C_{10} LAS, short-chain SPC (carboxylic sulfo-phenyl acids), and NPEC (nonyl-phenoxy-monocarboxylates).

The sorption process is associated with the hydrophobic nature of compounds such as:

i. more polar AS were noticed in the dissolved phases;

ii. less polar CS (cationic surfactant) and NS notified in the particulate phases where their transport is associated with suspended solids [35].

Through WWTP, the primary degradation of surfactants occurs by the activity of microorganisms to decrease toxic effects on living organisms. Microbes can use surfactants as their energy source and growth requirements during degradation processes. The efficiency of biodegradation of surfactant compounds in the environment is affected by many factors such as the chemical composition of analytes and physic-chemical parameters such as temperature, light, presence of oxygen, and salinity. On the other hand, some of the compounds (e.g., LAS, ditallow dimethyl ammonium chloride (DTDMAC)) may be persistent under anaerobic conditions [8, 35, 36]. Quiroga et al. [37] discovered that salinity adversely affects sodium dodecyl sulphate (SDS) degradation, while temperature increases the degradation process. Also, sediment enhances the biodegradation rate by gathering both surfactants and bacteria together. Cserhati et al. [20] reported that the adsorption of surfactants on sediment leads to stimulate the bacteria to attach them and cause biodegradation of these compounds. Manzano et al. [38] noticed that APE degradation was increased by temperature; at 7°C, about 68% of surfactant degraded while 96% at 25°C. While the degradation of APEs was slow down with light [36].

The presence of surfactants in the water networks leads to their natural decomposition (half-life time of hours to a few days) according to their characteristics and environmental parameters. These surfactants can also be subjected to either adhesion to suspended solids or accumulation in sediments. In environments with a shortage of oxygen (starting at a depth of a few cms), only anaerobic pathways can degrade surfactants. Although processes in the anaerobic state are slower or not detected (e.g., DTDMAC), and pollutants in the sediment are stay longer time [8, 35]. In experimental studies, researchers found that the acceptable degradation percent of LAS with the use of anoxic marine sediments (up to 79% in 165 days) [34, 35].

3.6 Surfactants in sediment

The adsorption of surfactants on sediments depends on their charge, which is a significant factor. The CA can undergo sorption in deposit much faster. In contrast to AS like sodium dodecylbenzene sulfonate (SDBS) [39]. Factors that increase the ability of CS to adsorb on sediment particles are pH, organic carbon contents, charge of its head, and surface sediment charge. So SDBS shows lower sorption than the CS due to the negative charge of SDBS as compared to CS that have a positive charge [39]. Thus the adsorption of surfactant types in sediment can be put in the order as Cationic > Nonionic > Anionic. Marcomini et al. [40] showed a significant temporal difference in LAS concentration and nonyl-phenol ethoxylates (NPEOs) in Lake Venice. They have explained high temperature that affects biodegradation. Temporal variation had for some surfactants in Glatt River (Switzerland) [30]. The surfactants are nonyl-phenol, lipophilic nonylphenol monoethoxylate and nonyl-phenol diethoxylate.

4. Materials

4.1 Equipments and instruments

The types of equipment and instruments used in the current study as below (**Table 3**)

4.2 Chemicals materials

See **Table 4**.

5. Methods

5.1 Study area

This research was done along the Tigris River within Baghdad city in Iraq for four specific locations for the period from February 2017 to February 2018. Samples (water and sediment) had collected per month for five months of the wet season and eight months of the dry season. The collection of samples had done between 8.30 am to 2.30 pm hours. Four sites had chosen to cover from north to south of Baghdad city. Sample locations are; Al-Muthanna bridge (Site 1-upstream), Al-Sarrafia Bridge (Site 2-midstream), Al-Shuhada Bridge in the north of Baghdad city (Site 3-midstream) and Al-Dora Bridge in the south of Baghdad city (site 4-downstream); (**Figure 4**). Global Positioning System (GPS) (**Table 5**), locations of the research sites were determined.

5.2 Water sampling and analysis

Duplicate water samples (1 liter) had collected from the surface layer (depth 20–30) in stopper fitted clean polyethylene bottles pre-washed with distilled water. The polyethylene bottles were rinsed several times before filling with water samples from the river. The physical and chemical properties for water samples were measured directly after collection. Air temperature (AT), electrical conductivity (EC), turbidity (Tur.), water flow (WF), water temperature (WT), pH were all measured in the field. At the same time, other parameters were analyzed directly in the Environmental Research Center at the University of Technology, Iraq. Such as salinity (S‰), total dissolved solids (TDS), total suspended solids (TSS), nitrite (NO_2), nitrate (NO_3), phosphate (PO_4), biological oxygen demand (BOD), chemical oxygen demand (COD), dissolved oxygen (DO), organic matter percent (OM%), and total organic carbon percent (TOC%). All tests had done by the standard methods [41].

Item	Devices	Company/Origin
1	Temperature, pH, Salinity, EC (portable meter H19811)	WTW/Germany
2	Incubator	Memmer/Germany
3	High-Performance Liquid Chromatography-Tandem Spectrometry (HPLC 8040)	SyknmS1122/Germany
4	COD Meter	Lovibond/Germany
5	Distillation device	Waterpia/Korea
6	Multiparameter photometer/C99	Hanna/Romania
7	Multiparameter photometer/HI83200	Hanna/Romania
8	Turbidity meter	Lovibond/Germany
9	Oven	Memmer/Germany
10	Sensitive balance	Phoenix/Korea
11	Vacuum pump	China
12	Ekman Grape Sampler	BDH/Germany
13	Photolab S12 (PHD)	WTW/Germany
14	Filter paper 0.45 μm	Whittman/UK
15	Ultrasonic Bath	ISOLAB/Germany
16	GPS device	GPS Map 78 s Germin/Tiwan

Table 3.
List of equipments and instruments in this study.

Item	Material	Company/Origin
1	$MnSO_4.H_2O$	Fluka/Germany
2	KOH	Fluka/Germany
3	KI	Sigma-Aldrich/Germany
4	NaN_3	Fluka/Germany
5	NaOH	Fluka/Germany
6	Na_2, NO_3, PO_4 Kit	Hanna/Romania
7	COD	Lovibond/Germany
8	$K_2Cr_2O_7$	Fluka/Germany
9	$Fe(NH_4)_2(SO_4)_2.6H_2O$	Fluka/Germany
10	H_2SO_4	Fluka/Germany
11	H_3PO_4	Fluka/Germany
12	NaF	Fluka/Germany
13	HCL	Fluka/Germany
14	Formaldehyde 37–40%	Romil/UK
15	Methanol	Romil/UK
16	Ethylacetate	Fluka/Germany
17	Dichloromethane	Romil/UK
18	Acetic acid	Fluka/Germany
19	Acetonitrile	Fluka/Germany
20	Anionic surfactant Kit	WTW/Germany
21	Nonionic surfactant	WTW/Germany

Item	Material	Company/Origin
22	4-dodecylbenzene sulfonic acid mixture of isomersm ≥95%	Sigma-Adrich/Germany
23	4-nonylphenyl-polyethylene glycol non-ionic	Sigma-Adrich/Germany
24	$Na_2S_2O_3.5H_2O$	Fluka/Germany

Table 4.
List of chemical materials in this study.

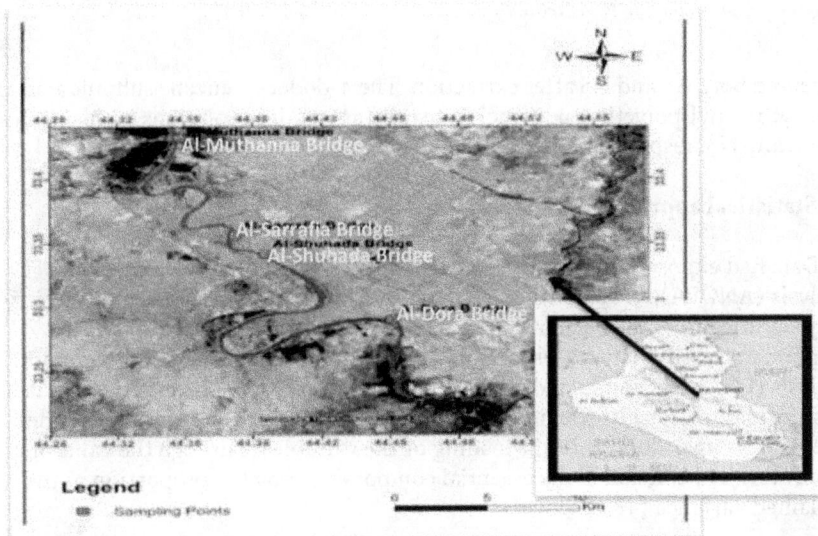

Figure 4.
The study sites in the Tigris River, Baghdad city-Iraq.

5.3 Sediment sampling, extraction, and analysis

Sediment samples had collected using an Ekman Grab (n = 3) for each site to 5 cm depth from the river. Excess water drained and added an adequate 10% formalin volume to submerge the sediment for storage as glass jars and transported to the laboratory. Aluminum foil cleaned with methanol was put over the pot's mouth and then put into the cap to avoid sample contamination. Laboratory sediment samples had reserved at approximately 4°C before surfactant analysis [42].

For 16 hours at 80°C in the oven, the sediment sample for AS dried. After excluding large stones and grit from the dry sediments, surfactant compounds were then extracted (10 gm) with methanol at 50°C (240 V, 3A, 50 Hz) by Ultrasonic water bath (ISOLAB/Germany). Three 10 min extractions (50 ml and 2 x 40 ml) had been done, and then by centrifugation. It then concentrated the combined extract to 2 ml [42].

For NS, sediment samples were homogenized before extraction by sieving with a 2 mm stainless steel sieve. Also, in the same method above, extraction of surfactant compounds was done with Ultrasonic water bath by using a mixture of methanol-dichloromethane (7:3, v/v). The final elutes evaporated afterward with a gentle stream of nitrogen gas and reconstituted with 1 ml of methanol [33].

The photometer photo lab S12 (PHD) and the High-Performance Liquid Chromatography system configuration (HPLC) (Syknm-S1122- Germany)

Site number	Site name	Coordinates	
		Longitude (E)	Latitude (N)
1	Al-Muthanna Bridge	44°34'55.50"	33°42'83.22"
2	Al-Sarrafia Bridge	44°37'36.01"	33°35'37.53"
3	Al-Shuhada Bridge	44°38'79.03"	33°33'79.59"
4	Al-Dora Bridge	44°45'02.84"	33°28'96.82"

Table 5.
The geographical positions (GPS) of the study sites.

measured both AS and NS after extraction. The 4-dodecylbenzene sulfonic acid and 4-nonyl-phenyl-polyethylene glycol were used as standard solutions in this study for AS and NS, respectively.

5.4 Statistical appraisal

Data had exposed to descriptive statistical analysis and one-way variance analysis (ANOVA). Probabilities less than 0.05 ($P < 0.05$) have been used statistically significant. Also, Cluster Analysis (CA) had used the Statistical Release 7 program to classify data, cases, or objects or clusters. The principal component analysis (PCA) was conducted as a series of irrelevant variables to retrieve critical information. Results provided plotting graphs in which the elements of the forecasts groups, along with the loading of the variables. Through the value of Eigenvalue had concluded the essential component or by the proportion of the explained variance [43, 44].

6. Results

6.1 Physicochemical parameters of Tigris River water

The results of the physicochemical parameters of the Tigris River water samples had compared with the Iraqi Maintaining System Law [45] and the Canadian Council of Ministers of the Environment (CCME) [46, 47] guidelines illustrated in **Table 6**.

Air temperature (AT) had ranged between 12.55–43.73°C. In comparison, the water temperature (WT) showed a noticeable seasonal trend with a minimum value of 10.36°C recorded in the wet season and a maximum value of 30.11°C in the dry season.

An EC in this study had ranged between 580.50 and 1108.75 μs/cm in dry and wet seasons, respectively, indicated levels higher than the limit standards.

The concentration of salinity (S‰) varied from 0.2 to 0.48‰, respectively, in the dry and wet seasons. The TDS ranged from 362.75 mg/L during the dry season to 711.75 mg/L during the wet season, but these high concentrations for S‰ and TDS are within the limited value.

The lowest value of total suspended solids (TSS) in the dry season was 3.00 mg/L, and the highest level in the dry season was 84.50 mg/L, while turbidity varied from 10.61 to 193.75 NTU in the wet and dry seasons, respectively.

Water flow (WF) had recorded the highest value in the dry season (0.71 m/s) **(Table 6)**.

Parameters	Range		Mean	Standard Deviation	Standard values	
	Minimum	Maximum			Law 25/1967	CCME
Physical Parameters						
Air temperature (AT) (°C)	12.55 (w)	43.73 (d)	26.27	±9.64	—	—
Electrical conductivity (EC) (μs/cm)	580.50 (d)	1108.75 (w)	876.27	±148.05	0.5–1.0	—
Salinity (S) ‰	0.20 (d)	0.48 (w)	0.35	±0.08	—	—
Total dissolved solids (TDS) (mg/L)	362.75 (d)	711.75 (w)	563.87	±105.47	1000	500
Total suspended solids (TSS) (mg/L)	3.00 (d)	84.50 (d)	18.58	±22.05	60	—
Turbidity (Tur) (NTU)	10.61 (w)	193.75 (d)	67.83	±65.36	5	5
Water flow (WF) (m/s)	0.31 (w)	0.71 (d)	0.47	±0.13	—	—
Water temperature (WT) (°C)	10.36 (w)	30.11 (d)	21.59	±6.83	>35	15
Chemical parameters (standard values)						
pH	7.43 (d)	8.25 (w)	7.75	±0.22	6–9.5	6.5–9
Nutrients (mg/L)						
Nitrite (NO_2)	0.01 (d)	0.45 (d)	0.11	±0.12	0.06	0.06
Nitrate (NO_3)	0.64 (d)	8.97 (d)	4.18	±2.82	15	13
Phosphate (PO_4)	0.07 (w)	1.52 (w)	0.66	±0.43	0.4	0.1
Organic						
Biological oxygen demand (BOD) (mg/L)	0.53 (w)	3.67 (d)	1.5	±0.79	>5	—
Chemical oxygen demand (COD) (mg/L)	3.75 (w)	88.25 (d)	36.73	±34.9	>100	—
Dissolved oxygen (DO) (mg/L)	4.63 (d)	11 (w)	6.18	±2.06	‹5	5.5–9
Organic matter (OM) (%)	0.43 (d)	5.55 (d)	1.7	±1.44	—	—
Total organic carbon (TOC) (%)	0.27 (d)	2.24 (w)	0.88	±0.56	—	—

- = not applicable, w = wet season, d = dry season.
Law 25/1967 = Iraqi River Maintaining System Law.
CCME = Canadian Council of Management of the Environment.

Table 6.
Physicochemical characteristics of Tigris River for wet and dry seasons.

The pH value was between 7.43 in the dry season and 8.25 in the wet season.
Nutrients include nitrite (NO_2), nitrate (NO_3), and phosphate (PO_4).
Concentration ranges recorded in dry and wet seasons were 0.01–0.45 mg/L

for NO_2, 0.64–8.97 mg/L for NO_3, and 0.07–1.52 mg/L for PO_4, respectively. Naturally occurring ions in water as part of the nitrogen cycle are NO_2 and NO_3. Concentrations were remarkably higher for all three nutrients in the wet season than those in the dry season except for PO4 in the wet season. NO_2 displayed higher concentrations than dry season requirements, while NO_3 had declined significantly compared with acceptable values for both dry and wet seasons (**Table 6**).

In this study, the measured organic materials are biological oxygen demand (BOD), chemical oxygen demand (COD), dissolved oxygen (DO), percent organic matter (OM%), and total organic carbon (TOC%). Ranges observed in wet and dry seasons were 0.53–3.67 mg/L for BOD, 3.75–88.25 mg/L for COD, while DO values ranged from 4.63–11.00 mg/L respectively in dry and wet seasons. In this study, the DO is within the allowed limit (**Table 6**).

OM% ranged from 0.43 to 5.55% in the dry season in the present study, although the TOC% in dry and wet seasons ranged from 0.27% to 2.24%, respectively. In this study, the highest values had been registered in dry season for OM% and in wet season for TOC% (**Table 6**).

6.2 Soil texture

The findings of the soil texture differed among the sites of the study as follows (**Table 7**): in S1 it was clay loam (40.4% silt, 30.6% sand, and 29% clay), in S2 it was clay loam (37.7% silt, 27.5% clay and 34.8% sand), S3 it was clay loam (38.6% silt, 33% clay and 28.4% sand) and silty clay loam (51.6% silt, 37% clay and 44.4% sand) at S4.

6.3 Descriptive analysis

Descriptive analysis for anionic surfactant (AS) and nonionic surfactant (NS) concentrations in sediment from the Tigris River sites during the study period by using photo lab S12 (PHD) and HPLC had demonstrated in **Table 8**.

The concentrations of AS using PHD ranged from 41.83 µg/g to 72.05 µg/g during the wet season. In comparison, the minimum NS levels in the wet or dry season were not measurable while the maximum concentrations in the dry season had registered 56.19 µg/g. HPLC results showed concentrations of AS in dry season ranging from 10.80 µg/g to 135.74 µg/g. During the wet season the minimum NS level was not measurable, and the maximum concentration in the dry season was 467.31 µg/g.

The two measurement methods (PHD and HPLC) have been compared by using a T-test analysis. The results revealed no significant variations between the tests

Site	Soil percentage			Soil class
	Clay%	Silt%	Sand%	Soil texture
S1	29	40.4	30.6	Clay loam
S2	27.5	37.7	34.8	Clay loam
S3	33	38.6	28.4	Clay loam
S4	37	51.6	11.4	Silty clay loam

Table 7.
Soil texture of sediment samples.

Parameters	Range		Mean	Standard deviation	Standard values	
	Minimum	Maximum			Law 25/1967	CCME for rivers
Phtolab (µg/g)						
Anionic Surfactant (AS)	41.83 (w)	72.05 (w)	52.85	9.88	—	—
Nonionic Surfactant (NS)	ND (w&d)	56.19 (d)	17.12	26.73	—	—
HPLC (µg/g)						
Anionic Surfactant (AS)	10.80 (d)	135.74 (d)	34.15	35.32	—	—
Nonionic Surfactant (NS)	ND (w)	467.31 (d)	163.80	147.38	—	—

- = not applicable, w = wet season, d = dry season.
Law 25/1967 = Iraqi River Maintaining System Law.
CCME = Canadian Council of Management of the Environment.

Table 8.
Surfactant concentrations in the sediment of the Tigris River for wet and dry seasons.

obtained by the two instruments for measuring AS at P‹0.05 (t = 0.088), while considerable discrepancies for measuring NS had been obtained at P‹0.05 (t = 0.004).

6.4 Cluster analysis

Two clusters diagram shows (**Figure 3a**) during the wet and dry season. Whereas two highest values (72.05 and 69.71 (µg/g) for AS by PHD, (ND) for NS by PHD) had recorded, whereas during the wet season they were 14.83 and 18.20 µg/g for AS (HPLC) and 56.17 and 55.03 µg/g for NS (HPLC). No detectable (ND) for NS (by HPLC) was recorded in dry season, particularly in June and August 2017 and May 2017, which indicates a marked variation in the season.

During the study period the cluster diagram (**Figure 6a**) shows two clusters. The first consisted of two sub-clusters; first, the pair of S3-S4: HPLC (37.5 and 235.79 µg/g, respectively) had reported specifically high concentrations of AS and NS in these sites. However, measurement with PHD did not show any detectable NS levels. Second, the pair of S1-S2: PHD (57.88 and 53.17 µg/g) had registered at S1, the closest highest values for AS. In comparison, S2 registered the similarly highest by PHD values (34.32 and 34.17 µg/g) for NS.

6.5 Correlation matrix

Table 9 shows the correspondence matrix of the results that recorded the following strong correlations (P‹0.05): Significant negative correlation of AT with NS (PHD) (r = −0.997) and a strong positive correlation with AS (HPLC) (r = 0.999) obtained. The correlation matrix also shows a strong degree of correlation between NS determined by PHD and TSS (r = 0.998), COD (r = 0.998), and NO_2 (r = 0.999).

6.6 Correlation between soil texture and surfactants in sediment

Table 10 demonstrated the correlation matrix between soil texture and occurrence of AS and NS at Tigris River deposit. Nevertheless, the results revealed no clear correlation.

Parameters	AT (°C)	TSS (mg/L)	COD (mg/L)	NO$_2$ (mg/L)
AS (µg/g) PHD	−0.778	−0.170	−0.294	−0.206
	P = 0.432	P = 0.892	P = 0.810	P = 0.868
NS (µg/g) PHD	−0.997	−0.697	−0.783	−0.723
	P = 0.050	P = 0.509	P = 0.428	P = 0.485
AS (µg/g) HPLC	0.999	0.720	0.803	0.745
	P = 0.029	P = 0.488	P = 0.407	P = 0.465
NS (µg/g) HPLC	0.787	0.998	0.998	0.999
	P = 0.423	P = 0.036	P = 0.045	P = 0.012

The correlation marked is significant at P<0.05.

Table 9.
Correlation between physicochemical and surfactants in Tigris River sediment during the study period.

Parameters	Clay %	Silt %	Sand %
AS (µg/g) PHD	0.269	0.191	−0.232
	P = 0.73	P = 0.81	P = 0.77
AS (µg/g) HPLC	0.940	0.807	−0.897
	P = 0.06	P = 0.19	P = 0.10
NS (µg/g) HPLC	0.389	0.824	−0.678
	P = 0.61	P = 0.18	P = 0.32

he analysis had done according to temporal variation for all four sites.

Table 10.
Correlation between soil texture and surfactants in sediment.

7. Discussion

7.1 Physicochemical parameters of Tigris River water

All the data for AT and WT during the study period was within the permissible limit of weather rates for Baghdad city during wet and dry seasons (**Table 6**) [48]. This result is consistent with the previously reported role of AT for the heat budget of the Tigris River [49].

The excellent indicator assessment for total dissolved solids (TDS) in the water of the aquatic ecosystem is electrical conductivity (EC) [41]. During the wet season, such high concentrations of major ions have recorded as those of the dry season may be attributed to increased surface runoff, the flow of irrigation water return, soil salinity, and increased human activities [50].

High values of S‰ and TDS parameters may result from increased surface runoff, river geological erosion, increased evaporation rate, and increased human activity, all of which may result in increased ion concentrations [50].

The factors that can lead to an increase of the TSS level are silt, decaying plant and animal matter, industrial wastes, and sewage. But such high TSS values will cause many stream health and aquatic life problems [51, 52]. River sediments represent suspended solids that are reliant on discharge [41]. Turbidity found values above the allowable limits, with turbid water evident by the eye, probably due to the presence of organic, inorganic matter, bacteria, silt, algae, etc. [53].

WF determines the degree and type of deposition and, thus, the nature of sediment [54]. Water flow is an essential factor that moves the pollutants into regions far from their origin. The reason that might cause an increase in flow rate is melting snow in the summer season; this explains much lower levels of sediment-measured pollutants [55].

The values of pH indicating river water is typically alkaline slightly and within the permissible amount for aquatic living [56].

In the environment, the NO_3 had known to be more stable than NO_2 [57]. The microbial activity, especially during the summer season because most biological processes take place during this season, is one of the reasons that led to a decrease in the nitrate concentration in the Tigris River due to the uptake process by these microorganisms [58]. For the concentration of PO_4, it was higher than the permitted level of aquatic life for both seasons, and the highest value had recorded in the wet season. In the water body, the PO_4 is an important nutrient, and only the soluble form, inorganic phosphorus, can be directly utilized by aquatic biota [59]. The presence of phosphorus in the environment is either through the natural or activities of humans. Natural phosphorus sources include atmospheric precipitation, natural rock, and mineral dissolution, weathering of inorganic soluble minerals, biomass decomposition, runoff, and sedimentation. The anthropogenic source, by comparison, contains detergents, animal wastes, fertilizers, wastewater, and effluent from the septic tank, and industrial discharge [60].

BOD is a function of the amount of oxygen the bacteria consume, which decomposes organic matter into both surface water and waste [58]. The BOD in this study is within the permissible limits [45]. COD is a measure of the number of chemicals, usually organics, that consume dissolved oxygen [61]. All values in wet and dry seasons were within the permissible limits, and they agree with those found by previous investigations [62]. One of the parameters that maintain biological life in water is DO, and its variations depend on temperature and the presence of algal communities [63]. Raising a water body's flow rate would increase the amount of dissolved oxygen in the water, due to the flow rate increases the atmospheric oxygen diffusion and movement from and into the water. Organic matter (quantitly) in the water impacts the dissolved oxygen levels by decreasing it [64].

TOC% estimate in sediment and soil samples is an essential criterion for determining environmental quality. In the ecosystem, organic matter exists in components of soil, ground, water, and sediment. The presence of these compounds in the sediment results in their interaction with metal ions allowing soluble or insoluble complexes to form. Such complexes, in effect, associate with minerals in the sediment to form particles capable of absorbing them into other pollutants [65]. TOC indicated river pollution because of the proportionality between TOC content and organic matter, which has an affinity for trace anionic and nonionic surfactant contaminants [66].

Urano et al. [67] showed that the sediment's adsorption potential tends to be independent on the residual surface area but is more related to the organic carbon content. Also found adsorption of AS and NS values on the microbiota equal to their sediment adsorption values. Organic matter (quantity) in the sediment supplies matrices for the adsorption of hydrophobic compounds and disposed them incomplete slowly [33].

7.2 Soil texture

The soil texture differed among the sites in this study (**Table 7**) from clay loam at S1, S2, S3, and silty clay loam at S4 [68, 69].

7.3 Descriptive analysis

Table 8 illustrates the descriptive analysis for anionic surfactant (AS) and nonionic surfactant (NS) concentrations in the sediment of the Tigris River. There is no standard has found, whether Iraqi or international, about the permissible limits of the presence of the surfactants in river sediments to compare with data of the present study [70].

The main important factor to absorption surfactants compound on sediment is the different charges of these compounds, so the cationic surfactants (CA) can undergo sorption in sediment much more significant in contrast to AS such as sodium dodecylbenzene sulfonate (SDBS) [39]. The coefficient of equilibrium distribution for CS is twice more significant than that for SDBS. Also, low organic carbon content and neutral pH provide CS with more excellent sorption capability to the sediment, due to the electrostatic interactions between the positively charged ammonium groups ($[(CH_3)_3NR]^+$), forming the heads of CS and the overall negatively charged sediment surface. SDBS shows lower sorption than CS, because the negative charge of SDBS compared to CS, which has positively charged [39]. And the adsorption of surfactant forms in sediment as Cationic > Nonionic > Anionic may be classified in the sequence. No measurable NS concentrations were reported using PHD in this study during 9 of the 13 months of the study period (3 in the wet season and 6 in the dry season), as opposed to higher AS concentrations. Lif and Hellsten [71] have shown that the NS has an amide group, comprising a small portion of the total volume of NS. However, their development and use are growing due to excellent chemical stability with rapid biodegradation and relatively simple processes of manufacture based mainly on renewable raw materials.

7.4 Cluster analysis

The seasonal variation illustrates in **Figure 5b**. In comparison to the wet season with lower temperatures, concentrations of NS and AS (by HPLC) observed the highest values in the dry season. The explanation could be either to lower pollutant (surfactant) inputs into the river or to more effective biodegradation of compounds studied in the dry season in the river water. Marcomini et al. [40] observed a significant seasonal variation in LAS and nonyl-phenol ethoxylates (NPEOs) concentrations at Lake Venice. Mainly due to increased biodegradation at temperatures above 20°C (late spring and summer). In 1994, coworkers observed the similar seasonal variation of some compounds in the Glatt River (Switzerland) [30]. Such as nonyl-phenol (NP), lipophilic nonyl-phenol monoethoxylate, and nonylphenol diethoxylate (NP_1EO, and NP_2EO).

A significant difference had observed for NS (HPLC) in which the two highest values at S1 and S4. At the same time, the lowest NS (HPLC) levels were at both S2

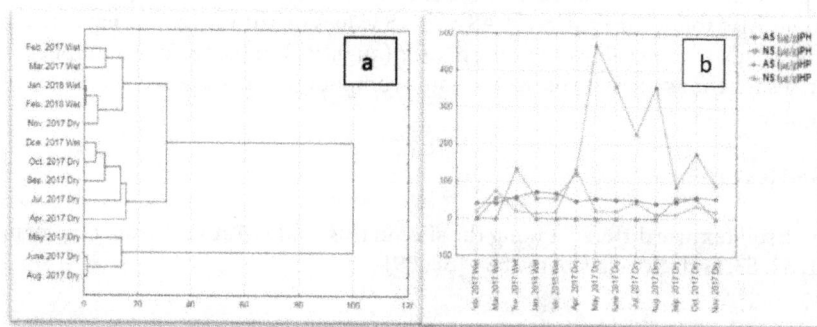

Figure 5.
a- Cluster diagram of temporal of the sampling period for wet and dry seasons, b- temporal variation of surfactants in sediment for wet and dry seasons.

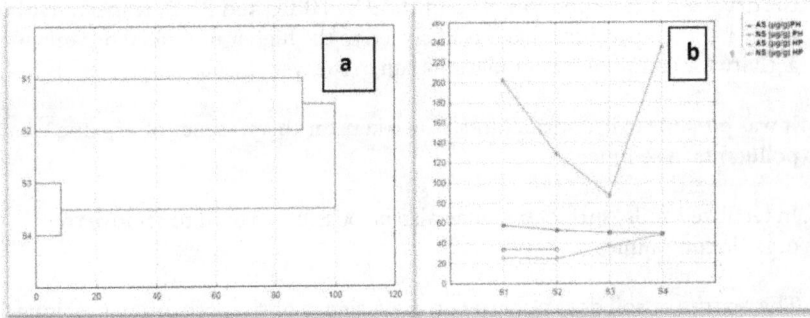

Figure 6.
a-cluster diagram of spatial clustering of sampling sites for the wet and dry season, a b-spatial variation of surfactants in sediment for dry and wet seasons.

and S3. Nevertheless, during the study period, no apparent difference was found for other surfactants (**Figure 6b**), which suggests that S1and S4 reported the highest levels of pollution with NS (by HPLC) relative to the lowest in S2 and S3, indicating these sites as hot spots for point sources of municipal and industrial discharges.

7.5 Correlation matrix

The correlation matrix between Physicochemical and surfactants in Tigris River sediment during the study period (**Table 9**) shows strong correlations ($P^‹0.05$) of (AT, TSS, COD, and NO_2) with AS and NS, which means that only these parameters affect on the presence and degradation of anionic and nonionic surfactants either positive and negative correlation.

7.6 Correlation between soil texture and surfactants in sediment

Cano and Dorn [72] and Brownawell et al. [73] reported that the alcohol ethoxylate surfactant sorption is better associated with the sediment's clay content than to its organic carbon content (**Table 10**). This study found no clear correlation.

8. Conclusions

Several conclusions that drown based on the findings of this study:

1. Two measuring techniques had applied with similar efficiency to measure anionic (AS) in river sediments while HPLC was more efficient for nonionic surfactants (NS)

2. The sediment serves as a sink for the sedimentation of AS and NS, the residual in the river water.

3. Four environmental parameters (air temperature, total suspended solids, chemical oxygen demand, and nitrite) were more critical factors impact on surfactants.

4. Nonionic surfactants and, most likely, their degradation products, nonylphenol, are significant contaminants because of toxic impacts in the aquatic environment, especially during the dry season. The highest concentration in

the dry season recorded (467.31 and 56.19 by HPLC and PHD, respectively) due to the high temperature that has likely led to the higher microbial organism's activity for compound degradation than in the wet season.

5. It was possible to use surfactants as markers for the presence of organic pollutants in sediments.

6. In fertilized soils, surfactants may also be possible to used to remove the organic compound.

7. The texture of soil does not affect the efficiency of the adsorption precipitation of AS and NS on river sediment.

Acknowledgements

The authors would like to acknowledge all laboratories in the Department of Environmental Pollution/ Environmental Research Center/University of Technology-Baghdad-Iraq and Department of Biology/College of Science for Women/Bagdad University-Baghdad-Iraq for conducting the experiments.

Also, the authors would like to extend thanks and appreciation to the members of the Ministry of Interior/Baghdad Operations Command/River Polices for protecting the collection of samples.

Author details

Rana R. Al-Ani[1], Fikrat M. Hassan[1,2*] and Abdul Hameed M. Jawad Al-Obaidy[1]

1 Environment Research Center, University of Technology, Baghdad, Iraq

2 Department of Biology, College of Science for Women, University of Baghdad, Baghdad, Iraq

*Address all correspondence to: fikrat@csw.uobaghdad.edu.iq

IntechOpen

References

[1] Schmitt TM. Analysis of surfactants. 2nd ed. Revised and Expanded. Madison Avenue: New York; 2001. 540p.

[2] Olkowska E, Polkowska Z, Namie'snik J. Analytics of surfactants in the environment: problems and challenges. Chem. Rev. 2011; 111 (9): 5667-5700.

[3] Kreisselmeier A, Durbeck HW. Determination of Alkylphenols, alkylphenolethoxylates and linear alklbenezenesulfonates in sediments by accelerated solvent extraction and supercritical fluid extraction. J. Chromatogr. A. 1997; 775: 187-196.

[4] White R, Jobling S, Hoare SA. Environmentally persistent alkylphenolic compounds are estrogenic. Endccrinol. 1994; 135: 175-182.

[5] Madsen T, Boyd HB, Nyle'n D, Pedersen AR, Petersen GI, Simonsen F. Environmental and Health Assessment of substances in Household Detergents and Cosmetic Detergent products. Environ. Pro. Miljoprojekt. 2001; 11: 615.

[6] Olkowska E, Ruman M, Polkowska Z. Occurrence of surface active agents in the environment. J. Anal. Methods Chem. 2014; 15. DOI: 10.1155/2014/769708.

[7] Pielou EC. Freshwater. Chicago. University of Chicago Press. 1998; 286p.

[8] Ying GG. Fate, behavior and effects of surfactants and their degradation products in the environment. Environ. Int. 2006;32: 417-431.

[9] Lechuga M, Fernandez-Serrano M, Jurado E, Nunez-Olea J, Rios F. Acute toxicity of anionic and nonionic surfactants to aquatic organisms. Ecotoxico. Environ. Saf. 2016; 125: 1-18.

[10] Rozas F, Castellote M. Electrokienetic remediation of dredged sediments polluted with heavy metals with different enhancing electrolytes. Electrochim. Acta. 2012; 86: 102-109. DOI: 10.1016/j.electacta.2012.03.068.

[11] Issa MJ, Al-Obaidi BS, Muslim RI. Evaluation of some trace elements pollution in sediments of the Tigris River Wasit governorate, Iraq. Baghdad Science Journal. 2020; 17(1): 9-22. DOI: http://dx.doi.org/10.21123/bsj.2020.17.1.0009.

[12] Inaba K, Koji A. HPLC determination of linear alkylbenzenesulfonate (LAS) in aquatic environment, seasonal changes in LAS concentration in polluted lake water and sediment. Intern. J. Environ. Anal. Chem. 1988; 34: 203-213.

[13] Aboul-Kassim TA, Simoneit BRT. Detergents: A review of the nature, chemistry, and behavior in the aquatic environment. Part I. Chemical composition and analytical techniques. Critical Reviews in Environmental Science and Technology. 2009; 23(4): 325-376.

[14] Okpokwasili GO, Nwabuzor CN. Primary biodegradation of anionic surfactants in laundry detergents. Chemosphere, 1988; 17: 2175-2182.

[15] Smulders E, Von Rybinski W, Sung E, Rahse W, Steber J, Wiebel F, Nordskog A. Laundry detergents. (In Ullmanns Encyclopedia of Industrial Chemicals), Bohnet M, Brinker CJ, Clemens H, Cornils B, Evans TJ, Greim H, Hegedus LL, Heitbaum J, Herrmann WA, Karst U *et al* Eds. Wiley-VCH GmbH, Weinheim: Germany; 2009.

[16] Tadros TF. Applied surfactants: principle and applications. Wiley-VCH, Weinheim: Germany; 2005.

[17] Haigh SD. A review of the interaction of surfactants with organic contaminants in soil. Sci Total Environ. 1996; 185: 161-70.

[18] Olkowska E, Ruman M, Kowalska A, Polkowska Z. Determination of surfactants in environmental samples Part II. Anionic compounds, Ecological Chemistry and Engineering S. 2013; 20 (2): 331-342.

[19] Rosen MJ, Dahanayake M. Industerial utilization of surfactants: principles and practice. American Oil Chemist's Society, Champaign III: USA; 2000.

[20] Cserhati T, Forgacs E, Oros G. Biological activity and environmental impact of anionic surfactant. Environ Int. 2002; 28: 337-48.

[21] McDonell G, Russel AD. Antiseptics and disinfectants: activity, action and resistance. Clin Microbial Rev. 1999; 12: 147-79.

[22] Cserhati T. Alkyl ethoxylated and alkylphenol ethoxylated nonionic surfactants: Interaction with bioactive compounds and biological effects. Environ Health Perspect. 1995; 103: 358-64.

[23] Holt MS, Waters J, Comber MHI, Armitage R, Morris G, Newberry C. AIS/CESIO environmental surfactant monitoring programe. SDIA sewage treatment pilot study on linear alkylbenzene sulphonate (LAS). Water Res. 1995; 29: 2063-71.

[24] Ivankovic T, Hrenovic J. Surfactants in the environment. Arth Hig Rada Toksikol. 2010; 61: 95-110.

[25] Pravilnik O. Croatian regulations on boundry limits of hazardous novine. 1999; 40.

[26] Tozum-Calgan SRD, Atay-Guneyman NZ. The effects of an anionic and a nonionic surfactant on growth and nitrogen fixing ability of a cyanbacterium *Gloeocapsa*. J Environ Sci Health Part A. 1994; 29: 355-70.

[27] Scott MJ, Jones MN. The biodegradation of surfactants in the environment. Biochimica et Biophysica Acta. 2000; 1508(12): 235-251.

[28] Gonzaler S, Petrovic M, Barcelo D. Advanced liquid chromatography-mass spectrometry (LC-MS) methods applied to wastewater removal and the fate of surfactants in the environment. Trends in Analytical Chemistry. 2007; 26 (2): 116-124.

[29] Berna JL, Ferrer J, Moreno A, Prats D, Ruiz Bvia F. The fate of LAS in the environment. Tenside Surfactants Deterg. 1989; 26(2): 101-107.

[30] Ahel M, Giger W, Koch M. Behaviour of alkylphenol polyethoxylate surfactants in the aquatic environment-I. occurance and transformation in sewage treatment. Water Research. 1994; 28(5): 1131-1142.

[31] Petrovic M, Barcelo D. In emerging organic pollutants in wastewaters and sludges. Barcelo D ed. Springer Heidelberg: Germany; 2004.

[32] Garcia MT, Campos E, Ribosa I. Biodegradability and ecotoxicity of amine oxide based surfactants. Chemosphere. 2007; 69: 1574-8.

[33] Petrovic M, Fernandez-Alba AR, Borrull E, Marce RM, Mazo EG, Barcelo D. Occurrence and distribution of nonionic surfactants, their degradation products, and LAS incoastal waters and sediments in Spain. Environmental Toxicology and Chemistry. 2002; 21(1): 37-46.

[34] Lara-Martin PA, Gomez-Parra A, Gonzalez-Mazo E. Reactivity and fate of synthetic surfactants in aquatic

environments. Trends in Analytical Chemistry. 2008A; 27(8): 684-695.

[35] Lara-Martin PA, Gomez-Parra A, Gonzalez-Mazo E. Sources, transport and reactivity of anionic and nonionic surfactants in several aquatic ecosystems in SW Spain: a comparative study. Environmental Pollution. 2008B; 156(1): 36-45.

[36] Mann RM, Boddy MR. Biodegradation of a nonylphenol ethoxylate by the autochthonous microflora in lake water with observations on the influence of light. Chemosphere. 2000; 41(9): 1361-1369.

[37] Quiroga JM, Sales D, Gomez-Parra A. Experimental evaluation of pollution potential of anionic surfactants in the marine environment. Water Res. 1989; 23: 801-8.

[38] Manzano MA, Perales JA, Sales D, Quiroga JM. The effect of temperature on the biodegradation of a nonylphenol polyethoxylate in river water. Water Res. 1999; 33: 2593-600.

[39] Pan G, Chengxia J, Dongye Z, Chun Y, Hao C, Guibin J. Effect of cationic and anionic surfactants on the sorption and desorption of perfluorooctane sulfonate (PFOS) on natural sediments. Environmental Pollution Journal. 2009; 157: 325-330.

[40] Marcomini A, Pavoni B, Sfriso A, Orio AA. Persistent metabolites of alkylphenol polyethoxylates in the marine environment. Mar. Chem. 1990; 29: 307-323.

[41] APHA (American Public Health Association), Standard Methods for the Examination of Water & Wastewater. 22nd Edition, Washington, DC, USA. 2012; 1200.

[42] Gordon WJ, Muller N, Gysman SJ, Marshall CJ,

Sparham SM, O'Connor MJ, Whelan. Effect of laundry activities on in-stream concentrations of linear alkylbenzene sulfonate in a small rural South African river. Sci Total Environ. 2009; 407(15): 4465-4471. DOI:10.1016/j.scitotenv.2009.04.023

[43] McGarial K, Cushman S, Stafford S. Multivariate statistics for wildlife and ecology research. Springer: New York; 2000. 81-128p. DOI:10.1007/978-1-4612-1288-1

[44] Mckenna G. An enhanced cluster analysis program with bootstrap significant testing for ecological community analysis. Environmental Modeling and Software. 2003; 18: 205-220.

[45] Rivers Maintaining System and General Water from Pollutin (RMS and GWP) No 25, (1967). Iraqi Official Gazette; No. 1446 on 16 July. 1967; 2: 108.

[46] CCME, (Canadian Council of Ministers of the Environment), Canadian water quality guidelines for the protection of aquatic life: Canadian water quality index 1.0 technical report. In Canadian Environmental Quality Guidelines. Winnipeg; Manitoba. 2000; 5. http://www.ccme.ca/assets/pdf/wqi usermanualfctsht e.pdf.

[47] Lumb A, Doug H, Tribeni S. Application of CCME water quality index to monitor water quality: A cases of the Mackenzie River Basin Canada. Environ. Monit. Assess. 2015; 113: 411-429.

[48] Ministry of Transportation/Iraqi Meteorological organization and Seismology at Baghdad International Airport (MT/IMOS). 2018.

[49] Al-Lami AA, Kassim TI, Dulymi AA. A limnological study on Tigris River, Iraq. The Sci. J. of Iraqi Atomic Commission. 1999; 1: 83-98.

[50] Miller RL, Bradford WL, Peters NE. Specific conductance: Theoretical cosiderations and application to analytical quality control, in US. Geological Survey Water-Supply Paper; Retrieved from. 1988: 8-23. http://pubs.usgs.gov/wsp/2311/report.pdf.

[51] Al-Shujairi SOH, Sulaiman SO, Najemalden MA. Variations of major ionic composition and salinity of Tigris River within Iraq. SJES. 2015; 1(2): 64-70.

[52] EPA, 5.5 Turbidity. In water: Monitoring & Assessment. 2012 Retrieved from http://water.epa.gov/type/rsl/monitoring/vms55.cfm.

[53] Daphne LHX, Utomo HD, Kenneth LZH. Correlation between turbidity and total suspended solids in Singapore Rivers. JWS. 2011; 1(3): 313-322.

[54] Wetzel RG, Likens GE. Limnology analyses. 3nd ed; Springer. 2000; 432p.

[55] Weiner ER. Application of environmental chemistry. Lewis Publishers, London: New York, 2000; 347p.

[56] Al-Ani RR, Al-Obaidy AMJ, Badri RM. Assessment of water quality in the selected sites on the Tigris River, Baghdad-Iraq. IJAR. 2014, 2(5): 1125-1131. http://www.journalijar.com

[57] ICAIR. Life systems, Inc. Drinking water criteria document on nitrate/nitrite. Washington, DC, United State Environmental Protection Agency, Office of Drinking water. 1987; 5.

[58] WHO, World Health Organization, Nitrate and Nitrite in drinking water: Background document for development of WHO Guidelines for drinking water quality. 2011. WHO/SDE/WSH/07.01/16/Rev/1, World Health Organization, 20 Avenue Appia, 1211 Geneva 27, Switzerland.

[59] Wetzel RG. Limnology. 3nd. Edition. Academic Press: California; 2001. 729.

[60] Fadiran AO, Dlamini SC, Mavuso A. A comparative study of the phosphate levels in some surface and ground water bodies of Swaziland. Bull; Chemical Society of Ethiopia. 2008; 22(2): 197-206.

[61] Salah KF, AbdulGhafoor KF, Abdalwahab EM. Assessment of pollution level of Habbaniya Lake, Iraq, using organic pollution indicators. IJLR. 2014; 7(1): 25-36.

[62] Al- Hiyaly SAK, Ma'alah WN, Al-Azzawi MN. Evaluating the effects of medical city wastewater on water quality of Tigris River. Eng. & Tech. J., Second conference on environment and sustainable development 28-29-Oct-2015, 2016; 34(3): 405-417

[63] Lashari K, Korai A, Sahato G, Kazi T. Limnological studies of Keenjhar Lake (District Thatta), Sindh, Pakistan. PJAEC. 2009; 10: 39-47.

[64] WHO, World Health Organization, Rolling revision of the WHO guidelines for drinking-water quality, Draft for review and comments. 2000; 77p. http://www.HIs.Gov.bc.ca/protect/pdf/WHO-V2-2000.

[65] Laze A, Lazo P, Arapi V. Determination of total organic carbon (TOC) in sediment of Mat River. JIEAS. 2011; 6(5): 699-703.

[66] Laza L, Gomoiu MT, Boicenco L, Vasiliu D. Total organic carbon (TOC) of the surfaces layer sediments covering the seafloor of the Romanian Black Sea Coast. J. Geo-Eco-Marina. 2012; 18: 121-132.

[67] Urano K, Saito M, Murata C. Adsorption of surfactants on sediments. Chemosphere. 1984; 13(2): 293-300.

[68] Salman JM, Hassan FM, Baiee MA. Practical methods in Environmental and pollution laboratory. National library: Baghdad; 2017. 119-120p.

[69] USDA, National resources conversation service. 2018. http://www. nrcs.usda.gov/wps/potal/nrcs/detail/ soils/survey/?cid=nrcs142p205416

[70] Hassan FM, Al-Obaidy AMJ, Al-Ani RR. Detection of detergents (surfactants) in Tigris River-Baghdad Iraq. IJEW. 2017; 6(2): 1-15.

[71] Lif A, Hellsten M. Nonionic surfactants containing an amide group, in van Os NM,ed, Nonionic surfactant Science Series, 72. Marcel Dekker: New York; NY; USA; 1998. 177-22p.

[72] Cano ML, Dorn PB. Sorption of an alcohol ethoxylate surfactant to natural sediments. Environ. Toxicol. Chem. 1996; 15: 684-690.

[73] Brownawell BJ, Chen H, Zhang W, Westall JC. Sorption of nonionic surfactants on sediment materials. Environ. Sci. Technol. 1997; 31: 1735-1741.

Chapter 5

Oil-Mineral Flocculation and Settling Dynamics

Andrew J. Manning, Leiping Ye, Tian-Jian Hsu,
James Holyoke and Jorge A. Penaloza-Giraldo

Abstract

In recent decades, oil spill contamination has tended to occur more commonly in deltaic and estuarial systems. The management of oil spillages has been a major challenge in the surrounding deltas due to the highly sensitivity nature of deltaic ecosystems. Many deltas have an abundance of clay minerals that can flocculate, and these play an important role in determining the transport of spilled oil contamination and its eventual fate, particularly given that suspended sediment and microbial activities are often prevalent and diverse in natural environments. The primary work presented here focuses on laboratory experimental studies that help develop improved parameterizations of flocculation processes for oil-sediment-biogeochemical modeling. Oil-mineral flocs (OMA) have been successfully created from a series of laboratory flocculation experiments. A floc video instrument LabSFLOC-2 has been adopted for the first time to study the settling dynamics of OMAs. Experimental results reveal OMAs can easily form in any oil, cohesive sediment, and seawater mixtures. However, Kaolin and Bentonite forms dramatically different OMA structures, which leads to their variable characteristics. In the Bentonite clay cases, the oil flocs tend to be much larger and with higher densities than those in Kaolin clay cases, resulting in significant variability of flocs settling velocities.

Keywords: cohesive sediments, flocculation, settling velocity, floc measurement, oil-mineral floc, sediment dynamics, oil contamination

1. Introduction

1.1 Scope of the problem

The rise in anthropogenic activities and industrialized related development in coastal and marine environments, and their ensuing recovery, from contamination by oil spillages, have been a considerable challenge [1]; especially in coastal deltaic regions where they have high vulnerability to aquatic ecosystem and associated public health issues [2]. During the past decades, contamination from oil spill incidents has gravitated to occur more frequently in vulnerable deltaic waters [3, 4]. To illustrate the global scale of the problem, a few examples of delta regions from around the world are presented.

In Africa, oil spill sites are a common phenomenon in the Niger Delta region. Between 1976 and 1996, a total of 4647 incidents resulted in the spilling of approximately 2,369,470 barrels of oil into the environment [5]. Of this quantity, an estimated 1,820,410.5 barrels (77%) were not recovered [5]. They tend to originate from a variety of sources including: leaks during processing, corrosion of oil pipes, poor maintenance of infrastructure, and deliberate acts of vandalism or theft of crude oil from pipes.

The Nigerian coastline is about 853 kilometers in length from the border with the Republic of Benin in the west to the Republic of Cameroon in the east. Encompassing an area of approximately 70,000 km², the Niger Delta is one of the largest wetlands in the entire world. The region comprises: sandy and muddy morphological features, swamp forests (both seasonal and permanent), low-land rain forests, and mangroves—both saline and brackish [6]. The region is topographically characterized by numerous rivulets, creeks, canals, and rivers of various sizes. The coastal line borders the Atlantic Ocean and is subjected to direct tidal interaction. In contracts, the mainland experiences episodes of flooding from the riverine networks, dominated by the River Niger. The soils underlying the Niger delta are generally characterized as soft, highly compressible, organic, and inorganic silty clays overlying fine sands at great depths [7].

Historically, the largest oil spill in Nigeria occurred during January 1980, when an offshore well blew out and oil spread throughout Nigeria's Atlantic coastline. This spill caused extensive damage to nearly 340 hectares of mangrove [5] and was equivalent to approximately 200,000 barrels of crude oil (this is equivalent to 8.4million US gallons). Although not fully verified, over the past half century, it is estimated that oil spill incidents account for nearly 546 million gallons of oil [8] into the Niger Delta environment. This is comparable to approximately 11 million gallons per year [9]. UNEP [10] reported that many communities in the Niger Delta, in particular Ogoniland region, are continuing to live with a persistent state of pollution from oil spills.

Oil spills have blighted a myriad of the Nigeria Delta environment including: the ambient air, ground and surface waters, and crops (via bioaccumulation). Ordinioha and Brisibe [8] disclose that associated contaminants can include (but not limited to): trace metals; hydrocarbons, such as carcinogens, namely polycyclic aromatic hydrocarbon and benxo (a) pyrene; and naturally occurring radioactive substances.

For example, the Bodo oil spill in the Niger Delta region in 2008, which was caused by operational problems, recorded approximately 4000 barrels of oil spill a day for a 10-week period. The scale of the spill is likened to the Exxon Valdez 1989 disaster in Alaska, where 10 million gallons of oil destroyed the remote coastline [5].

The Niger Delta has ecologically sensitive wetlands, which makes the impact of oil spills to be widely felt by the people in the region. The people depend on the environment for traditional livelihood (farming and fishing) and subsistence; thus the contamination of water, land and air quality continuously erode their capacity to support the lifestyle and economy survival [9, 11]. When spills occur, whether on lands or waters caused by operational error, sabotage and theft, equipment failure, and aging pipeline, it contaminates water use for domestic purpose, fishing and farming activities with severe impact on the local people who are left with no alternative.

In the United Kingdom, since 1924, the Firth of Forth has received effluent from Scotland's major petrochemical industry and refinery in addition to hydrocarbon inputs from many other sources [12]. This has created a large residue of hydrocarbon contaminants within its sediments. The major inputs have produced localized lethal effects, in that productive estuarine intertidal habitat has been lost or its functioning altered. The changes are primarily due to organic enrichment with only very near-field toxic effects. Some of the effects result from historical contamination, but existing discharges continue to have a deleterious effect. New petrochemical discharges in both firth and estuary appear to have a minimal impact, although the historical contamination makes this partly inconclusive [12].

Of particular note, the largest oil spill event in human history (at the time of publication) occurred in 2010 and was the Deepwater Horizon (DWH) disaster. This occurred in the vicinity of numerous coastal delta zones, and a significant net influence on: human activities, ecosystem contaminations, and environmental changes. Of the habitats recurringly oiled throughout the DWH spill event, Martin [13] reported that salt marshes were among the most frequently affected (45%), with remedial activities occurring on less than 9% of the affected territories [14]. The resilience of these salt marsh habitats is vital to the persistence of the native fish species that populate these marsh regions, together with the vivacity that is redirected to pelagic food webs [15]. The vulnerability of estuarine ecosystems and their fauna to oil released from DWH has been illustrated by numerous studies (e.g. [16–19]).

DWH released approximately 4.9 million barrels (or 779 million L) of crude oil into the Gulf of Mexico [20, 21]. Michel et al. [14] estimated that the DWH spill directly affected approximately 1773 km of the shoreline habitat (much of this deltaic). In addition to the bleak ecological impacts the spill site imposed on neighboring shoreline, there was an unanticipated sedimentation of oil-associated marine snow (often referred to as Marine Oil Snows - MOS) throughout the water column, extending down to the seafloor; this gave rise to a supplementary prolonged impact on benthic zones [22–24].

Once oil is spilled into an aquatic environment, it tends to float as it is less dense (i.e., more buoyant) than the surrounding water, and this connection with the overlying atmosphere can facilitate the natural weathering processes on the oil contaminants. Having said that, a number of studies [23, 25] observed that a significant amount of the floating oil droplets readily aggregated with suspended matter already present in the water column, including (but not limited to): fecal pellets, detritus, sediments, and phytoplankton. In addition, both Passow and Alldredge [26] and Malpezzi et al. [27] found that this aggregation process was markedly enhanced by the presence of biologically secreted substances specifically Transparent Exopolymer Particles (TEP) or the more generic Extracellular Polymeric Substance (EPS; e.g. [28–30]). These MOS would eventually settle and deposit on the seafloor as they were transported through the water column [22, 31].

Following an oil spill, chemical dispersants are frequently chosen to remedy visible surface contaminants [32–34]. However, in the wake of dispersing an oil slick into smaller constituent oil droplets, there is a likelihood that these diminutive oil droplets agglomerate with other cohesive suspended materials present in the water column to further augment MOS flocculation and settling [35, 36]. Many deltaic regions, including estuarine and coastal water systems with energetic flows [37–39], are often dominated by muddy sediments comprising clay-based minerals and EPS (such as those exuded by phytoplankton). When oil is introduced into the sedimentary matrix, Zhao et al. [40] report that they can freely flocculate into oil-particle aggregates (OPAs). Ye et al. [41] concluded that this interchange between cohesive minerals, oil contaminants, and biological EPS collectively perform a pronounced role in the fate of oil spills in a natural ecosystem. Thus, an improved understanding of the formation, physical characteristics, and settling dynamics of OPAs is extremely advantageous when assessing how they will affect a subaquatic ecosystem [42–44].

Apart from the large amount of spilt oil floating at the water surface, there is still a considerable portion of oil being transported within the water column and settling to the seafloor after/during flocculating with natural sediment (especially clay mineral in deltaic regions) and biological materials (including EPS). And the interactions of oil and aquatic particles (sediment and biological materials) can be one of the keys for the spilt oil bio-degradation, oil-weathering or geological deposition, and resuspension processes. However, the flocculation processes of the different types of natural clay minerals when combined with oil are still poorly understood; especially their resultant settling dynamics.

From the 2010 Deepwater Horizon event, for example, round 3–5% spilt oil settling into the seafloor and still 11–25% spilt oil still uncounted for transporting or settling in the ocean water column (e.g., [20, 21]). The aggregation and settling processes can potentially lead to oil eventually residing in a quasi-preserved status within the seafloor sedimentary deposits, and these oil-contaminated sediments having an extended residence time. As such, this problem triggered this study, which not only addresses the improvement of our poor understanding of the multiple oil-mineral structures but also the oil-mineral flocculation processes and its resultant settling dynamics under certain turbulent flow conditions.

Through a series of well-controlled laboratory experiments, our goal is to improve our understanding of the settling velocity and floc properties associated with the flocculation of oil-contaminated clay minerals. This laboratory research is part of the Consortium for Simulation of Oil-Microbial Interaction in the Ocean (CSOMIO) with an overarching goal to advance the numerical model predictions for the future oil spill mitigation.

1.2 Overviews of oil-mineral aggregates (OMAs) studies

Oil-mineral aggregates (OMAs) are oil droplets stabilized by fine mineral particles in water. Since the mineral particles and oil interactions in aquatic system were first mentioned by Poirier and Thiel in 1941, for the few past decades numerous studies have been examining how oil hydrocarbons absorb onto mineral particles at a molecular level, (e.g., [45–47]). More recently, most oil-mineral aggregates (OMAs) studies have been focused on the: formation mechanisms [48, 49], characterizations [50], and "influence factors" such as salinity [51], temperature [50], oil types [50], and temporal turbulence [52]. Most of these past OMAs studies were predominantly based on laboratory experiments in order to better quantify the flow conditions and particle/oil concentration. The complex nature of OMAs has resulted in several challenges to simulate/mimic their formation and fate in numerical models, especially when they interact with the cohesive particles present in natural deltaic aquatic systems.

The first laboratory experiment related to minerals and oil droplets was presented by Delvigne [53], which suggested that oil types, mineral types, turbulence energy, and water salinity were the main influence on the aggregation process. Subsequently, numerous OMAs laboratory studies have been conducted and most indicate that spilt oil bio-degradation can be enhanced by increasing oil-mineral flocculation in a natural water column [54–56]. Floch et al., [51] quantified the amount of oil incorporated into OMAs and suggested that the extent of OMA formation is not significantly different from that of seawater for salinity values as low as 1.5–0.15 (1/20 to 1/200 of pure seawater). Omotoso et al., [48] found that the degree of oil-mineral interactions should be dependent on the viscosity of the crude oil and the type of mineral present, because the oil droplet surface changes with dynamic viscosity parameter.

A series of shaker jar tests by Khelifa et al., [50] indicated that although droplet size and shape were not correlated to oil viscosity, the concentration of oil droplets present during the flocculation process became extremely sensitive to the oil viscosity, temperature, and asphaltenes-resins content (ARC). Cloutier et al., [57] conducted an annular flume experiment to determine critical shear stress removing oil from a surface by resuspension and the effect of suspended sediment concentration (SSC) on the oil erosion processes. They found that SSC at 200–250 mg/l was observed to give maximum erosion efficiency and is therefore suggested as the optimal concentration for erosion and elimination of heavy crude oil at a water temperature of 13°C.

A more recent laboratory study was conducted by Sun et al., [58] to investigate the kinetics of oil-sediment aggregates formation under various mixing intensities using marine sediments (standard reference material 1941b) and Arabian heavy crude oil. Sun et al., [58] noticed that the formation of oil suspended particles aggregates increased exponentially throughout the mixing duration and peaked during a 5-hour period. Optimum trapping effectiveness of oil rose from 24 to 47%, and the necessary shaking (i.e. physical agitation) time lessened from 4.5 to 1.2 hours as both the mixing level and sediment concentration intensified. The maximum oil-to-sediment ratio reached 240–680 mg of oil per gram of sediment during the aforementioned 5-hour period. Observations of revealed that during more quiescent conditions, the oil-sediment aggregates formed were predominantly of a solid type and single droplet aggregates. While higher turbulent agitation resulted in more multidroplet type aggregates.

Moreover, many OMA studies have been focused on oil-mineral aggregates formation and their appearance using microscopic and imaging analysis in laboratory. Stoffyn-Egli and Lee [49] used laboratory protocols and microscopic methods to detect and classify the OMA types. They identified three types of OMAs: droplet, solid, and flake aggregates, and these are controlled by mineral types, oil surface properties, viscosity, and oil/mineral ratios. Moreover, they also report clear evidence that turbulence greatly enhances the OMA formation. Delvigne [59] used natural sediment in the laboratory experiment to study the physical appearance of oil contaminated sediment. Through his quantitative investigations, it was concluded that the division of oil in the different phase is affected by the oil-sediment interaction process, oil type, and oil concentrations.

Delvigne's [59] microscopic observations of oil-contaminated sediment samples lead to three possible phases of presence of oil: oil droplets, oil-coated sediment particles, and "oil patches." Therefore, both Stoffyn-Engli and Delvigne found the existence of consistent OMA types through their laboratory studies, including ① mineral particles attaching around the surface of big oil droplet, ② small oil droplets coated by sediments, and ③ flake-shaped oil-mineral aggregates. Khelifa et al., [50] reported more detailed OMAs study by using an epi-fluorescence microscopy to analyze the shape, size distribution, and oil concentration within OMAs. These results also show these types of OMAs with further quantification on more specific oil components. Furthermore, O'Laughlin et al. [60] used a high-resolution imagery method to record settling OMAs to analyze the particle size, density, and settling velocity. Their investigation revealed the settling velocities of artificially formed OMAs to be on the order of $0.1 \sim 0.4$ mm·s^{-1}. The OMA formation increases with suspended sediment concentration (>50 mg·l^{-1}).

Following the path led by all the abundant laboratory experiments of OMAs for the last decades (mentioned previously), numerical models have been adapted to predict the oil-mineral aggregation. Fitzpatrick et al., [61] presented a simple algorithm for oil-particle aggregates with the algorithm representing oiling conditions in a natural river delta, after the formation of oil-particle aggregates within the water column. Three groups of oil-particle aggregates have been included in the model, in recognition that there are likely multiple sizes and densities of oil globules and oil-particle aggregates in the riverbed. The groups range from 2 mm single oil globule with a 10 μm silt coating to more complex aggregates with multiple smaller globules and diameters of 31 μm and 100 μm. Densities range from just greater than the density of freshwater for the large oil globule with silt coating (1.034 g·cm^{-3}) to somewhat heavier and close to the density of organic matter for the oil-particle aggregates (1.511 g·cm^{-3}). Settling velocities range from 0.2 to 20 mm·s^{-1}, depending on the amount of oil relative to the size of the aggregate. A major assumption is that the oil-particle aggregates stay intact (i.e., they do not break up or disaggregate), during model runs. The simplified transport algorithm for oil-particle aggregates was suggested to be a start for future modeling of oil-particle aggregates formation, transport, and deposition. In addition, Zhao et al., [62] presented a new conceptual-numerical model, A-DROP, to predict oil amount trapped in oil-particle aggregates.

A new conceptual formulation of oil-particle coagulation efficiency has been introduced to account for the effects of oil stabilization by particles, particle hydrophobicity, and oil-particle size ratio on oil-particle aggregates formation. A-DROP was able to closely reproduce the oil trapping efficiency reported in experimental studies. The model was then used to simulate the oil-particle aggregates formation in a typical nearshore environment. Modeling results indicate that the increase of particle concentration in the swash zone would speed up the oil-particle interaction process; but the oil amount trapped in oil-particle aggregates did not correspond to the increase of particle concentration. It is suggested that the developed A-DROP model could become an important tool in understanding the natural removal of oil and developing oil spill countermeasures, by means of oil-particle aggregation.

1.3 Justification for this study

From the extensive literature review presented in the previous related studies, we deduce that flocculation of the oil-mineral aggregates is one of the keys to improve our understanding of the fate of spilt oil in coastal and deltaic aquatic regions. Most OMA studies focus on: the formation mechanism, physical appearance, and their influencing factors. However, extremely few researchers have systematically investigated the OMA characteristics in terms of how they affect and influence their resultant settling dynamics during the flocculation process. This presents a knowledge gap in predicting the amount and fate of oil throughout the water column and extending all the way to the seafloor, particular in muddy deltaic regions. The study presented here aims to fill this knowledge gap to better understand the settling dynamics and environmental impacts from the oil sediment flocculation in coastal and deltaic ecosystems, with the expectation of improving numerical modeling to predict the spilt oil fate in a more accurate, efficient, and reliable way.

OMAs have been proven both in laboratory and field to generate easily under turbulent flow environments in natural aquatic systems (including deltas) where there is a presence of cohesive materials, including minerals and biological matters. Because of the complex structure of natural OMAs, laboratory tests can provide more details on how specific natural materials (e.g., clay mineral) influence OMA flocculation and settling dynamics. Besides the mixing devices (magnetic stirrer and reciprocal shaker) to generate the OMAs and high-resolution digital microscopy camera to observe the details of OMAs, we also utilize a Vectrino to evaluate the generated turbulence and a state-of-the-art novel LabSFLOC-2 video camera system to observe the OMAs' characteristics and most importantly their settling velocities. Collectively, a more robust and systematical study of OMAs has been conducted.

The chapter is divided as follows. Section 2 describes the experimental methodology and instrumentation utilized, with the main results reported in Section 3. A comparison of the OMA characteristics and a discussion of the implications to oil sediment transport and flocculation modeling are presented in Section 4, followed by some key points and concluding remarks summarized in Section 5.

2. Technical approach

2.1 Experimental design and instrumentations

A series of magnetic stirring jar (**Figure 1a**) and reciprocal shaking jar (**Figure 1b**) experiments have been conducted at the Center for Applied Coastal Research, University of Delaware, USA. Pure Kaolin clay, Bentonite clay, Xanthan gum powder (a proxy of biological Extracellular polymeric substance), and raw Texas crude oil (Dynamic viscosity: 7.27 cP) with various proportions have been used to generate different types of mineral flocs and Oil-Mineral Aggregates (OMAs) samples, including pure kaolin flocs, pure bentonite flocs, mixed kaolin-bentonite flocs, oil-kaolin flocs, oil-bentonite flocs, oil-kaolin-bentonite flocs, and oil-kaolin-bentonite-EPS flocs (mineral clay: 100–200 mg/l; oil: 50 mg/l). Artificial seawater (salinity ≈ 35‰) has been made from mixing clean water and pure salt. Magnetic stirring speeds have been set up to 500 rpm (Device range: $0 \sim 1000$ rpm) while reciprocal shaking speeds were at 120 rpm, 160 rpm, and 200 rpm for comparing the turbulence influence on the OMAs generating in jar tests. Both stirring and shaking for each experimental run last up to 2 hours and

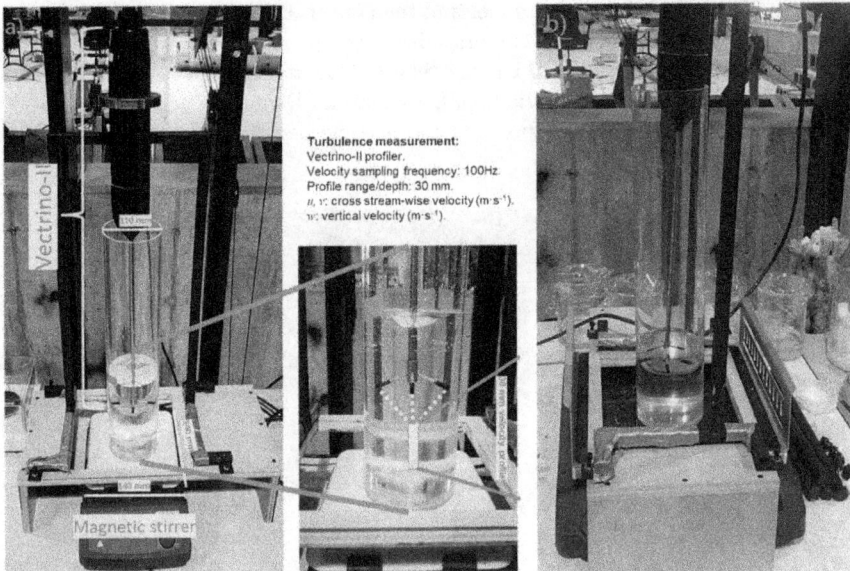

Figure 1.
Jar experimental setup: a) shows the magnetic stirrer and b) shows the reciprocal shaking jar. A Vectrino was mounted above to monitor and record the three-way flow velocity.

settling down for overnight, and it has been believed being long enough to generate mature OMAs for further studies.

A Nortek produced Vectrino (https://www.nortekgroup.com/products/vec trino) (Profiling Acoustic Doppler Velocimeter, sampling frequency in this work used 100 Hz) was used in stir-jar test to measure three-way velocities and calculate the statistics of turbulence. The Vectrino is a high-resolution acoustic velocimeter used to measure 3D water velocity fluctuations within a very small sampling volume and at sample rates of up to 200 Hz generally. It has a 30 mm profiling data collecting zone starting from 4 cm below the mid of sensor probes. It works by sending out a short acoustic pulse from the transmit element. The Vectrino was mounted on the shelf above the magnetic stirrer, and the sensor probes were in the water column (see **Figure 1**). The time series data set was collected in the same conditions (up to 500 rpm in stirrer and 200 rpm in shaker, artificial seawater and 110 mm diameter jar) with all the stir jar experiments with oil-mineral mixtures. Therefore, the turbulence of the mixing water column can be evaluated and provided as a constant value/pattern.

The mass settling dynamics of OMAs were observed using the low intrusive LabSFLOC-2 system (the second version of Laboratory Spectral Flocculation Characteristics instrument; [63, 64]) (**Figure 2**). This instrument was originally developed by Manning [65] originally, and it measures an entire floc population for each sample being assessed. LabSFLOC-2 utilizes a low-intrusive 2.0 MP Grasshopper monochrome digital video camera with a Sill TZM 1560 Telecentric (F4, with a magnification of 0.66 or 1:1.5) lens to optically observe all individual settling flocs (e.g., [66]) nominally 0.075 m above the base of a Perspex settling column (0.1 m square by 0.35 m high) within a 1 mm depth of field in the center of the column. Each pixel was 5 μm, with a 0.6% maximum distortion. More details can be found in [67].

A high-resolution digital microscope system (**Figure 3**) has been used to obtain the floc detailed structures and number statistical analysis. All the floc samples were directly collected from the running experiment in real time and using wide mouth (> 2 mm) plastic pipettes (least original floc disturbance) transferring from the mixing jar to the microscope slides and then observing under 10-times zoom-in screen on a DELL laptop via the camera software provided by AmScope.

A Beckman Coulter LS 13320 Particle Size Analyzer (**Figure 4**) located in the Advanced Materials Characteristic Lab, University of Delaware, can be used to determine the clay mineral components and particle size distributions of the floc or

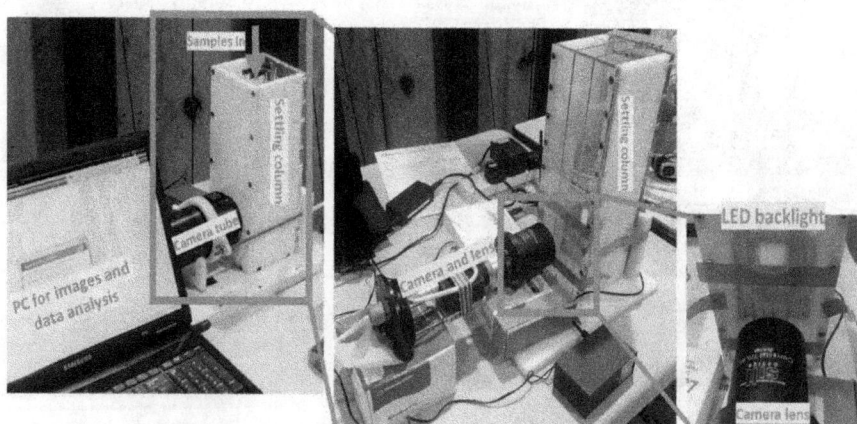

Figure 2.
The LabSFLOC-2 setup on the desk beside the stir jar system for real-time samplings.

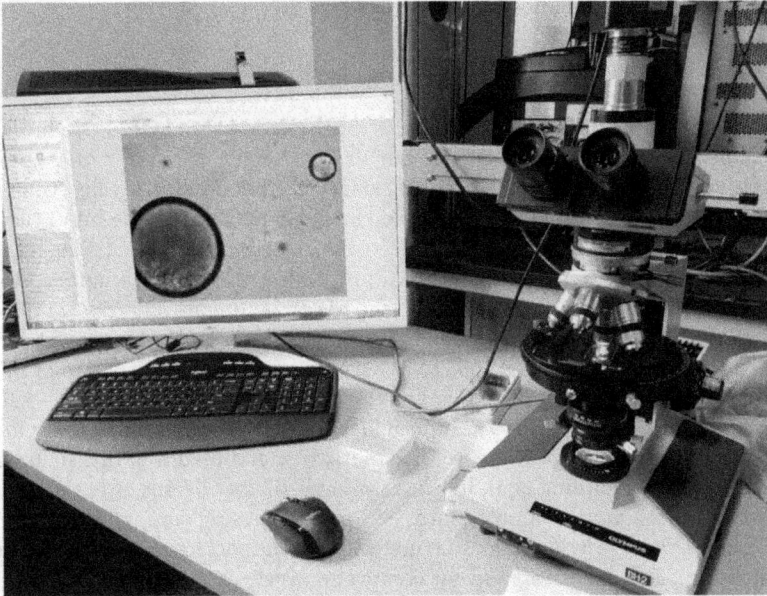

Figure 3.
High-resolution digital camera microscope.

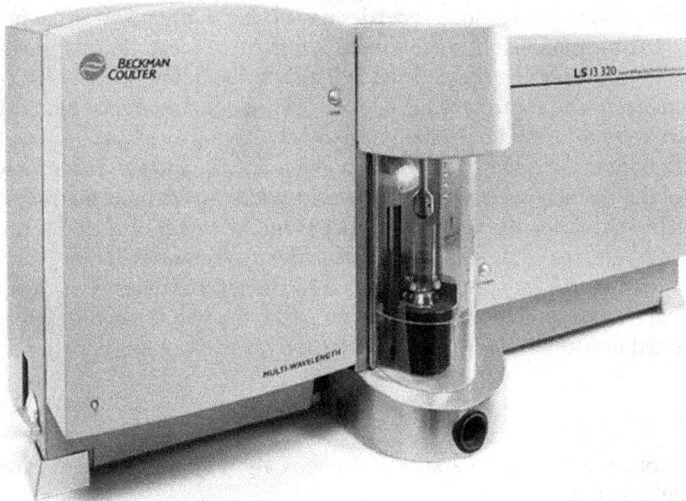

Figure 4.
Beckman coulter LS 13320 particle size analyzer.

sediment samples. This LS 13320 MW high-resolution device observes individual particles (across the entire size range) using laser diffraction. The resultant sizes are accurately determined using the principles of Mie and Fraunhofer theories of light scattering.

Considering the potential influences of oil and mineral clay addition to the flocculation initiation and floc structures, both adding oil before and after mineral clay into artificial seawater in the mixing jar have been tested and flocs images have been compared in Section 3.

2.2 Data processing and analysis

2.2.1 LabSFLOC-2 camera floc data

As one of the most novel floc video instruments, the LabSFLOC-2 produces not only visible individual floc images, but can also enable the estimation of essential quantitative floc properties (including floc size, effective density, settling velocities, etc.) through postprocessing of the raw floc image data. The recorded AVI files of streamed floc settling videos are not Codec compressed, so they can be analyzed with Matlab software routines. During postprocessing, the HR Wallingford Ltd. *DigiFloc software–version 1.0* [68] is then used to semiautomatically process the digital recording image stack to obtain floc size and settling velocity spectra. A modified version of Stokes Law [69] then permits an accurate calculation of individual floc effective density [70].

Additionally, LabSFLOC-2 system provides the following supplementary individual floc information: floc porosity, floc mass, fractal dimension, floc shape, and mass settling flux. Manning et al. [64, 71] provides further details of both the LabSFLOC-2 floc acquisition procedures and postprocessing computations, respectively. LabSFLOC-2 provides data that covers many important aspects of flocculation; these floc data are necessities for comprehensively assessing and characterizing Oil-Mineral-Microbial settling dynamics and for improving the parameterization and calibration of numerical models.

In order to investigate the general spatial variation in the floc properties, a selection of the sample mean and Macrofloc:microfloc (using mean floc size 160 μm demarcation) parameterized floc properties have been summarized at the nominal LabSFLOC-2 acquisition height. A complete summary of mean floc parameters for all the samples in this report has been listed in **Table 1**, including floc size (D_{Mean}), mean effective density, mean settling velocity (Ws_{Mean}), and fractal dimensions (f_n).

Floc sizes were measured from the image by overlaying an ellipse on each floc, which yields both major and minor axes of a given floc: D_x and D_y. This provides an indication of the floc shape in terms of the height/width aspect ratio. Settling velocity is determined by measuring the vertical distance from the center of each floc travels between two frames; the time step size between the two frames is known, which means that the floc settling velocity can be calculated by the ratio of distance to the time step size. To aid in the interpretation of the floc size data, the two orthogonal dimensions were converted into a spherical equivalent floc diameter, D_n, using equation:

$$D_n = (D_x.D_y)^{0.5} \tag{1}$$

As mentioned, a modified version of Stokes Law [69] permits an accurate estimate of floc effective density:

$$\rho_e = (\rho_f - \rho_w) = \frac{W_s 18\mu}{Dn^2 g} \tag{2}$$

where g is the acceleration due to gravitational, and μ is dynamic molecular viscosity.

International Equation of State of Sea Water, 1980 [72] was employed to establish the relevant water density from measurements of water temperature and salinity. Floc effective density is the difference between ambient water density (ρ_w) and the bulk density floc (ρ_f). Eq. (2) enabled simultaneously measured individual floc settling velocities (W_s) and D to be directly related to the corresponding individual floc effective density.

Samples	Kaolin			Bentonite			Kaolin-Bentonite		
Demacation	Mass	microfloc	Macrofloc	Mass	microfloc	Macrofloc	Mass	Microfloc	Macrofloc
Floe number	2631	2128	503	1705	681	1024	2998	2420	578
Mean floe size (µm)	120	101	199	185	105	238	104	76	225
Mean effective density (kg·m^{-3})	315	336	224	127	187	87	410	461	200
Mean settling velocity (mm·s^{-1})	2.41	1.82	4.89	2.00	1.08	2.61	1.97	1.27	4.90
Fractal dimension (nf)	2.54	2.54	2.55	2.30	2.30	2.30	2.57	2.59	2.53

Samples	Oil-Kaolin			Oil-Bentonite			Oil-Kaolin-Bentonite		
Demacation	Mass	microfloc	Macrofloc	Mass	microfloc	Macrofloc	Mass	Microfloc	Macrofloc
Floe number	3102	2696	406	1592	580	1012	2610	1975	635
Mean floc size (µm)	102	86	204	198	115	246	120	87	222
Mean effective density (kg·m^{-3})	249	269	113	127	167	104	408	446	290
Mean settling velocity (mm·s^{-1})	1.21	1.03	2.41	2.53	1.07	3.36	3.33	1.94	7.63
Fractal dimension (nf)	2.40	2.40	2.36	2.33	2.29	2.35	2.57	2.57	2.58

Table 1.
Summary for microfloc and macrofloc of various types of oil-mineral flocs.

To assist in data interpretation, the general trend of each floc population could be depicted by the calculations of various sample average floc parameters. The bimodality of each floc population [73] could be investigated by calculating the Macrofloc and smaller microfloc fractions [74, 75] as $D_n > 160$ µm and $D_n < 160$ µm, respectively [74].

2.2.1.1 Examples of LabSFLOC-2 floc size vs. settling velocity distributions

The previous section provided an overview of how floc properties can be obtained through LabSFLOC-2 measurements and subsequent postprocessing analysis on the floc images. We now illustrate in more detail the typical trend of individual floc sizes vs. flocs settling velocities. The scatterplots in **Figure 5** depict individual spherical-equivalent dry mass weighted floc sizes (x-axis) plotted against their corresponding settling velocities (y-axis) of each experimental sample collected and analyzed by LabSFLOC-2 camera system.

Figure 5a shows the resultant flocs from a mixed Kaolinite and Bentonite suspension having floc size range from 15 to 508 microns. The plot shows a group of small-sized flocs (<100 microns) with low settling velocities (<0.2 mm/s) formed; based on comparisons, this fraction is lacking from floc populations composed of both pure Kaolin and pure Bentonite clays. Kaolinite-Bentonite flocs consist of a significant portion of small microflocs (81% of the 2998 flocs). Although the settling velocities range from 0.1 to 11 mm/s, the higher portion of microflocs causes a net reduction of averaged settling velocity to ~2 mm/s.

When oil is added to Kaolinite-Bentonite mixture (**Figure 5b**), although the maximum floc size is 90 microns smaller than the corresponding non-oil suspension (**Figure 5a**), we observe more high-density Macroflocs with settling velocities greater than 10 mm/s (peaking at 20 mm/s). We also observe 5% more Macroflocs (by population), but their effective density is quite widely spread; from very porous Macroflocs with a density close to water (i.e., below the red line), to those Macroflocs of a higher density—approaching the density of a quartz particle flocs (i.e., the pink line).

2.2.2 Vectrino data (turbulence)

The Vectrino data recorded 5–10 minutes continuously three-way velocities in the mid of jar water column as an interval of 30 mm from the 4 cm below its

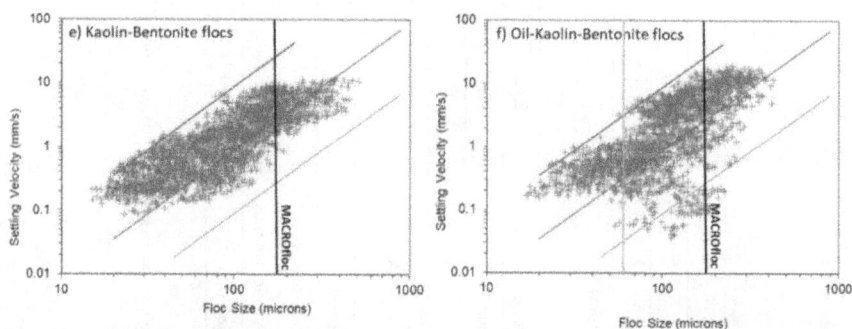

Figure 5.
Plots of floc sizes vs. settling velocities of each type of mineral clay and oil mixtures. The three diagonal lines represent contours of Stokes-equivalent constant effective density (i.e., floc bulk density minus water density): Pink = 1600 kg·m^{-3} (equivalent to a quartz particle), green = 160 kg·m^{-3}, and red = 16 kg·m^{-3}. Solid black vertical line shows the separation of microflocs (<160 microns) and macroflocs (>160 microns); gray vertical line shows the average oil droplets size (57 microns).

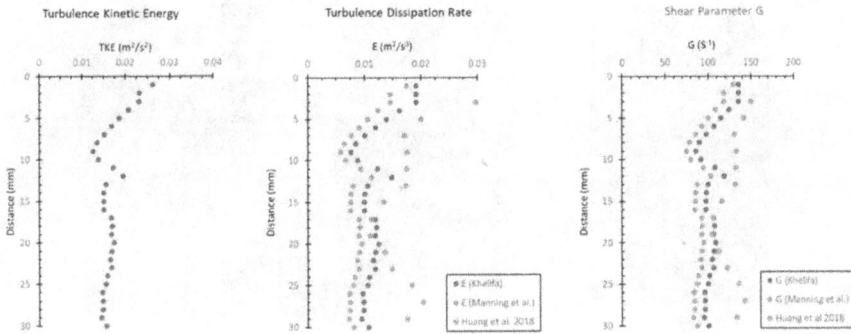

Figure 6.
Turbulence kinetic energy (TKE), dissipation rate (ε), and the flow shear parameter G calculated by approaches from Khelifa [71, 76, 77], respectively.

transducers. The raw data file.DAT has been input into Matlab for statistical analysis. Three approaches [71, 76, 77] have been utilized to calculate the turbulence kinetic energy (TKE), dissipation rate (ε), and the flow shear parameter G and for comparing the reliability and accuracy (**Figure 6**).

In Khelifa's [77] study, a constant l_t ($\approx$ 47 mm) has been assumed, which represents characteristic velocity and length scales of turbulence in a specific jar. Therefore, turbulence dissipation rate (ε) has been calculated by $\frac{1}{0.047}\left(\frac{2}{3}TKE\right)^{3/2}$. On the other hand, Manning et al. [71] used the shear parameter G (units of s^{-1}), which is the root-mean-square of the gradient in the turbulent velocity fluctuations, and turbulent shear stress τ to evaluate the turbulence [78]. These are related through the shear velocity u* by $G = \left(\frac{u^{*3}\xi}{\kappa\nu z}\right)^{1/2}$, where κ is the von Karman's constant (taken as 0.40), ν is kinematic viscosity of the water, z is height above the bed, h is water depth, and ξ = 1 − z/h. Different from the dependence on specific conditions of the former two approaches, Huang [76] calculated dissipation rate (ε) by common frequency spectrum (φ(f)) using Taylor's frozen turbulence hypothesis [79] to give: $\varphi(f) = \left(\frac{U}{2\pi}\right)^{2/3}\alpha\varepsilon^{2/3}f^{-5/3}$, where f is the frequency, and U is the mean velocity. Thus, εcan be calculated by $\frac{2\pi}{U}\alpha^{-3/2} < f^{5/2}\varphi^{3/2}(f)>$.

2.2.3 Microscope images analysis

The images collected from the digital microscope of each floc sample (e.g., **Figure 7**) have been number counted and shape analyzed individually for further statistical analysis, and particles flocculation stickiness rate discussions. The temporal development of flocs has been analyzed by using the evolutionary changes in the floc numbers observed during the microscope experiments; this technique was also utilized to represent the stickiness of flocs in time series (see [80]). For each mineral clay flocculation experimental run of the jar tests, floc samples have been collected in real time and screen observed under microscope at the time interval of 1 minute, 2 minutes, 4 minutes, 8 minutes, 12 minutes ... 140 minutes. Then all the individual floc numbers have been counted manually for each microscope image (e.g., **Figure 8**). The numbers of flocs formed in temporal indicate the development of the flocculation, which has been used to represent the floc particles stickiness rate parameter (result will be presented in Section 3.3).

Additionally, averaged oil droplets size has been analyzed by averaging the oil droplet size from the microscope images (e.g., **Figure 9**). Delvigne et al., [53, 81, 82]

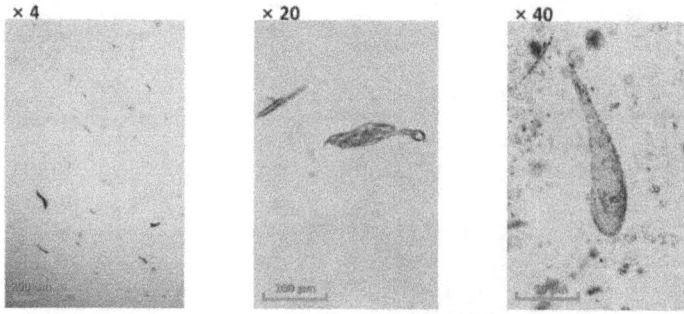

Figure 7.
Example of oil-mineral aggregates images viewed under high-resolution microscopy.

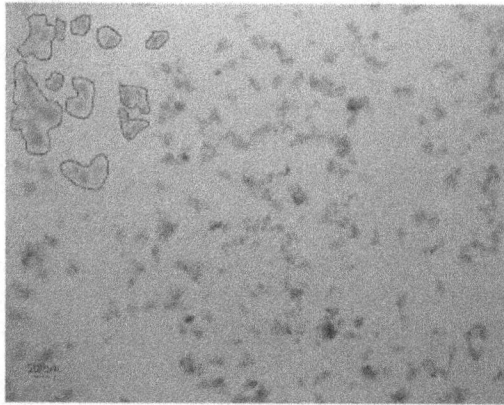

Figure 8.
Floc number manually counted from the microscope images. The red part showing the individual floc can be recognized manually for further statistical analysis.

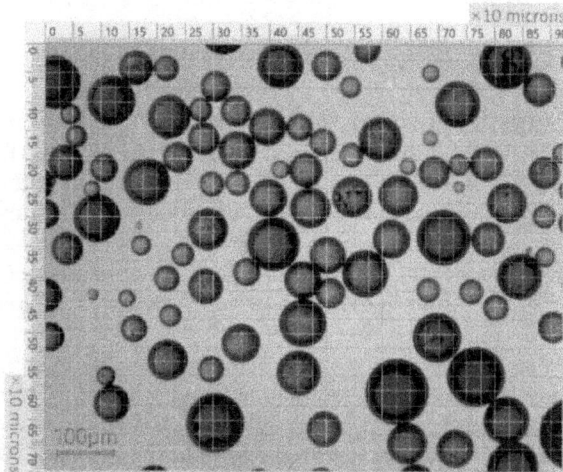

Figure 9.
Pure oil droplets samples and image from microscope for manually counting the averaged droplet size.

report that the level turbulent activity and the viscosity of oil both heavily control oil droplet sizes in turbulent flows. Observations of the distributions of dispersed oil droplets sizes (D) were seen to follow a pattern corresponding to $N \sim D^{-2.3}$; N is the number concentration of oil droplets of size D. This empirical interrelation was valid for oil dispersion due to breaking waves during the short period plunging phase [82]. Furthermore, pouring experiments conducted by Delvigne and Hulsen [83] showed that the oil dispersion coefficient was not affected by oil viscosity when the oil viscosity was <1 cm ^{2}/s.

In contrast, at higher oil viscosities, the dispersion decreases considerably with viscosity. Fraser and Wicks [84] have discussed how the concept of critical Weber number [85] may be used to estimate the maximum size of stable oil droplets dispersed in sea. Li and Garrett [86] have investigated theoretically the effect of oil viscosity on the size of droplets, assuming that viscous shear is the mechanism for droplet breakup. Their research showed that the ratio $(\mu_d/\mu_c)^n$ was comparable with the largest droplet present, whereby μ_c is the viscosity of the uninterrupted (i.e., continuous) phase, μ_d is the viscosity of the oil droplet. The superscript term "n" is equivalent to 3/8 when the oil droplet size exceeds 50% of the Kolmogorov length scale [87], or otherwise it reverts to 1/8.

3. Results

3.1 Oil droplet size

The pure oil droplet sizes have been statistically analyzed from 10 selected microscope images. The results show the median size of the mass oil droplets is 60 microns, mean size is 57 microns, minimum size is 12 microns, maximum size is 120 microns, and 90% of the entire oil droplets observed are <80 microns, and 99% are <110 microns. Compared with the previous oil droplets size studies mostly for the dispersed oil droplets size, these are slightly larger. Lunel [88] also estimated that 90% of the oil droplets (50% of the oil volume) were < 45 μm in diameter, and that 99% of the oil droplets (80 ~ 90% of the oil volume) were < 70 μm. In addition, Khelifa et al. [50] suggest that considering the mechanism of droplet breakup is controlled by oil and continuous phase properties and the energy dissipation rate εby turbulence.

3.2 Floc structure

Both Kaolin and Bentonite cases samples have been analyzed using a micro-scope, and the high-resolution (10-times zoomed in scale) images provided the details of the floc structure for each mixture. Three basic floc types have been identified: i) mineral flocs/aggregates, ii) oil droplets attaching/combining Kaolin aggregates, and iii) stringy/flake-shaped oil-Bentonite aggregates.

3.2.1 Kaolin floc and oil-kaolin aggregates structures

Figure 10a shows a representative sample of settled pure kaolin clay floc (Kaolin clay: 100 mg/L), which is about 250 microns in length. The floc was generated under a 2-hour constant turbulent condition (turbulence dissipation rate: 0.015 ~ 0.02 m^2/s^2). With the addition of 100 mg/L Texas crude oil in the mixing jar (oil added into mineral floc mixtures), the oil droplets can be observed being attached or half embraced within the kaolin clay structures (such as **Figure 10b** and **c**). Basically, we observed mineral flocs of Kaolin clay attached or combined with

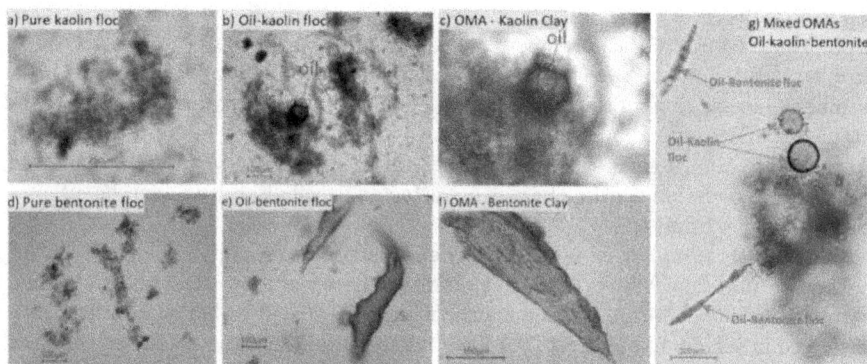

Figure 10.
Floc images from the high-resolution digital microscope camera. a) the pure kaolin clay floc structure; b) & c) oil and kaolin clay mixed floc structure; d) the pure bentonite clay formed flocs structure and e) & f) oil and bentonite clay formed flocs, & g) mixed OMAs with oil-kaolin-bentonite flocs.

the oil droplets (surface) and developed lager oil-Kaolin aggregates. This is consistent with one of the OMA types shown in the previous studies (such as [49, 50]). They found that droplet OMA (or oil coated by sediment aggregates) appears as oil (spheres) with mineral particles or flocs attached to their surface only. In this case, the droplets do not enclose mineral particles. The quantity of minerals attached to a droplet is highly variable.

3.2.2 Bentonite floc and oil-bentonite aggregate structures

The Bentonite clay cases (Bentonite clay: 100 mg/L), generated with same turbulent dissipation rate, are shown in **Figure 10d-f**. Pure Bentonite clay particles can be much more attachable than the Kaolin clay particles (more quantitative comparison on flocculation rate will be given later). The Bentonite flocs are long stringy-shaped and up to 100 ~ 200 μm in width and several hundred microns in length (**Figure 10d**). However, largely different from the oil-Kaolin case, the oil-Bentonite mixture flocs show an entire reshaped oil structure. The sphere-shaped oil droplets have now disappeared and formed more randomly stringy/flake-shaped oil-Bentonite aggregates, up to hundreds of microns in size as shown in **Figure 10e** and **f**.

Compared with the previous studies, the oil-Bentonite aggregates observed here belong to the category of flake OMA. Flake aggregates have the appearance of membrane structures, usually floating or neutrally buoyant, which can attain hundreds of microns in length. They have been observed in several laboratory studies (such as [49, 50]). From inspection of the microstructure of oil-Bentonite aggregates, they appear extremely orderly with a "feather-like" or "dendritic" appearance. These flakey structured aggregates tend to collapse under fast or extended agitation during high shear stress episodes. The resultant crumpled flake aggregates are less buoyant and more compact and often depicted as a solid OMA. However, they can be distinguished from solid OMA, by their favorable mineral pattern organization [49].

3.2.3 Mixed oil-kaolinite-bentonite aggregate structures

After mixing equal amounts of Kaolinite and Bentonite clay together with oil in the mixture, the resultant comprises both droplet OMAs and flake OMAs present in the same floc population (**Figure 10g**). From inspection, the large flake-shaped OMA observed in **Figure 10g** has similar floc sizes to those in the oil-Bentonite case (see **Figure 10e-f**). In the next sections, we will investigate different mineral flocs and OMAs settling velocities and discuss their relationship to floc structures.

3.3 Floc characteristics

3.3.1 Mean floc size, macroflocs and microflocs distributions

In terms of the results from the jar tests, the comparisons between the results in **Table 1** reveal the oil, kaolinite, and bentonite components' influence on the OMA floc characteristics, respectively. A systematic comparison is given next.

3.3.1.1 Floc characteristics between pure kaolin, bentonite flocs, and mixed kaolin-bentonite flocs

A comparison between Kaolinite and Bentonite flocs shows that the Kaolinite flocs have around 50% higher floc number and 50% smaller floc size in total. Interestingly, the larger number of flocs in Kaolinite is only due to microflocs. Simply in terms of Macrofloc number, Kaolinite has only half of that of Bentonite. Overall, the Kaolinite floc effective density is around 2.5 times higher than that of Bentonite flocs, which results in approximately 20% quicker settling velocity of Kaolinite flocs than that of Bentonite flocs. Finally, the fractal dimension of Kaolinite flocs is about 2.4, which is higher than that of Bentonite flocs of around 2.0. With these floc characteristics, we can conclude that the Kaolinite clay has lower flocculation rate than that of Bentonite, and this also suggests that Kaolinite clay shows a lower relative cohesivity than Bentonite clay.

After mixing equal amounts of Kaolinite and Bentonite clay (50% and 50%), the resultant mixed floc sample shows slightly (14%) higher floc numbers in total than that of Kaolinite floc sample and significantly more flocs (1.75 times) than the total present in the corresponding Bentonite floc sample. Meanwhile, the mixed floc sample has the smallest averaged floc size (104 microns) and settling velocity (1.97 mm/s), but the largest averaged total effective density (410 kg/m^{-3}) when comparing with those of the two pure clay samples.

Noticeably the Macroflocs in the mixed Kaolinite-Bentonite sample shows medium averaged floc size and averaged effective density of 225 microns and 200 kg/m^{-3}, respectively; that places the mixed Kaolinite-Bentonite sample midway between pure Kaolinite (199 microns, 224 kg/m^{-3}) and the pure Bentonite (238 microns, 87 kg/m^{-3}) cases. And the averaged Macrofloc settling velocity (4.9 mm/s) of mixed Kaolinite-Bentonite is similar to that of pure Kaolinite (4.89 mm/s), but much faster than that of pure Bentonite (2.61 mm/s). This indicates a signal that within the mixed clay case, the Macroflocs development is somewhat enhanced by the more cohesive bentonite component, but the entire mixed floc characteristics are still dominated by the effect of the less cohesive Kaolinite, especially via the microfloc population. This seemingly subtle point is raised here because it may play a more important role when interacting with oil droplets (see the next section).

3.3.1.2 Floc characteristics between pure kaolin, bentonite clay flocs and oil-kaolin, oil-bentonite flocs, respectively

The addition of a crude oil component into the Kaolin floc sample, oil-Kaolin floc numbers increased by around 18% in total, while the total averaged floc size decreased by around 18%, and the average total effective density also decreases by 20%. As a result, we obtain a reduction of averaged total settling velocity by half. Therefore, the oil addition decreases the Kaolinite flocculation rate (i.e., it produces smaller flocs) and the lower density of oil droplets further reduces the OMA density. Noticeably, for Kaolinite, its OMAs show a significant reduction in settling velocity. This is particularly demonstrated in the Macrofloc fraction: adding oil reduces the number of Macroflocs, and increases mean Macrofloc size and fractal dimension very slightly. It is the significant reduction of mean effective density (reduce by half) that causes a twofold slowing of the Macrofloc OMAs settling velocity.

For the Bentonite case, adding an oil component decreases the total number of flocs by around 7%, while the mean total floc size increases by around 7%. Although the averaged total effective density is in the same level, the averaged total settling velocity increases by 25%, which may be caused by the increase of floc size. In Macroflocs, although the mean floc number, size. and fractal dimension stay in same levels, their mean effective density and settling velocity tend to be higher after added oil. Therefore, for the entire floc group, adding oil to the matrix slightly increases flocculation in the oil-Bentonite case, and for its Macroflocs, effective density and settling velocity both rose by adding oil. Here, we can see that the response of Kaolinite and Bentonite to the addition of oil is distinctly different, and this is related to the very different OMA structures illustrated in **Figure 10**.

3.3.1.3 Floc characteristics between mixed kaolin-bentonite flocs and oil-kaolin-bentonite flocs

In total, floc number shows a reduction of around 18% by adding oil into a mixed Kaolinite and Bentonite suspension, and the averaged total floc size contrarily increases by around 15%, which may indicate an increase of floc cohesivity due to the addition of oil droplets. However, a more careful observation suggests that the changes due to the addition of oil show a reversed trend between the microfloc and Macrofloc fractions. Namely, by adding oil the microfloc flocculation rate is increased, which causes an increase in the microfloc mean size. In contrast, by adding oil to the suspension, the Macrofloc flocculation rate decreases and translates to a net rise in floc density. Significantly, by adding oil, the mean settling velocity quickens by almost 70%; this is due to the increase in the resultant average microfloc size, plus a marked rise in floc density within Macroflocs.

Overall, adding oil to the mixed Kaolin-Bentonite suspension reveals a more unexpected response. Our results show that in terms of the floc numbers, size, density, and settling velocity, the addition of oil to a Kaolin-Bentonite mixture suggests a slightly more exacerbated change in floc property patterns, when compared with the same floc characteristics from adding oil into pure Bentonite flocs. Therefore we deduce, in Kaolin-Bentonite mixtures, oil is selectively actively interacting noticeably more with Bentonite rather than with Kaolinite clay minerals (under similar clay content conditions), respectively.

3.3.2 Size classes analysis

3.3.2.1 Settling velocity

The averaged settling velocity of each size class is shown in **Figure 11**. For the Kaolin case, the pure mineral flocs show a marked quickening of the settling velocities as the floc sizes grew through the 32–512 micron subfractions. By adding oil to Kaolinite, although the settling velocities initially still show the same increase

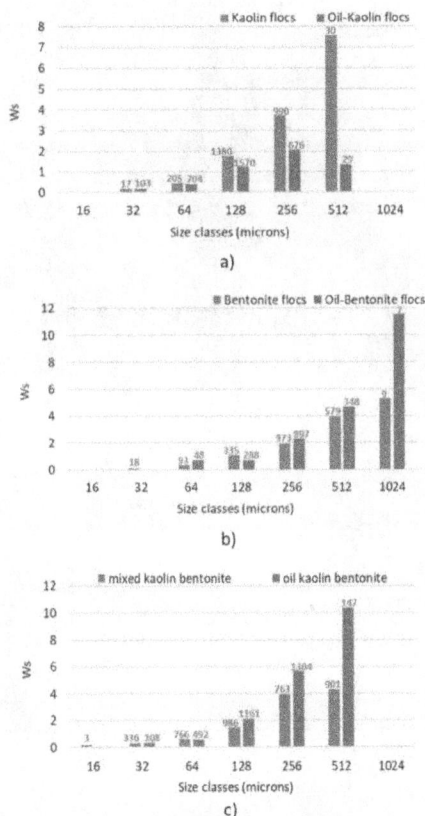

Figure 11.
Averaged settling velocity of each size classes flocs. a) the kaolin and oil-kaolin cases, and b) the bentonite and oil-bentonite cases and c) mixed kaolin-bentonite and oil-kaolin-bentonite cases. The numbers on the top of the bars show the floc numbers in each class.

through the 32–256 microns flocs, on reaching the 512 micron class, the settling velocity abruptly reduced by half.

Bentonite clay suspensions depict a very different flocculation trend (when compared with the corresponding Kaolinite flocs) for both pure mineral and oil-Bentonite cases. Bentonite produces a rise in settling rates with the growing flocs, especially in the larger size fractions from 512 to 1024 microns. The settling velocity doubles by adding oil into Bentonite flocs.

Combined Kaolin and Bentonite mixtures again show an increasing settling velocity with growing floc sizes trend for both for the mixed minerals flocs and oil-Kaolin-Bentonite flocs. The dramatic rise in of settling velocity by adding oil occurs in the 512-size class; the 1024 class present for Bentonite-only flocs has now disappeared.

3.3.2.2 Effective density

For a clearer idea of floc effective density trends, seven size groups have again been utilized as shown in **Figure 12**. Both Kaolin and Bentonite flocs show the decreasing propensity of the effective density through increasing size classes, and a similar trend once oil is added. Effective density decreases at a quicker rate for pure Bentonite when compared with pure Kaolinite (**Figure 12b**). However, the addition

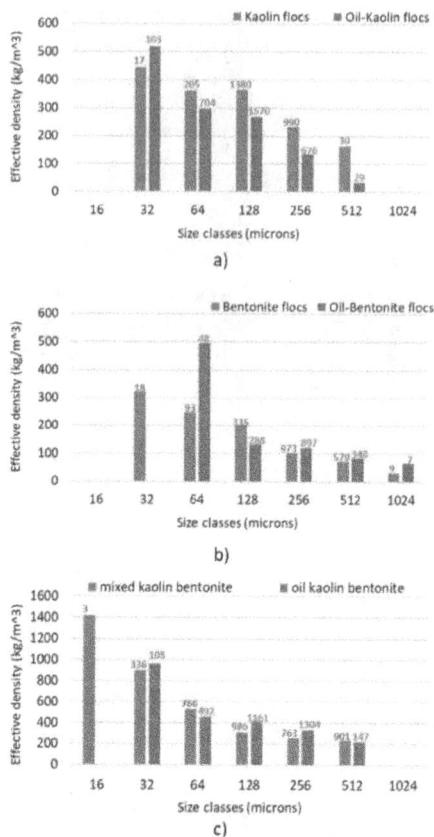

Figure 12.
Averaged effective density of each size classes flocs. a) the kaolin and oil-kaolin cases, b) the bentonite and oil-bentonite cases and c) mixed kaolin-bentonite and oil-kaolin-bentonite cases. The numbers on the top of the bars show the floc numbers in each class.

of oil into these both types of mineral clay flocs saw changes predominantly in size class 256–512 microns for Kaolin case and 32–64 microns for Bentonite case, where the oil concentrations are most probably more abundant. More specifically, adding an oil component tended to reduce the floc effective densities, especially dramatically in the 256–512 micron class; the main exception being the 16–32 micron class, which shows slight increases in effective densities (**Figure 12a**).

On the other hand, an oil addition to the sedimentary matrix instigated an increase in the effective densities in oil-Bentonite clay flocs, especially in the 32–64 micron class; the only exception being the 64–128 micron class. Therefore, the most noticeable density decrease was due to the addition of oil to Kaolinite within the Macrofloc fraction (256- and 512-micron classes). The most significant increase of density when oil is added to Bentonite is at the microfloc class 32–64 micron, as well as Macroflocs of 1024 micron. Noticeably, at 16–32 micron class, there may well be no oil-floc (for both Kaolinite and bentonite) due to oil droplet sizes.

3.3.2.3 Fractal dimension

The fractal dimension changes in size classes of both Kaolin and Bentonite cases also reveal that the oil addition decreased *fn* largely in the 256–512 microns floc size range for Kaolin flocs, and 32–64 microns for Bentonite cases (**Figure 13**).

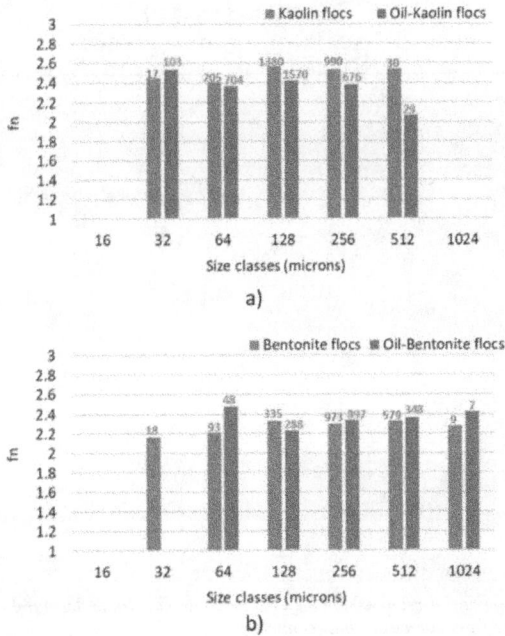

Figure 13.
Averaged fractal dimensions of each size classes flocs. a) the kaolin and oil-kaolin cases, b) the bentonite and oil-bentonite cases. The numbers on the top of the bars show the floc numbers in each class.

On average, Kaolinite flocs have larger fractal dimensions (*fn* between 2.4 and 2.6) than those of Bentonite flocs (fn from 2.2 to 2.3). The addition of oil decreased the fractal dimensions in Kaolinite case, particularly showing a substantial reduction in the 256–512 micron class, with the exception of the 16–32 micron class where it rose slightly.

In contrast, the oil-Bentonite flocs exhibit larger fractal dimensions than corresponding Bentonite flocs particularly in the 32–64 micron range the exception being the 64–128 micron class. Therefore, the consequence of fractal dimension changes after adding oil is primarily in the Macroflocs (256–512 micron) for the Kaolinite case, and more so in smaller microflocs (32–64 micron) for the Bentonite case.

3.3.2.4 Mineral flocculation stickiness rate development as a time series

Flocculation occurs and trends to be stable in mineral clay particles with the mixing time under a constant turbulent condition. With the flocculation development, smaller mineral aggregates can stick together becoming newer lager flocs, which leads to a number decay with the same total primary particles condition. **Figure 14** shows the Kaolin (orange solid circles) and Bentonite (blue solid circles) clay particles flocs were developing with the decreasing of total floc numbers. From the first 10 minutes, the floc numbers reduced dramatically in both Bentonite and Kaolinite flocs. Between 10 and 30 minutes, the floc numbers demonstrate much more stability, but still a low level in fluctuation. After 1 hour, the trend in floc numbers seems to have become more stable.

It is considered that this widespread decay rate in floc numbers can be attributed to the mineral flocculation stickiness rates between each temporal scale. Both the Kaolin and Bentonite floc development showed similarity, although the Kaolin floc

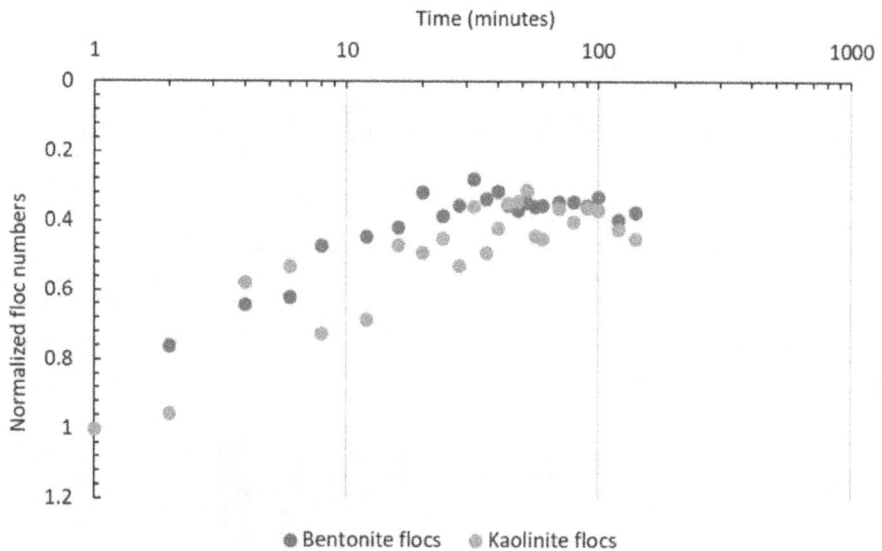

Figure 14.
Statistical result of the temporal development of both bentonite and kaolinite floc numbers, derived from the averaged size of mass flocs from microscope images analysis.

numbers tended to reduce dramatically within the first 10 minutes, there was much less fluctuation in the floc numbers after the 20-minute period. The floc numbers keep stabilized from around 30 minutes and until to the end of the test.

Overall, the floc number decay rate, which can represent the stickiness of mineral clay flocs, occurred more rapidly for Bentonite clay floc, and was much more stable than that of Kaolinite flocs. It is recommended that a higher frequency of temporal data is required for further detailed analysis, especially for the initial 10-minute phase.

4. Discussions

The laboratory experimental results presented above provide sound information of the multitypes of pure mineral clay, oil-mineral flocs characteristics, and their respective flocculation dynamics under a certain constant flow turbulent condition and seawater environment; these can be representative of the deltaic environments (including coastal and estuarine). The two clay mineral types, Kaolin and Bentonite, formed into distinctly different OMAs. Their implication for future contributions to oil-contaminated sediment transport modeling is discussed in the following sections.

4.1 Comparison of oil droplets, oil-kaolin and oil-bentonite aggregates

The oil droplets formed under the turbulent mixing in the experimental jar tests resulted in a mean size of 57 microns, with a minimum and maximum droplet size of 12 and 120 microns, respectively. 90% of the entire oil droplets observed were < 80 microns, and 99% were < 110 microns. The induced turbulence ($\varepsilon \approx 0.015$ m^2/s^3 and $G \approx 120$ s^{-1}, **Figure 6**) was characteristic of a highly turbulent region, when compared with modeling predictions of the estuarine flow conditions [89].

Under the consistent experimental conditions and concentrations, microscopy analysis reveal for the oil-Kaolin aggregates, the clay mineral particles/flocs adhere on the oil droplets surface, and the mineral particles act as a web structures surrounding the oil droplet, thus preventing it attaching to other oil droplets or further rebonding to oil slicks. Since the Kaolin mineral particles can be attached together as a much larger structure than an individual oil droplet, the oil can be observed being attached or even embraced within the Kaolin flocs (**Figure 10a-c**). And the oil droplet can be safely preserved with the Kaolin clay flocs after the entire oil droplet surface area has been occupied by mineral particles. This type of oil-mineral aggregate has been proven to be the same with previous OMAs studies, such as Zhao et al. [62].

On the other hand, with the equivalent concentration of oil and mineral clay particles and under the same mixing turbulent conditions, Bentonite clay particles tend to reshape the oil droplets into large (~ as large as 900 microns floc has been observed) stringy oil-Bentonite aggregates. From the beginning, the Bentonite particles form low density and large size predominantly stringy-configuration mineral flocs. With the addition of oil droplets, the Bentonite mineral flocs start attaching and combining with the oil. After a number of hours of mixing and flocculating, the oil component can be absorbed in the high porosity Bentonite flocs and adopt the characteristics we observe of oil-Bentonite aggregates (**Figure 10d-e**).

In terms of mixed kaolin-bentonite flocs, the results indicate a preference for kaolinite-like or bentonite-like properties within different size classes. Ye et al. [41] noted pronounced kaolinite-like features in the smaller microflocs, whereas the larger macrofloc size fraction produced bentonite-like features. These are important factors when considering the mineral interaction with oil droplets.

A conceptual structure of both oil-Kaolin and oil-Bentonite aggregates are illustrated in **Figure 15a** and **b**, respectively. Previous research has revealed that the quantity of oil droplets stabilized by oil particle aggregates formation can be strongly influenced by the ratio of oil to number of particles, plus the individual aggregate sizes and shapes [90–92]. But more significantly, the original cohesive sediment type has been proven as a key of oil-mineral aggregates structure, which can directly influence all other physical characteristics of the OMAs.

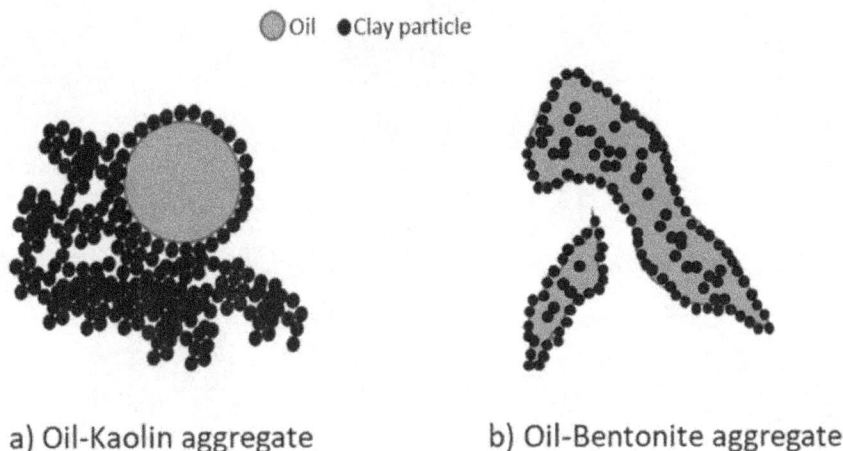

a) Oil-Kaolin aggregate   b) Oil-Bentonite aggregate

Figure 15.
The illustration showing the conceptual structure of kaolin clay formed oil aggregates and bentonite clay formed oil aggregates with difference.

4.2 Implications to the oil sediment transport and flocculation modelings

The oil-Kaolin flocculation has been proven to be mineral clay particles/flocs attaching or embracing original oil droplet together until the oil surface area fully occupied by clay particles to develop new OMAs and settling or transport through the water column. From the mass population statistical analysis, floc numbers slightly increased after adding the oil into the Kaolin mixtures because of the contribution of oil droplets to the floc entire numbers. Although the averaged floc size and effective density slightly reduced by adding oil, the average settling velocity reduced by half with the oil addition, from 2.4 mm/s in pure kaolin flocs to 1.2 mm/s with oil-Kaolin flocs. This indicates that the oil addition has a low-level influence on the Kaolin floc size and numbers, but clearly decreases the effective density and settling velocity in the mass populations. However, the plots cluster overlap area highlighted in **Figure 16** also reveals that the main difference between pure Kaolin flocs and oil-Kaolin flocs is located in the upright area of the population cluster plots where the oil-Kaolin floc settling velocities are unchanged, whilst the pure Kaolin flocs depict a trend of quickening settling velocities with the growing floc size. Clearly, the difference between the oil con-taminated flocs and pure mineral flocs is within the macrofloc fraction (floc size >160 microns).

Therefore, it can be accepted that taking the macrofloc subpopulation analysis into consideration is key to a clearer understanding of oil-Kaolin flocculation, and can be more representative, even though the macrofloc fraction comprises less than 20% of the total population mass. From the statistical results of the macrofloc subpopulation, the reduction in both their effective density and settling velocity is larger than those in the entire population. This suggests that the oil addition is most likely influencing these particular floc characteristics more sensitively, and it is more accurate to compare with pure mineral flocs and oil-mineral flocs within this subpopulation. These factors can be used to enlighten the present oil sediment and

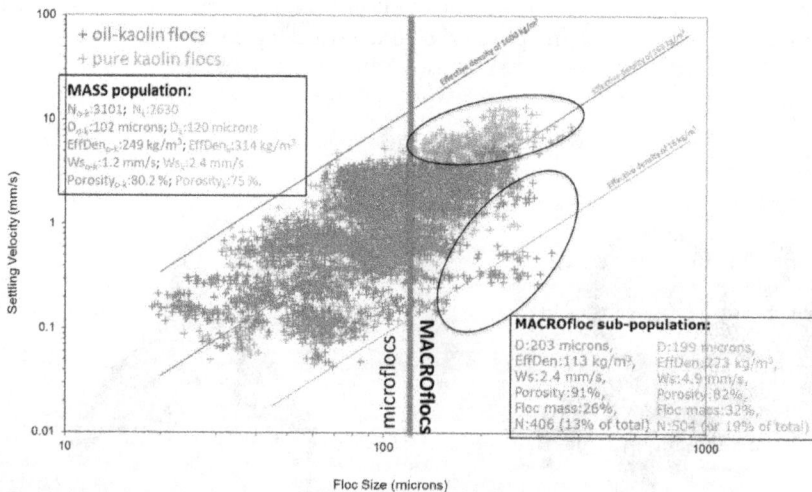

Figure 16.
Plots of floc sizes vs. settling velocities of kaolin flocs (light blue) and oil-kaolin mixing flocs (red). The information in the boxes shows the mass population and MACROfloc subpopulation statistically results, respectively. The three diagonal lines represent contours of Stokes-equivalent constant effective density (i.e., floc bulk density minus water density): Pink = 1600 $kg \cdot m^{-3}$ (equivalent to a quartz particle), green = 160 $kg \cdot m^{-3}$, and red = 16 $kg \cdot m^{-3}$. Solid blue vertical line shows the separation of microflocs (<160 microns) and macroflocs (>160 microns).

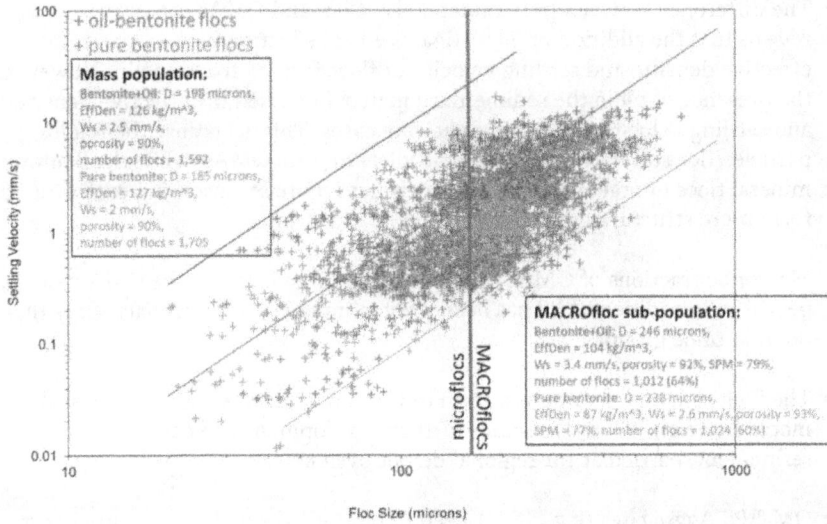

Figure 17.
Plots of floc sizes vs. settling velocities of bentonite flocs (light blue) and oil-bentonite mixing flocs (red). The information in the boxes shows the mass population and macrofloc subpopulation statistically results, respectively. The three diagonal lines represent contours of Stokes-equivalent constant effective density (i.e., floc bulk density minus water density): Pink = 1600 kg·m^{-3} (equivalent to a quartz particle), green = 160 kg·m^{-3}, and red = 16 kg·m^{-3}. Solid blue vertical line shows the separation of microflocs (<160 microns) and macroflocs (>160 microns).

flocculation modeling work, essentially because the present oil-particle aggregates models are primarily based on the assumptions of the structures of oil-Kaolin flocculation process (such as, [61, 62]).

Conversely, for mass populations, the oil-Bentonite flocs show almost no difference with Bentonite flocs statistically in terms of their floc numbers, effective density, settling velocity, and porosity (**Figure 17**). However, the macroflocs show a nominal 30% rise in effective density and settling velocity in Oil-Bentonite flocs, when compared with pure Bentonite flocs; this reveals an opposing situation to that of the Kaolin case.

This can be explained by the unique structure of oil-Bentonite aggregates, which radically differ from oil-Kaolin aggregates (as discussed in an earlier section). Rather than the OMAs forming from simply attaching oil droplets and mineral particles/flocs (as with Kaolin), the Bentonite flocs and oil droplets become mutually absorbed. Therefore, the oil-Bentonite aggregates tend to be more robust in structure and larger in size than their corresponding oil-Kaolin aggregates, which leads to an increase in both effective density and settling velocity for the Bentonite types. This more complex Bentonite flocculation structure can pose additional issues from the numerical modeling perspective in cohesive sediment flocculation. However, since the Bentonite clay is one of the most common mineral particles in natural environments, this tends to be a necessity for the future cohesive sediment transport and flocculation modeling improvement within muddy deltaic regions.

5. Summary of key points

- There is a vast difference in OMAs structures between Kaolin and Bentonite minerals.

- The difference between pure mineral clay flocs and OMAs characteristics reveals that the addition of oil instigates a net reduction in the floc size, effective density, and settling velocity of flocs formed from Kaolin. However, the presence of oil in the sedimentary matrix increased the effective density and settling velocity of flocs in Bentonite cases. This is because the Kaolin particles/flocs directly attach to oil droplets to form OMAs, while the Bentonite mineral flocs interact with the oil component more by absorbing each other to form more structurally complex oil flakes and strings.

- Macrofloc fractions of OMAs can be focused on to improve the OMA models, because the addition of oil has been proven more sensitive to macroflocs than the microfloc group.

- The inclusion and parameterization of OMA data for the use in numerical modeling (e.g., [93, 94]) require further development and continued refinement, particular for regional deltaic systems.

- OM-EPS Aggregates (e.g., [80]) need to be studied extensively to predict the natural oil contamination and environmental changes after oil spills. 3D assessment techniques such as those demonstrated by Zhang et al. [95] and Spencer et al. [96] may provide further enlightenment.

Acknowledgements

The research reported in this chapter was primarily part of the Consortium for Simulation of Oil-Microbial Interactions in the Ocean – CSOMIO that was funded by GoMRI - Gulf of Mexico Research Initiative (Grant no. SA18–10), together in part by the National Science Foundation (NSF) OCE-1924532. AJM's contribution toward this chapter was also partly assisted by the TKI-MUSA project 11204950-000-ZKS-0002, HR Wallingford company research FineScale project (Grant no. ACK3013_62), and NSF grant OCE-1736668. LY's contribution was also partly supported by Southern Marine Science and Engineering Guangdong Laboratory (Zhuhai) (No. 311020003). The datasets presented in this study can be found in online repositories GRIDDC and can be found at: https://data.gulfresearchinitiative.org.

Conflict of interest

The authors declare no conflict of interest.

Author details

Andrew J. Manning[1,2,3,4,5,6,7]*, Leiping Ye[2,8], Tian-Jian Hsu[2], James Holyoke[2,9] and Jorge A. Penaloza-Giraldo[2]

1 HR Wallingford Ltd, Coasts and Oceans Group, Wallingford, UK

2 Department of Civil and Environmental Engineering, University of Delaware, Center for Applied Coastal Research, Delaware, USA

3 University of Hull, Environment and Energy Institute, Hull, UK

4 University of Florida, Gainesville, FL, USA

5 Stanford University, Stanford, CA, USA

6 TU Delft, Delft, Netherlands

7 University of Plymouth, Plymouth, Devon, UK

8 Sun Yat-Sen University, School of Marine Sciences, and Southern Marine Science and Engineering Guangdong Laboratory, Zhuhai, China

9 Department of Civil, Architectural and Environmental Engineering, University of Texas at Austin, Austin, TX, USA

*Address all correspondence to: a.manning@hrwallingford.com

IntechOpen

References

[1] Doshi B, Sillanpää M, Kalliola S. A review of bio-based materials for oil spill treatment. Water Research. 2018; **135**:262-277. DOI: 10.1016/j.watres.2018. 02.034

[2] Barbier EB, Hacker SD, Kennedy C, Koch EW, Stier AC, Silliman BR. The value of estuarine and coastal ecosystem services. Ecological Monographs. 2011; **81**:169-193

[3] Peterson CH, Rice SD, Short JW, Esler D, Bodkin JL, Ballachey BE, et al. Long-term ecosystem response to the exxon Valdez oil spill. Science. 2003; **302**:2082-2086. DOI: 10.1126/science. 1084282

[4] Radoviæ JR, Domínguez C, Laffont K, Díez S, Readman JW, Albaigés J, et al. Compositional properties characterizing commonly transported oils and controlling their fate in the marine environment. Journal of Environmental Monitoring. 2012;**14**:3220-3229. DOI: 10.1039/c2em30385j

[5] Agbonifo P. Oil spills injustices in the Niger delta region: Reflections on oil industry failure in relation to the united nations environment programme (unep) report. International Journal of Petroleum and Gas Exploration Management. 2016;**2**(1):26-37

[6] Nwilo PC, Badejo OT. Oil spill problems and management in the Niger Delta. International Oil Spill Conference. American Petroleum Institute. 2005;**2005**(1):567-570. DOI: 10.7901/2169-3358-2005-1-567

[7] Ijeoma BIC, editor. Nigerian Coastal Erosion and Subsidence. Technical Report No. 1. Prepared for EEC/ Nigerian Coastal Erosion Research Project. 1991. p. 137

[8] Ordinioha B, Brisibe S. The human health implications of crude oil spills in the Niger Delta, Nigeria: An interpretation of published studies. Nigerian Medical Journal. 2013;**54**(1): 10-16

[9] Amnesty International. UN Confirms Massive Oil Pollution in Niger Delta. 2011. Available online at: http://www. amnestyusa.org/news/news-item/un-confirms-massive-oil-pollution-in-niger-delta

[10] UNEP. Environmental Assessment of Ogoniland. United Nations Environment Programme Report. UNEP, SMI (Distribution Services) Limited. 2011. pp. 1-25, 262. ISBN: 978-92-807-3130-9, Job No.: DEP/ 1337/GE

[11] Barry FB. Environmental Injustices: Conflict and Health Hazards in the Niger Delta. Washington, DC: Substantial Research Paper; 2010. pp. 1-73

[12] Elliott M, Griffiths AH. Contamination and effects of hydrocarbons on the forth ecosystem, Scotland. Proceedings of the Royal Society of Edinburgh, Section B: Biological Sciences. 2011;**93**(3–4) The Natural Environment of the Estuary and Firth of Forth, 1987:327-342. DOI: 10.1017/S0269727000006783

[13] Martin CW. Avoidance of oil contaminated sediments by estuarine fishes. Marine Ecology Progress Series, Theme Section: Response of nearshore ecosystems to the Deepwater Horizon oil spill. 2017;**576**:125-134. DOI: doi.org/ 10.3354/meps12084

[14] Michel J, Owens EH, Zengel S, Graham A, Nixon Z, Allard T, et al. Extent and degree of shoreline oiling: Deepwater horizon oil spill, Gulf of Mexico, USA. PLoS One. 2013;**8**:e65087. DOI: 10.1371/journal.pone.0065087

[15] Peterson GW, Turner RE. The value of salt marsh edge vs interior as a habitat for fish and decapod crustaceans in a Louisiana tidal marsh. Estuaries. 1994; 17:235-262

[16] Fodrie, F.J., Able, K.W., Galvez, F., Heck, K.L. Jr and others (2014). Integrating organismal and population responses of estuarine fishes in Macondo spill research. Bioscience, 64, 778–788.

[17] Pezeshki SR, DeLaune RD. United States Gulf of Mexico coastal marsh vegetation responses and sensitivities to oil spill: A review. Environments. 2015; 2:586-607

[18] Rozas LP, Minello TJ, Miles MS. Effect of Deepwater horizon oil on growth rates of juvenile penaeid shrimps. Estuar Coast. 2014;37:1403-1414

[19] Silliman, B.R., Van De Koppel, J., McCoy, M.W., Diller, J. and others (2012). Degradation and resilience in Louisiana salt marshes after the BP–Deepwater horizon oil spill. Proceedings of the National Academy of Science USA 109: 11234–11239

[20] Atlas RM, Hazen TC. Oil biodegradation and bioremediation: A tale of the two worst spills in us history. Environmental Science & Technology. 2011;45:6709-6715. DOI: 10.1021/es2013227

[21] Crone TJ, Tolstoy M. Magnitude of the 2010 gulf of Mexico oil leak. Science. 2010;330:634-634. DOI: 10.1126/science.1195840

[22] Daly KL, Passow U, Chanton J, Hollander D. Assessing the impacts of oil associated marine snow formation and sedimentation during and after the deep water horizon oil spill. Anthropocene. 2016;13:18-33. DOI: 10.1016/j.ancene.2016.01.006

[23] Passow U, Ziervogel K. Marine snow sedimented oil released during the Deepwater horizon spill. Oceanography. 2016;29:118-125. DOI: 10.5670/oceanog.2016.76

[24] Romero IC, Toro-Farmer G, Diercks A-R, Schwing P, Muller-Karger F, Murawski S, et al. Large-scale deposition of weathered oil in the gulf of Mexico following a deep-water oil spill. Environmental Pollution. 2017;228: 179-189. DOI: 10.1016/j.envpol.2017.05.019

[25] Passow U, Ziervogel K, Asper V, Diercks A. Marine snow formation in the aftermath of the Deepwater horizon oil spill in the Gulf of Mexico. Environmental Research Letters. 2012;7: 035301. DOI: 10.1088/1748-9326/7/3/035301

[26] Passow U, Alldredge AL. A dye binding assay for the spectrophotometric measurement of transparent exopolymer particles TEP. Limnology and Oceanography. 1995;40: 1326-1335. DOI: 10.4319/lo.1995.40.7.1326

[27] Malpezzi MA, Sanford LP, Crump BC. Abundance and distribution of transparent exopolymer particles in the estuarine turbidity maximum of Chesapeake Bay. Marine Ecology Progress Series. 2013;486:23-35. DOI: 10.3354/meps10362

[28] Hope JA, Malarkey J, Baas JH, Peakall J, Parsons DR, Manning AJ, et al. Interactions between sediment microbial ecology and physical dynamics drive heterogeneity in contextually similar depositional systems. Limnology and Oceanography. 2020;65(10):2403-2419. DOI: 10.1002/lno.11461

[29] Parsons DR, Schindler RJ, Hope JA, Malarkey J, Baas JH, Peakall J, et al. The role of biophysical cohesion on subaqueous bed form size. Geophysical

Research Letters\Early View. 2016;**43**: 1-8. DOI: 10.1002/2016GL067667

[30] Tolhurst TJ, Gust G, Paterson DM. The influence of an extracellular polymeric substance (EPS) on cohesive sediment stability. Proceeding in Marine Science. 2002;**5**:409-425. DOI: 10.1016/ s1568-2692(02)80030-4

[31] Passow U, Sweet J, Francis S, Xu C, Dissanayake AL, Lin Y-Y, et al. Incorporation of oil into diatom aggregates. Marine Ecology Progress Series. 2019;**612**:65-86. DOI: 10.3354/ meps12881

[32] Kujawinski EB, Kido Soule MC, Valentine DL, Boysen AK, Longnecker K, Redmond MC. Fate of dispersants associated with the Deepwater horizon oil spill. Environmental Science & Technology. 2011;**45**:1298-1306

[33] Lubchenco J, McNutt MK, Dreyfus G, Murawski SA, Kennedy DM, Anastas PT, et al. Science in support of the Deepwater horizon response. Proceedings. National Academy of Sciences. United States of America. 2012;**109**:20212-20221

[34] McNutt MK, Camilli R, Crone TJ, Guthrie GD, Hsieh PA, Ryerson TB, et al. Review of flow rate estimates of the Deepwater horizon oil spill. Proceedings. National Academy of Sciences. United States of America. 2012;**109**:20260-20267

[35] Fu J, Gong Y, Zhao X, O'reilly SE, Zhao D. Effects of oil and dispersant on formation of marine oil snow and transport of oil hydrocarbons. Environmental Science & Technology. 2014;**48**:14392-14399. DOI: 10.1021/ es5042157

[36] Kleindienst S, Paul JH, Joye SB. Using dispersants after oil spills: Impacts on the composition and activity of microbial communities. Nature Reviews. Microbiology. 2015;**13**: 388-396. DOI: 10.1038/nrmicro3452

[37] Manning AJ, Dyer KR. Mass settling flux of fine sediments in northern European estuaries: Measurements and predictions. Marine Geology. 2007;**245**: 107-122

[38] Manning A, Schoellhamer D. Factors controlling floc settling velocity along a longitudinal estuarine transect. Marine Geology. 2013;**345**:266-280. DOI: 10.1016/j.margeo.2013.06.018

[39] Prandle D, Lane A, Manning AJ. Estuaries are not so unique. Geophysical Research Letters. 2005;**32**(23). DOI: 10.1029/2005GL024797

[40] Zhao L, Torlapati J, Boufadel MC, King T, Robinson B, Lee K. V-drop: A comprehensive model for droplet formation of oils and gases in liquids incorporation of the interfacial tension and droplet viscosity. Chemical Engineering Journal. 2014;**253**:93-106. DOI: 10.1016/j.cej.2014.04.082

[41] Ye L, Manning AJ, Hsu T-J. Oil-mineral flocculation and settling velocity in saline water. Water Research. 2020;**173**:115569. DOI: 10.1016/j.watres.2020. 115569

[42] Fisher CR, Montagna PA, Sutton TT. How did the Deepwater horizon oil spill impact deep-sea ecosystems? Oceanography. 2016;**29**:182-195. DOI: 10.5670/oceanog.2016.82

[43] Kingston PF. Long-term environmental impact of oil spills. Spill Science and Technology Bulletin. 2002; 7:53-61. DOI: 10.1016/s1353-2561(02) 00051-8

[44] Teal JM, Howarth RW. Oil spill studies: A review of ecological effects. Environmental Management. 1984;**8**: 27-43. DOI: 10.1007/bf01867871

[45] Button DK. The influence of clay and bacteria on the concentration of dissolved hydrocarbon in saline

solution. Geochimica et Cosmochimica Acta. 1976;**40**:435-440

[46] Karickhoff SW, Brown DS, Scott TA. Sorption of hydrophobic pollutants on natural sediments. Water Research. 1978;**13**:241-248

[47] Meyers PA, Oas TG. Comparison of associations of different hydrocarbons with clay particles in simulated seawater. Environmental Science and Technology. 1978;**12**:934-937

[48] Omotoso OE, Munoz VA, Mikula RJ. Mechanism of crude oil-mineral interactions. Spill Science & Technology Bulletin. 2002;**8**(1):45-54. DOI: 10.1016/S1353-2561(02)00116-0

[49] Stoffyn-Egli P, Lee K. Formation and characterization of oil-mineral aggregates. Spill Science & Technology Bulletin. 2002;**8**(1):31-44. DOI: 10.1016/S1353-2561(02)00128-7

[50] Khelifa A, Stoffyn-Egli P, Hill PS, Lee K. Characteristics of oil droplets stabilized by mineral particles: Stabilized by mineral particles: Effects of oil type and temperature. Spill Science & Technology Bulletin. 2002;**8**(1):19-30. DOI: 10.1016/S1353-2561(02)00117-2

[51] Floch SL, Guyomarch J, Merlin F, Stoffyn-Egli P, Dixon J, Lee K. The influence of salinity on oil-mineral aggregate formation. Spill Science & Technology Bulletin. 2002;**8**(1):65-71. DOI: 10.1016/S1353-2561(02)00124-X

[52] Hill PS, Khelifa A, Lee K. Time scale for oil droplet stabilization by mineral particles in turbulence suspensions. Spill Science & Technology Bulletin. 2002;**8**(1):73-80. DOI: 10.1016/S1353-2561(03)00008-2

[53] Delvigne GA. Droplet size distribution of naturally dispersed oil. In: Fate and Effects of Oil in Marine Ecosystems. Dordrecht: Springer; 1987. pp. 29-40

[54] Lee K, Lunel T, Wood P, Swannell R, Stoffyn-Egli P. Shoreline cleanup by acceleration of clay-oil flocculation processes. In: Proceedings of 1997 International Oil Spill Conference. Washington, DC: American Petroleum Institute; 1997. pp. 235-240

[55] Lee K. Oil–particle interactions in aquatic environments: Influence on the transport, fate, effect and remediation of oil spills. Spill Science & Technology Bulletin. 2002;**8**(1):3-8

[56] Weise AM, Nalewajko C, Lee K. Oil–mineral fine interactions facilitate oil biodegradation in seawater. Environmental Technology. 1999;**20**:811-824

[57] Cloutier D, Amos CL, Hill PR, Lee K. Oil erosion in an annular flume by seawater of varying turbidities: A critical bed shear approach. Spill Science & Technology Bulletin. 2002;**8**(1):83-93. DOI: 10.1016/S1353-2561(02)00115-9

[58] Sun J, Zhao D, Zhao C, Liu F, Zheng X. Investigation of the kinetics of oil–suspended particulate matter aggregation. Marine Pollution Bulletin. 2013;**76**:250-257

[59] Delvigne GAL. Physical appearance of oil in oil-contaminated sediment. Spill Science & Technology Bulletin. 2002;**8**(1):55-63. DOI: 10.1016/S1353-2561(02)00121-4

[60] O'Laughlin CM, Law BA, Zions VS, King TL, Robinson B, Wu Y. Settling of dilbit-derived oil-mineral aggregates (OMAs) & transport parameters for oil spill modelling. Marine Pollution Bulletin. 2017;**124**(1):292e302

[61] Fitzpatrick FA, Boufadel MC, Johnson R, Lee KW, Graan TP, Bejarano AC, et al. Oil–Particle Interactions and Submergence from Crude Oil Spills in

Marine and Freshwater Environments: Review of the Science and Future Research Needs. USA: US Geological Survey (USGS); 2015

[62] Zhao L, Boufadel MC, Geng X, Lee K, King T, Robinson B, et al. A-DROP: A predictive model for the formation of oil particle aggregates (OPAs). Marine Pollution Bulletin. 2016;**106**: 245-259

[63] Manning AJ. LabSFLOC-2 – The Second Generation of the Laboratory System to Determine Spectral Characteristics of Flocculating Cohesive and Mixed Sediments. UK: HR Wallingford Ltd; 2015

[64] Manning AJ, Whitehouse RJS, Uncles RJ. Suspended particulate matter: The measurements of flocs. In: Uncles RJ, Mitchell S, editors. ECSA Practical Handbooks on Survey and Analysis Methods: Estuarine and Coastal Hydrography and Sedimentology. Cambridge: Cambridge University Press; 2017. pp. 211-260. DOI: 10.1017/9781139644426 ISBN 978-1-107-04098-4

[65] Manning AJ. LabSFLOC – A Laboratory System to Determine the Spectral Characteristics of Flocculating Cohesive Sediments. TR: HR Wallingford Technical Report; 2006. p. 156

[66] Manning AJ, Dyer KR. The use of optics for the in-situ determination of flocculated mud characteristics. Journal of Optics A Pure and Applied Optics. 2002;4(4):S71-S81

[67] Ye L, Manning JA, Hsu T, Morey S, Chassignet EP, Ippolito T. Novel application of laboratory instrumentation characterizes mass settling dynamics of oil-mineral aggregates (OMAs) and oil-mineral-microbial interactions. Marine Technology Society Journal. 2018;**52**(6): 87-90

[68] Benson T, Manning AJ. DigiFloc: The development of semiautomatic software to determine the size and settling velocity of flocs (HR Wallingford Report DDY0427-RT001). 2016

[69] Stokes GG. On the effect of the internal friction on the motion of pendulums. Transactions of the Cambridge Philosophical Society. 1851; **9**:8-106

[70] Manning AJ, Spearman JR, Whitehouse RJS, Pidduck EL, Baugh JV, Spencer KL. Laboratory assessments of the flocculation dynamics of mixed mud-sand suspensions. In: Manning AJ, editor. Sediment Transport Processes and their Modelling Applications. Rijeka, Croatia: InTech; 2013. pp. 119-164

[71] Manning AJ, Schoellhamer DH, Mehta AJ, Nover D, Schladow SG. Video measurements of flocculated sediment in lakes and estuaries in the USA. In: Proceedings of the 2nd Joint Federal Interagency Sedimentation Conference and 4th Federal Interagency Hydrologic Modeling Conference. Las Vegas: Joint Federal Interagency Conference; 2010

[72] Millero FJ, Poisson A. International one-atmosphere equation of state of seawater. Deep-Sea Research. 1981;**28**: 625-629

[73] Dyer KR, Cornelisse J, Dearnaley MP, Fennessy MJ, Jones SE, Kappenberg J, et al. A comparison of in situ techniques for estuarine floc settling velocity measurements. Journal of Sea Research. 1996;**36**:15-29

[74] Manning AJ. A study of the effects of turbulence on the properties of flocculated mud. [Ph.D. Thesis]. UK: Institute of Marine Studies, University of Plymouth; 2001. p. 282

[75] Eisma D. Flocculation and de-flocculation of suspended matter in

estuaries. Netherlands Journal of Sea Research. 1986;**20**:183-199

[76] Huang CJ, Ma H, Guo J, Dai D, Qiao F. Calculation of turbulent dissipation rate with acoustic Doppler velocimeter. Limnology Oceanography: Methods. 2018;**16**(2018):265-272

[77] Khelifa A, Fingas M, Brown C. Effects of dispersants on Oil-SPM aggregation and fate in US coastal waters. Final Report Grant Number: NA04NOS4190063. 2008

[78] Manning AJ, Dyer KR. A laboratory examination of floc characteristics with regard to turbulent shearing. Marine Geology. 1999;**160**(1–2):147-170. DOI: 10.1016/S0025-3227(99)00013-4

[79] Taylor GI. Diffusion in a turbulent air stream. Proceedings of the Royal Society of London, Series A: Mathematical and Physical Sciences. 1935;**151**(873):465-478

[80] Ye L, Manning AJ, Holyoke J, Penaloza-Giraldo JA, Hsu T-J. The role of biophysical stickiness on oil-mineral flocculation and settling in seawater. Frontiers in Marine Science. 2021;**8**: 628827. DOI: 10.3389/fmars.2021.628827

[81] Delvigne GAL, Sweeney C. Natural dispersion of oil. Oil and Chemical Pollution. 1988;**4**(4):281-310

[82] Delvigne GA. Natural dispersion of oil by different sources of turbulence. In: International Oil Spill Conference. American Petroleum Institute; 1993; **1993**(1):415-419

[83] Delvigne GA, Hulsen LJ. Simplified laboratory measurement of oil dispersion coefficient – application in computations of natural oil dispersion. Ottawa, ON, Canada: Environment Canada; 1994. pp. 173-187

[84] Fraser JP, Wicks M. Estimation of maximum stable oil droplet sizes at sea

resulting from natural dispersion and use of a dispersant. In: Arctic and Marine Oilspill Program Technical Seminar. Canada: Ministry of Supply and Services; 1995. pp. 313-316

[85] Hinze JO. Fundamentals of the hydrodynamic mechanism of splitting in dispersion processes. AIChE Journal. 1955;**1**(3):289-295

[86] Li M, Garrett C. The relationship between oil droplet size and upper ocean turbulence. Marine Pollution Bulletin. 1998;**36**(12):961-970

[87] Hinze JO. Turbulence. 2nd ed. New York: McGraw-Hill; 1975

[88] Lunel T. Dispersion: Oil droplet size measurements at sea. In: Proceedings 1993 Oil Spill Conference. Tampa, Florida: American Petroleum Institute; 1993. pp. 794-795

[89] Verney R, Lafite R, Brun-Cottan JC. Flocculation potential of estuarine particles: The importance of environmental factors and of the spatial and seasonal variability of suspended particulate matter. Estuaries and Coasts. 2009;**32**(4):678-693

[90] Ajijolaiya LO, Hill PS, Khelifa A, Islam RM, Lee K. Laboratory investigation of the effects of mineral size and concentration on the formation of oil–mineral aggregates. Marine Pollution Bulletin. 2006;**52**:920-927

[91] Payne JR, Claton JR Jr, McNabb GD Jr, Kirstein BE, Clary CL, Redding RT, et al. Oil–ice–sediment interactions during freeze-up and break-up. US Department of Commerce, NOAA, OCSEAP. Final Report. 1989;**64**:382

[92] Frelichowska J, Bolzinger MA, Chevalier Y. Effects of solid particle content on properties of o/w pickering emulsions. Journal of Colloid and Interface Science. 2010;**351**(2):348-356

[93] Cui L, Harris CK, Tarpley DRN. Formation of oil-particle-aggregates: Numerical model formulation and calibration. Frontiers in Marine Science. 2021;**8**:629476. DOI: 10.3389/fmars. 2021.629476

[94] Dukhovskoy DS, Morey SL, Chassignet EP, Chen X, Coles VJ, Cui L, et al. Development of the CSOMIO coupled ocean-oil-sediment-biology model. Frontiers in Marine Science. 2021;**8**:629299. DOI: 10.3389/ fmars.2021.629299

[95] Zhang N, Thompson CEL, Townend IH, Rankin KE, Paterson DM, Manning AJ. Nondestructive 3D imaging and quantification of hydrated biofilm-sediment aggregates using X-ray microcomputed tomography. Environmental Science & Technology. 2018;**52**:13306-13313. DOI: 10.1021/acs. est.8b03997

[96] Spencer KL, Wheatland JAT, Bushby AJ, Carr SJ, Droppo IG, Manning AJ. A structure–function based approach to floc hierarchy and evidence for the non-fractal nature of natural sediment flocs. Nature - Scientific Reports. 2021;**11**:14012. DOI: 10.1038/ s41598-021-93302-9

Chapter 6

Impact of the Jamapa River Basin on the Gulf of Mexico

María del Refugio Castañeda-Chávez
and Fabiola Lango-Reynoso

Abstract

The Jamapa River basin is located in the central region of the State of Veracruz, it is born in the Pico de Orizaba and connects with the Veracruz Reef System in the Gulf of Mexico, both protected natural areas. The lower part of the basin has the contribution of two important effluents, Arroyo Moreno, which is a protected natural area, strongly impacted due to municipal discharges from the metropolitan cities Veracruz-Boca del Río-Medellín. And the Estero, which is part of a complex aquatic system that discharges its waters from the Lagunar Mandinga system to the Gulf of Mexico. Currently, there is a diversity of chemical and biological compounds that the basin receives from different sources of freshwater pollution, such as industrial waste, sewage, agricultural and urban runoff, and the accumulation of sediments. The climatic seasons are the determining factors in the composition of its sediments, due to the force exerted on the bottom of the river by the increase in rainfall, the force of the winds mainly in the north wind season, where the greatest quantity of polluting materials.

Keywords: basin, Jamapa River, anthropogenic activities, reef system

1. Introduction

The basins have an altitudinal function, that is to say, being made up of territories that are at different altitudes; the problems of the higher parts may directly affect the lower parts, such as the mouth and deposition, this, by interconnecting the geographical spaces formed by the flow of water, matter and energy [1]. The Jamapa River basin is the link between three protected natural areas of great economic, social and environmental importance for the sustainable development of the state of Veracruz. These areas are the Pico de Orizaba National Park or Cilaltépetl with 19,750 ha, Arroyo Moreno Protected Natural Area (ANPAM) with 287 ha and the National Park of the Veracruz Reef System (PNSAV) with 65,516.47 ha [2].

Human settlements and the economic activities that take place in the surroundings have strongly impacted the basin, from the highest part wastewater is discharged without treatment or with poor treatment that allows all pollutants and nutrients to reach its main effluents. It is considered that only the large cities that are in this basin have wastewater treatments such as Veracruz, Boca del Río, Córdoba, Huatusco and Coscomatepec. However, there are more than a thousand rural agricultural, livestock, aquaculture and fishing communities that do not treat their wastewater. Due to the variety of pollutants that are constantly dumped into the basin and that converge in four important natural areas due to their ecosystem functions, it is a priority to know the interactions that take place in the different components of these systems in order to find solutions to this problem. The research question has its origin in knowing what is the impact that the Jamapa River basin receives from the Arroyo Moreno protected natural areas and lagoons connected with the basin, and whose final destination is the natural resources of the Gulf of Mexico?

Works such as that of Ortiz [3] who carried out a *"Modification in the provision of environmental services due to the change in environmental heterogeneity in the Jamapa River Basin, Veracruz, Mexico."* whose main objective was to analyze the modification in the provision of environmental services due to the effect of the change in environmental heterogeneity in the Jamapa River Basin.

The results of the study conclude that the anthropic activity throughout the basin, caused by the increase in agricultural and urban areas, is the cause of the reduction, fragmentation and detriment of primary coverage. These changes in the coverage of the basin have modified the provision of environmental services, a decrease in land areas provided by those associated with "Support" functions, and an increase in the percentage of land provided by those related to the functions of "Provision", mainly food.

Castañeda-Chávez et al., [4] carried out a water quality study in the lower basin of the Jamapa River, by analyzing the relationship between dissolved oxygen and temperature. The investigation showed that the dissolved oxygen levels in the different sampling sites and by season did not have significant differences; However, this parameter remained above that established in the national standards for water bodies, the temperature results showed significant differences in the north wind season. Salas-Monreal et al., [5] carried out the *annual variation of the hydrographic parameters at the confluence of the Río Jamapa and Arrollo Moreno (Mexico)*. They evaluated the quality of the aquatic environment at the mouth of the Río Jamapa, by monitoring the environmental parameters of dissolved oxygen, total nitrogen, chlorophyll, temperature and salinity in different climatic seasons. The results showed a variation of dissolved oxygen, total nitrogen, temperature and salinity in the different climatic seasons, while chlorophyll remained constant. Both authors point out the importance and need to carry out constant monitoring of these parameters in sites adjacent to the main sources of contamination, due to the possible effects of population growth and the increase in productive activities in the study area.

On the other hand, the impact that anthropic activities have on the aquatic environment of the Jamapa River basin is shown in the studies carried out in the Arroyo Moreno Protected Natural Area (ANPAM) by García-Villar et al., [6], where the temporal variation of the composition of fish species in the area, with the historical information collected with the fishermen and various statistical tools, they concluded that in the last two decades the richness, abundance and sizes of the fish species have decreased; particularly those used as a fishery resource, this effect is attributed to the use of this stream as a drainage of wastewater, which is a consequence of urban growth in the area and the lack of environmental management of this protected natural area.

The variation of species due to anthropogenic activity was confirmed by Rodríguez et al., [7]; evaluated the gross primary productivity (PPB) and plantonic respiration (PR) in the National Park of the Veracruz Reef System (PNSAV), characterized the area and identified the function of the organic metabolism of the ecosystem. The values indicated that the north zone had a difference with the south zone of the study area, being Playa Norte the most productive site in the system. The northern area presented a greater anthropogenic influence, due to a wastewater treatment plant, while the southern area is subject to the influence of the discharge from the Jamapa River during the rainy season. In contrast, the Cabezo reef was the least productive; this site is the farthest from the coast and therefore suffers less from the influence of the Jamapa River.

The impact that human activities have on water quality is not limited to surface waters, it also affects groundwater, as demonstrated by Landeros-Sánchez et al., [8] in their work entitled: *"Assessment of Water Pollution in Different Aquatic Systems: Aquifers, Aquatic Farms on the Jamapa River, and Coastal Lagoons of Mexico"* where the concentrations of nitrates, total coliforms (TC) and Vibrio sp., temperature, salinity, dissolved oxygen, and pH in groundwater of shallow wells in aquatic farms located along the river, and in lagoon systems, located in the state of Veracruz, Mexico. The results showed that agricultural effluents had total coliforms (TC) levels higher than 2419 NMP 100 mL^{-1} and the dissolved oxygen was at a minimum value of 1.7 mg L^{-1}, concentrations beyond those established in the Official Mexican Standards. They also identified the presence of Vibrio sp. in lagoon systems, for which they concluded that the impact of productive activities leads to health risks. However, derived from population growth and as a consequence the increase in the contribution of pollutants and nutrients from the Jamapa River to the Veracruz Reef National Park (PNSAV), it is vitally important to continue monitoring these pollutants in order to preserve the public health, since currently emerging pollutants such as antibiotics, hormones, microplastics, among others, are compounds that are discharged directly from this basin to the Gulf of Mexico.

With the above, it seeks to substantiate the importance of knowing the influence that the Jamapa River basin exerts on the center of the Gulf of Mexico, whose purpose is in the first instance the preservation of coastal marine ecosystems, the care of public health, as well as the proposal of possible mitigation measures with the development of environmentally friendly activities. With the aim of analyzing the impact of pollutants in the Jamapa River basin in the center of the Gulf of Mexico.

2. Study area

The Jamapa River Basin is located on the slope of the Gulf of Mexico, and occupies an area of 3, 918 km^2, and it is made up of the states of Veracruz Puebla; and the municipalities that comprise it are 31 from the state of Veracruz and 3 from the state of Puebla. Among the main cities are Córdoba, Huatusco, Coscomatepec, Atoyac, Cuitlahuac, Paso del Macho, Medellín, Soledad de Doblado, Fortín and Medellín. The eastern part of the basin is located on the southern Gulf coastal plain, and the western part is located on the neovolcanic axis (Mexican volcanic belt) [9].

The upper basin is the highest altimetric portion, with the steepest slope in the entire basin, with flow-erosive characteristics. The middle basin is a transition zone between the upper and lower basin, the slope is less steep than the upper basin. The lower basin is the deposition and discharge zone of the basin, the slope is softer or nul, is the exit area, composed of the flood plains [10].

The prevailing climates for 10 years are: **E (T)CH (w2)ig** (Very cold with rain regime in summer with ganges and isotherms type gait) It is located in the region of the peak of Orizaba; **Cb´(w2)igw** (Semi-cold with rains in summer with heatwave, Ganges-type gait and isotherms); **C(m)(f)igw** (Humid temperate with rains all year round with heat waves, Ganges-type gait and isotherms); **C(w1)ig** (Temperate with rains in summer with Ganges-type gait and isotherms); **C(w2)ig** (Temperate with rains in summer with ganges and isotherms, the most humid of the sub humid) [11].

To highlight the levels of sociocultural importance in the lower part of the Rio Cotaxtla and Rio Jamapa basins, with an area of 500km^2, lies in the location of 132 sites with monumental architecture of rammed earth, as testimony of pyramidal and monumental squares, they have been used to collect chronological information on archeological events, as an example, the Conchal Norte and La Joya are considered (**Figure 1**) [12].

2.1 Methodological description

A review and inclusion of the research results was carried out in the lower basin of the Jamapa River, to know what is the impact of anthropogenic activities that are having an impact on fresh and salt water, in addition to knowing what would be the impact of these activities on the aquatic organisms that are present in the lower Jamapa basin and the implications for the impact of pollution in the Gulf of Mexico. Some research results are also presented, where the presence of pollutants such as heavy metals and pesticides in the water is manifested, which is important to note

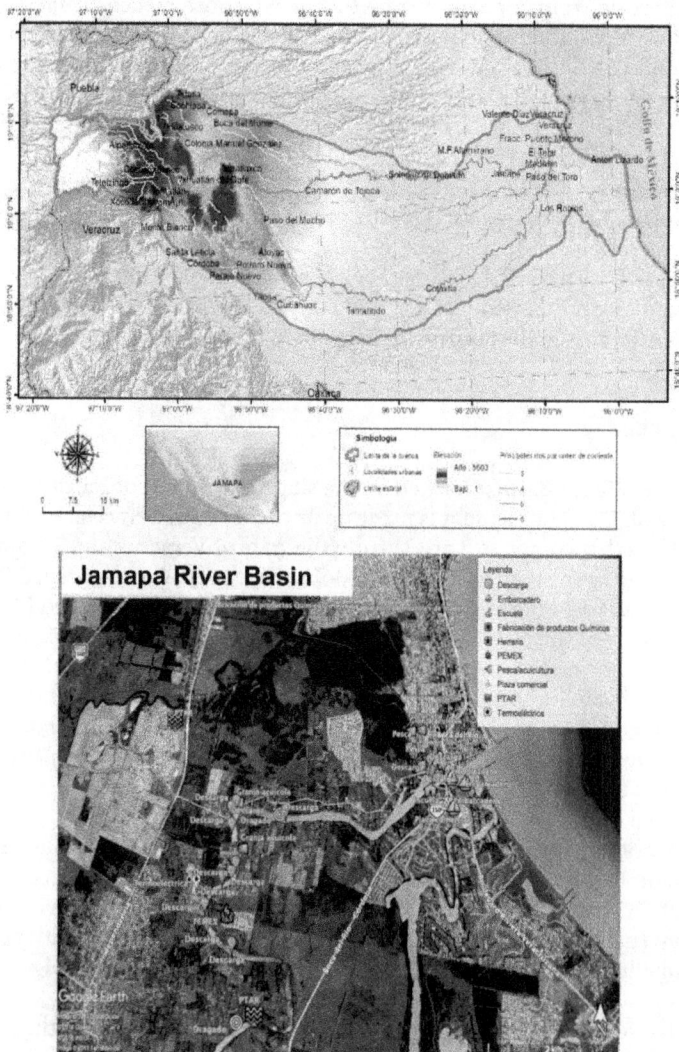

Figure 1.
Representative location of the lower basin of the Jamapa River.

that its importance lies in its massive use and recently introduction to the market of the agricultural sector. It was divided as main topics to address: Anthropogenic activities in the region; Contaminants in water, soil, sediment and organisms; Impacts on the Veracruz Reef System; Impact on the Gulf of Mexico.

2.2 Anthropogenic activities in the region

The character of the Jamapa River basin is heterogeneous, it encompasses present coverage, related to orography, and human activities in the region. That is, in the upper part of the basin, erosive flow conditions are associated, it links a vegetation of mountainous regions, with river slopes, associated with water flows. In the middle part of the basin, the vegetation and the slope (does not exceed 1%), allows the development of agricultural activities and pastures, that is, the development of the agricultural sector, begins to displace the primary vegetation and its alteration, otherwise In the lower part of the basin where anthropic activities occur in greater proportion as urbanization product of a softer or no slope, it is the area with the greatest deposition and discharge of the basin, an example of this is the metropolitan area of Veracruz and Boca del Rio [13]. The activities carried out in the coastal lagoons that influence the lower Jamapa basin, activities are carried out in areas with open and closed systems, in cages and ponds, for the production of marine and freshwater organisms [14]. In the upper part of the Jamapa river basin, rainfall that is between the ranges of 1200 to 1300 mm per year, represents 19.4% of the Jamapa basin, where pine forest communities such as oak, pine and the oyamel [15].

In development areas in the basin, the negative effects upstream will have direct impacts on the lower part of the basin, that is, they influence the coastal and marine ecosystems, they affect the capacity of the system on meteorological phenomena, increasingly violent and unpredictable. CONANP (National Commission of Protected Natural Areas) and its civil allies, companies and the settled population, carry out activities for the ordering, conservation and adaptation of productive activities, protection of natural resources, because, in terms of water, the shortage of 2.5 million people in the states of Puebla and Veracruz [16]. The extreme impacts of climate change in the Pacific and Gulf of Mexico, such as cyclones Ingrid and Manuel, are the product of the climatic variability of our environment, it is estimated that there is an area of high vulnerability in the north of the state of Veracruz, mainly where begins the slope towards the Gulf of Mexico (**Figures 2** and **3**) [17].

The anthropic activities that are carried out in the region of the Jamapa River basin could not only negatively impact the population, but also change the habitat of some species that nest in said area, such as the red-billed tropical bird (*Phaethon aethereus*) that feeds on various species of fish (*Loliolopsis diomedeae, Engraulidae, Clupeidae, Exocoetidae, Hemiramphidae, Carangidae, Gerreidae y Scombridae*) in the Gulf of Mexico [18]. Anthropic causes and the magnitude of climate change threaten the basic elements of human life, such as water supply, food production, health, land use, and the environment [19]. The Mandinga Lagoon System, located in the town of Mandinga; Alvarado; Veracruz is a body of water that receives water from the Jamapa River basin and the Gulf of Mexico tide, and its main activity is fishing; But the facts and social phenomena are articulated in different dimensions through the territory, the conception of humanity and nature have given rise to environmental and social problems, combined with historical aspects, such is the case of disputes between the conceived space and the local territory, which is more than a cultural aspect [20]. Since this system has direct influence from four municipalities Alvarado, Boca del Río, Medellín and Tlalixcoyan.

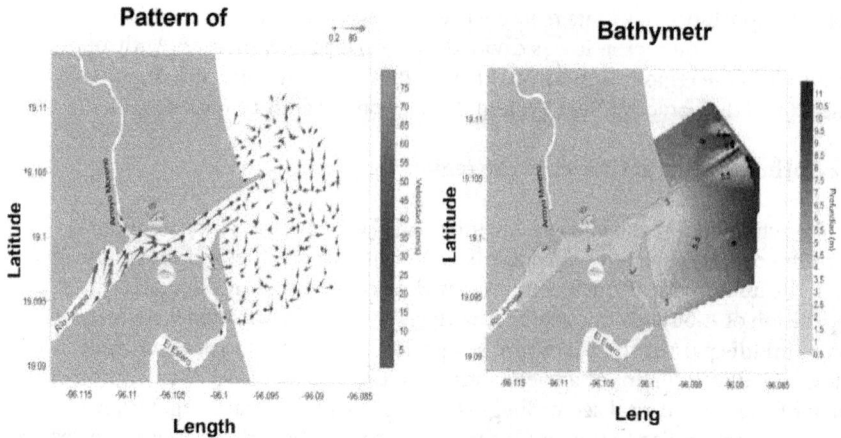

Figure 2.
Current pattern (left) and bathymetry (right), studies from august 2016.

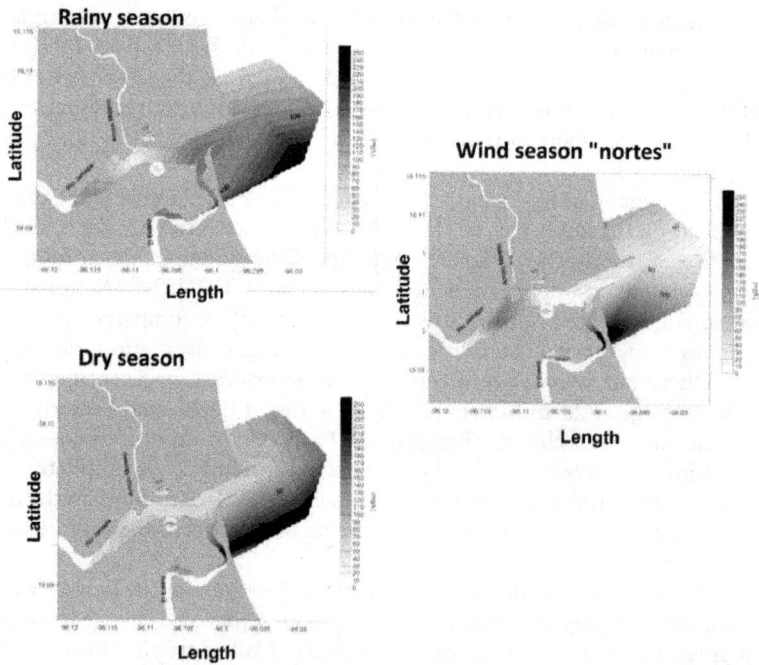

Figure 3.
Behavior of the suspended particles per season, in the water column at the mouth of the Jamapa River, Veracruz.

The Mandinga Lagoon System has been impacted by the high logging of the mangrove and fishing deterioration, by the decrease in some environmental services that the lagoon system provides to the inhabitants, such as artisanal fishing, in addition to the change of activities employment from construction, trucking and migration with a negative effect [21].

A study in three municipalities vulnerable to hydrometeorological phenomena in the state of Veracruz, showed that young people from Tlacotalpan, La Antigua and Cotaxtla are the ones who can become agents of change towards their families and the rest of the population, mainly because they know the uses and customs of the community and are proactive, this allows to identify natural leadership, to link intra-community and inter-community civil protection and strengthen solidarity and reciprocity (**Figure 4**) [22].

2.3 Pollutants in water, soil, sediment and organisms

In a study of surveys carried out among residents and users of the Jamapa basin, contamination was detected as the biggest problem at the municipal level, among which the contamination of the river, soil, improper handling and burning of garbage, factory waste stand out [23].

In **Table 1**, as part of the agricultural activities and watersheds that take place in the Jamapa River, neonicotinoid pesticide residues were found at different sampling points along the river route, the maximum values were 0.163 mg L^{-1} of thiamethoxam and mean values of 0.0417 mg L^{-1} of thiamethoxam, In the north wind season that begins in the month of November to February, the highest concentrations of this pesticide were recorded [24].

The Mandinga Lagoon System is associated with the Jamapa River basin, which is born with the melting of the Orizaba peak and travels 150 km, in this lagoon system there are different sources of point and diffuse contamination with the presence of *Vibrio cholerae*. The main factors of alteration and contamination of the lagoon are the change of land use, population increase and migration to the coast, the uncontrolled use of pesticides, and industrial waste (oil), little water treatment of waste, immoderate logging, desiccation of swamps for agricultural purposes. In 40 sampling sites of the lagoon system, a variation of the concentrations of *Vibrio cholerae*, and the highest concentrations were found in the round lagoon, higher than 1000

Figure 4.
River Jamapa Basin, Veracruz. A) Mouth of arroyo Moreno, B) arroyo Moreno, C) mouth of the Jampa River, D) mouth of the estuary of the Mandinga lagoon system and E) estuary of the Mandinga lagoon system.

Main riverbed	Slopes built-in	Activities / Establishments
Jamapa River	Jamapa River Cotaxtla River	Discharge of urban and industrial wastewater and transport of chemical pollutants in soils pesticides, pesticides, herbicides, heavy metals, emerging and microbiological pollutants.
Arroyo Moreno	Laguna Real, river and Channel	Domestic / industrial wastewater discharges La Zamorana cannel of the municipalities of Veracruz, Boca del Río- Medellin
Mandinga Lagoon System	Mandinga Lagoon	Discharges of Wastewater from El Dorado, Discharge of Wastewater from the domestic areas of the towns of Mandinga, El Conchal, Alvarado and Veracruz.

Table 1.
Main point and non-point sources in the Gulf of Mexico with an impact on the Jamapa River basin.

NMP/100 ml of water; Also, it was determined that the main source of contamination are discharges from homes and restaurants. Another important factor regarding the seasons, the greatest contamination is in the dry season, followed by the rainy season and the north wind season [25]. The presence of high densities of bacteria and organic matter generates the ideal characteristics for the development of parasites in wild populations of *Macrobrachium acanthurus* from the lower basin of the Jamapa River; in a diagnosis carried out in 2007 to 2008, five species of parasites were found: *Epistylis* sp., *Acineta* sp., *Lagenophrys* sp., gregarines and a ciliate (**Figure 5**) [26].

The dry river sub-basin, located in the center of Veracruz, is a part of the hydrological region "X Golfo Centro" and the Jamapa River basin, which is home to more than 200,000 people, who live in urban areas, and the 30% in rural areas, this river is used for agriculture, for the provision of drinking water and environmental support; at present it is heavily contaminated by organic matter, nitrogen and fecal matter [27]. Also, water erosion has been estimated in the Jamapa River Sub-basin,

Figure 5.
Point pollution sources in the Estero area, of the de Mandinga lagoon system, Veracruz.

with precipitation data of 10 years from 1990 to 2008, finding that is lost a total of 7787.8 ton/10 years and the month of July is when the highest average specific degradation in 10 years of 27.7 t. ha^{-1} [28].

In **Figure 6**, the concentrations of paraquat herbicide particles in water are shown in the lower basin of the Jamapa River in the rainy season, high residual concentrations were found mainly in the sampling site called Las Gualdras, these concentrations have as diffuse sources of contamination the crops established on the banks of the Jamapa River, such as crops of pineapple (*Ananas comosus*), watermelon (*Citrullus lanatus*), papaya (*Carica papaya* L.) and rice (*Oryza sativa*). On the other hand, the management of systemic insecticides with high mobility such as neonicotinoids, especially thiamethoxam, cause water pollution.

In **Figure 7**, it is shown that the Cotaxtla and Jamapa rivers bifurcate in the Gulf of Mexico, they present residuality of thiamethoxam exceeding the maximum residuality limit of the FAO of 0.01 mg/L, It was also observed that the highest concentrations were found in the Cotaxtla River, which crosses the municipalities of the center of Veracruz, such as Cotaxtla and Medellín de Bravo, a place of high agricultural activity.

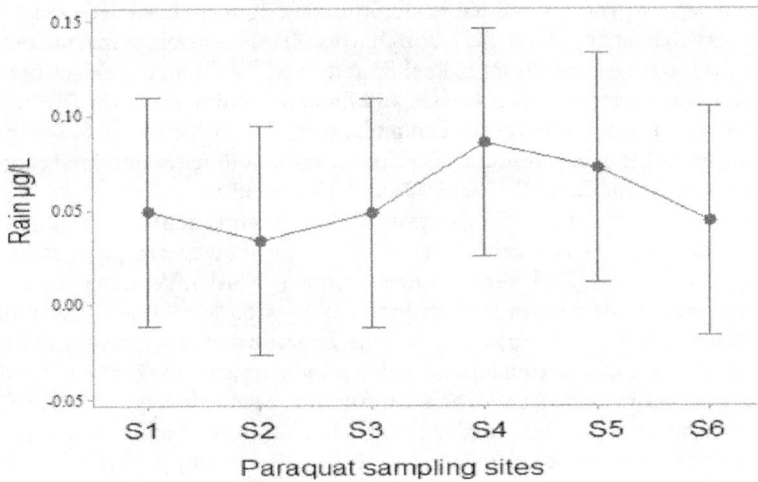

Figure 6.
Paraquat concentrations in the rainy season at six surface water sampling sites (ANOVA: α = 0.05), in the sub-basin of the port of Veracruz and Río Jamapa, S = wáter sampling site. S1(dos Bocas 1), S2(dos Bocas 2), S3(La Rayana), S4(las Gualdras), S5(arroyo Moreno), S6(La Bocana).

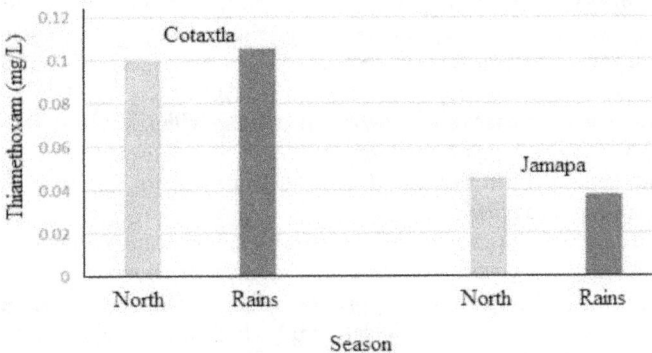

Figure 7.
Thiamethoxam concentrations in the Cotaxtla and Jamapa rivers in the north wind and rainy seasons.

2.4 Impacts on the Veracruz reef system

In the Veracruz Reef system, the variation in temperature was studied between the period of March 2011 to March 2012, it was found that the highest values in temperature occur in the months of August and September; With respect to the salinity of the water there is little variability because the concentrations are presented in a constant way. In the variation of oxygen, the values found were in a range of 2 to 5 ml L^{-1}, and nitrogen concentrations were observed low values in the months of August and September with 8.2 ml L^{-1} and high values greater than 9.0 ml L^{-1}, These variations are caused by the accumulation of sediments from the Jamapa River basin, in addition, a scenario is visualized by the microbiotic activity and an area of hypoxia is identified [29].

It is indicated that after the floods of the Jamapa River and the cold water intrusions that affected the Veracruz Reef System in the 1970s, the populations of Acropora palmara are recovering, because in 2007 and 2013 they have been found in 11 reefs north and on all southern reefs, mainly in shallow waters along the reef edge [30].

The presence of various pollutants of anthropic origin has been detected, Zamudio-Alemán et al., [31] in his research work: "Heavy metals in marine sediment of the National Park Veracruz Reef System (PNSAV)" had the objective of identifying the concentration of Cu, Cd and Zn in the sediments of the PNSAV, associated with the main sources of contamination, that influence the lower basin of the Jamapa River using atomic absorption spectrophotometry obtained concentrations of Cu,Cd and Zn of 0.1392;0.001 and 2.3606 mg kg^{-1}.

Montoya-Mendoza et al., [32] determined the concentrations of cadmium (Cd), lead (Pb), vanadium (V) and zinc (Zn) in the muscle of 30 specimens of *Pterois volitans*, in the National Park Veracruz Reef System (PNSAV), Veracruz, México, using spectrophotometry atomic absorption (AAS), after microwave digestion. Due to their proximity to the coast, and to the Veracruz port, they have suffered numerous severe impacts from human activity, industry, and the carry of sediments from the rivers of Jamapa, Papaloapan and Antigua; causing a set of disturbances by heavy metals, hydrocarbons and coliforms. Studies in 2007 of macrocrustaceans in this reef park, found the existence of more than 20 reef banks with different degrees in development and accretion, the presence of two groups of macrocrustaceans, located in the northern part and another in the southern part of the reef system, as a result of the presence of two oceanic gyres in front of the Jamapa River, which generate a physical barrier in this protected natural area [33].

The expansion works in the Veracruz Reef System National Park have generated negative changes in the ecosystems of the coral reefs; For example, in the Blanquilla reef, it was observed that the algae cover tends to decrease temporarily and increase the groups of invertebrates, also the reefs tend to increase like the Agaricia in the Blanquilla and in others the opposite happens like Agaricia in the Galician; In addition, there were diseases such as coral bleaching, which has been the disease with the highest percentage of mortality in 2017; On the other hand, the decrease in biomass and density of piscivores and fish of commercial interest is also attributed to the expansion of the port [34].

The environmental and socioeconomic problems currently being experienced in the PNSAV, requires a change in the work scheme that allows to recover the trust and participation of the different actors in favor of the maintenance of environmental services, since its declaration as a National Park, Ramsar site, Biosphere Reserve. So, from 1992 to date it has been working, without any management program since its name as ANP. On the other hand, despite the fact that there are several involved from researchers, authorities, with multidisciplinary scientific capacity and quality research institutions, a fisheries management program is required, in the area in synchrony with the park management program, to contribute the conflict facing the PNSAV, with economic activities including port activity, tourism and fishing, with impact work on the conservation of reefs, flora and fauna since its expansion of the port area to correct space problems [35].

2.5 Impact on the Gulf of Mexico

The pollution indices on the coast of the Gulf of Mexico, coming from the Jamapa River, found heavy metals as Al, V, Cr, Co, Ni, Cu, Zn, As, Cd and Pb, but the highest values are for Al2O3 higher than 120 µg g-1. The geoaccumulation index of metal concentrations in the area of the upper continental crust of the Jamapa are As (3.76), Cu (2.47), Zn (1.38), the determining factor of chemical concentrations is coming from anthropic activities and their distribution of the Jamapa River [36].

A study of eutrophication on the coast of the lower part of the Jamapa River, found in six sampling sites values of dissolved oxygen between the ranges of 3.41 to 6.19 mg L^{-1}, total nitrogen between ranges of 2.732 to 4.596 mg L^{-1} and temperatures of 27.67 to 30.30 °C, these values influence the water quality [37]. Local fishermen from the Jamapa river hydrological basin, in the Boca del Río municipality and field findings, indicate that species richness, abundance, fish sizes and their commercial use have decreased, as a result of the growth of the urban area, and the deterioration of the body of water and the negative effects on the fish population is also evident, in addition to that the characteristics of the water have been modified, mainly due to its poor quality [38].

The trend of the decrease in salinity in the large Mandinga lagoon between the city of Boca del Río and the town of Mandinga, in rainy or dry periods, is due to the contribution of the flow of the volume of water from the Jamapa River, which dilutes the concentration of salt; In the two-dimensional analysis carried out under low water conditions, the values ranged from zero to 0.5 km upstream from the border with the sea [39].

Matriz	Cd	Pb	Región	Referencia
Sedimento marino	1.0–13.9	0.9–37.7	Placa continental de Tabasco, Tamaulipas y Veracruz	[41]
Sedimento superficial	—	5.3–42.4	Isla de Sacrificios, PNSAV	[42]
Sedimento superficial	0.02–0.2	5.0–22.0	Costa central de Veracruz	[43]
Sedimento marino	<0.0–0.37	—	PNSAV	[44]
Sedimento superficial	—	53.1–107.3	PNSAV	[45]
Sedimento	0.01–5.27	0.02–2.3	Sistema Lagunar de Alvarado	[46]
Sedimento marino	—	2.90–16.6	Costa central de Veracruz	[31]
Núcleo de sedimento	<0.00–0.016	31.86–40.36	Sistema Lagunar de Alvarado	[47]
Sedimento marino	—	10.0–27.0	Costa de Tamaulipas	[48]
Núcleo de sedimento	—	0.1–26.2	Región sur del Golfo de México	[49]
Sedimento marino	2.9-33.33	0.0-11.72	PNSAV	[50]

Table 2.
Reported concentrations for cadmium and lead in sediments in the Gulf of Mexico. Concentrations in mg kg^{-1}.

There is evidence of metal contamination in water and sediment matrices in coastal and marine areas; such is the case of the presence of Lead in sediments (Pb), in the lagoons near Laguna Verde of 77.2 μg g^{-1}, Salada of 78.8 μg g^{-1} and the Mancha of 81.1 μg g^{-1}, It must be considered that lead is volatile and tends to be deposited in areas other than its origin, this could be influenced by the wind patterns that predominate in the Gulf of Mexico. In addition, in some metals the chromium in sediments present in the Laguna de Ostión with concentrations of 140.7 μg g^{-1} and in the Alvarado lagoon of 159.7 μg g^{-1} in the state of Veracruz, in addition, it tends to accumulate in sediments and increases its level in these areas. In the case of total Nickel (Ni) in the sediments of the coastal areas of the Gulf of Mexico, with a concentration pattern of 26.29 μg g^{-1} in the Mandiga lagoon, in the Jamapa and Actopan rivers, Papaloapan Veracruz, with concentrations below 100 μg g^{-1} [40]. **Table 2** shows the concentrations for cadmium and lead in sediments in the Gulf of Mexico.

3. Discussion

The presence of pollutants in the water in the Jamapa River basin, and the concentration of organisms that come from fecal matter such as total and fecal coliforms, have their origin from human activities, animal husbandry, agricultural activity and aquaculture, it was found that there is a variation in the chemical parameters (oxygen, nitrates and nitrites) in the quality of groundwater and surface water (lagoon water systems); An indicator of contamination in the Jampa River basin is the presence of *Vibrio* sp., there is also evidence of the negative impact of the extraction of water from the aquifer and of chemical and microbiological contaminants, such as the presence of the enteropathogenic bacteria *E. coli*, which is the cause of diarrhea syndromic diseases in adults and children, and some *E. coli* batteries that cause internal bleeding [51].

It is evident that all the discharges that the Jamapa River basin receives are the result of the runoff and infiltration processes of the lagoon systems that ultimately go to the Jamapa River, in addition to impacting the Gulf of Mexico, mainly the coral reef area, is a concentration of pathogenic bacteria from organic matter and anthropogenic activities among those identified are *Aurantimonas coralicidae, Halofolliculina corallasia, Helicostoma nonatum, Oscillatoria sp., Phormidium corallycticum, Phormidium valderianum, Serratia marcescens* and *Vibrio spp.* [52]. Among the indicator parameters of eutrophication on the lower coast of the Rio de Jamapa, they were found in the municipalities of Boca del Rio, Medellín and Alvarado in six analyzed sites (Medellín, Tejar, San José, Playa de Vacas, Arroyo Moreno, Boca del Rio) maximum temperatures of 30.3 °C, and dissolved oxygen of 5.83 mg L^1, and the maximum means and total nitrogen were 4.59 mg L^1 and 4.28 mg L^1, This causes alterations in nitrogen concentrations, an increase in temperature and high concentrations of nutrients in the water [53]. On the other hand, synthetic pollutants present in the Gulf of Mexico such as banned organochlorine pesticides such as Endosulfan, when concentrated in the water, can be present in aquatic organisms such as mollusks, crustaceans and fish; organisms in which they have been detected in maximum concentrations of 99.48 ± 16.21 ng g^{-1}, which could be a potential risk to public health and food trade in tourist areas such as Veracruz-Boca Río [54]. Among other organoclarated pesticides identified in sediments in the Alvarado lagoon system, some concentrations were aldrin (46.05 ng / g), dieldrin (22.13 ng / g) and endrin (21.23 ng/g), these concentrations are reported and originate from the agricultural activities [55].

Faced with this problem, it is important to generate agricultural, livestock and aquaculture production schemes with sustainable approaches, to mitigate the effects caused by pollution and greenhouse gases; For example, in the case of the management of pesticides of chemical and synthetic origin, combine production for pest control with products of plant or biological origin. Minimize the use of pesticides in agriculture and livestock, it is evident that aquaculture and fisheries production activities are having a negative impact. Ecosystems require sustainability strategies for the protection of their resources, the local participation of producers, the formation of networks, disseminating clear norms, building trust and credibility in transparent processes [56]. Consider that human settlements and conurbation areas need to strengthen wastewater treatment, improve solid waste treatment, and that municipal governments improve processing plants, prohibit construction on the banks of the Jamapa River [57].

4. Conclusion

The Jamapa River basin is the link of three protected natural areas, receives a significant impact from the sources that flow into the basin and which has a consequence in the Gulf of Mexico. Due to anthropogenic activities, many of the physical, chemical and biological pollutants are deposited in water, sediments and marine organisms, mainly in the reef area of the Gulf. It is necessary to seek strategies to mitigate the environmental impact generated by human activities, from the organization at the national, state and municipal level and the organization with the key actors participating in the different links of the agri-food chains, in addition to applying regulations and laws for the regulation and care of the environment.

In Mexico, 1471 hydrographic basins have been delimited that, for administrative purposes, the National Water Commission (CONAGUA) has grouped into 731 basins, which in turn make up 37 hydrological regions, again grouped into 13 economic-administrative regions.

The Jamapa River Basin is an example of the complexity that occurs in the basins of Mexico, where sustainable development faces the challenges of combining and harmonizing economic, social and environmental development in favor of growth and the preservation of different habitats to achieve an adequate quality of life and mitigate climate change.

Acknowledgements

The authors thank the Program for the professional development of teachers (PRODEP), the consolidated academic body ITBOR-CA-2 "Management of Coastal Resources and Environmental Sciences".

Conflict of interest

The authors certify that they have no conflict of interest during preparation of this chapter.

Author details

María del Refugio Castañeda-Chávez* and Fabiola Lango-Reynoso
National Technology of Mexico/Technological Institute of Boca del Río,
Veracruz, México

*Address all correspondence to: mariacastaneda@bdelrio.tecnm.mx

IntechOpen

References

[1] Garrido A., Pérez D.J.L. y Enríquez C. "Delimitación de zonas funcionales de las cuencas hidrográficas de México". En: Cotler H. (Coord.) Las cuencas hidrográficas de México. Diagnóstico y priorización. México: Instituto Nacional de Ecología/Fundación Gonzalo Río Arronte I.A.P. 2010. Disponible en: www2.inecc.gob.mx/publicaciones/consultaPublicacion.html?id_pub=639.

[2] CONANP, 2019. Decretos, programas de manejo CONANP, población total estimada: INEGI. Censo de población y vivienda 2010. Principales resultados por localidad (ITER), 2010. https://simec.conanp.gob.mx/ficha.php?anp=135&reg=5

[3] Ortiz L. J. A. Modificación en la provisión de los servicios ambientales por efecto del cambio en la heterogeneidad ambiental en la cuenca del Río Jamapa, Veracruz, México. Universidad Veracruzana. Instituto de ciencias marinas y pesquerías, maestría en ecología y pesquerías. 2013. 137 p.

[4] Castañeda C. M. del R., Sosa V. C. A., Amaro E. I. A., Galaviz V. I., Lango R. F. Eutrophication in the lower coastal basin of the Jamapa river in Veracruz, Mexico. International journal of research-Granthaalayah. 2017. 5. 12: 206-216. https://doi.org/10.29121/granthaalayah.v5.i12.2017.495

[5] Salas-Monreal, D., Díaz-Hernández, A., Áke-Castillo, J. A., Granados - Barba, A., & Riverón-Enzástiga, M. L. Variación anual de los parámetros hidrográficos en la confluencia del río Jamapa y arroyo Moreno (México). Intropica, 2020. 15. 1: 59-65. https://doi.org/10.21676/23897864.3402

[6] García V. A. M., Montoya M. J., Chávez L. R., 2019. Aproximación histórica de la composición de especies de peces en arroyo Moreno, Veracruz, México. Biocyt Biológica, Ciencia y Tecnología. 2019. 12. 48: 895-908.DOI: 10.22201/fesi.20072082.2019.12.72323.

[7] Rodríguez G. C. F., Aké C., Campos B. G., Productividad primaria bruta y respiración plantónica en el parque nacional sistema arrecifal veracruzano. Hidrobiológica. 2013. 23.2: 143-153.

[8] Landeros S. C., Lango R. F., Castañeda C. M. del R.., Galaviz V. I., Nikolskii G. I., Palomares G. M., Reyes V. C., Minguez R. M. Assessment of water pollution in different aquatic systems: Aquifers, aquatic farms on the Jamapa river, and coastal lagoons of México. Journal of Agricultural Science. 2012. 4. 7: 186-196. URL: http://dx.doi.org/10.5539/jas.v4n7p186.

[9] INECC-FGM. Plan de acción para el manejo integral de cuencas hídricas: Cuenca del río Jamapa. Proyecto: Conservación de cuencas costeras en el contexto del cambio climático. 2018. 151 pp.

[10] Ortiz L. J. A. Modificación en la provisión de los servicios ambientales por efecto del cambio en la heterogeneidad ambiental en la cuenca del Río Jamapa, Veracruz, México. Universidad Veracruzana. Instituto de ciencias marinas y pesquerías, maestría en ecología y pesquerías. 2013. 137 p.

[11] Reyes C. M. del P. Estructura y distribución del bosque de abies en la cuenca superior del Río Jamapa, Veracruz. Tesis de licenciatura en geografía. Universidad Nacional Autónoma de México. México, D. F. 2000. 111 p.

[12] Velasco G. J. E., Danneels V. A., Silva C. T., 2011. Patrones de macrodesgaste dental y diferenciación social en restos óseos del clásico en el centro de Veracruz. Estudios

de Antropología Biológica. 2011. 15:245-271.

[13] Ortiz L. J. A. Modificación en la provisión de los servicios ambientales por efecto del cambio en la heterogeneidad ambiental en la cuenca del Río Jamapa, Veracruz, México. Universidad Veracruzana. Instituto de ciencias marinas y pesquerías, maestría en ecología y pesquerías. 2013. 137 p.

[14] Salcedo G. M. G., Castañeda C. M. R., Lango R. F., Sosa V. C. A., Landeros S. C., Galaviz V. I. 2019. Influence of physicochemical parametrs on phytiplankton distribution in the lagoon system of Mandiga, México. Revista Bio Ciencias. 6,e427.Doi: https://doi.org/10.15741/revbio.06.e427.

[15] Reyes C. M. del P. Estructura y distribución del bosque de abies en la cuenca superior del Río Jamapa, Veracruz. Tesis de licenciatura en geografía. Universidad Nacional Autónoma de México. México, D. F. 2000. 111 p.

[16] Álvarez R. Conservación en las cuencas Jamapa y Antigua. Diario de Xalapa, Ciencia y luz. Edición Hernández-Gutiérrez E. 2017.

[17] Zerecero S. B. C., Ibarracán V. M. E., Gómez G. A., Hernández de la R. P., González G. M. de J., Escalona M. M. J., Sardiñas G. O., Rivera C., Toruño P., 2015. Modelo de indi-cadores de vulnerabilidad al cambio climático y su representación espacial en la región centro-Golfo de México. Revista iberoamericana de Bioeconomía y Cambio climático. 2015. 1.1: 149-184.

[18] Velarde E., Iturriaga L. J., Meiners C., Jiménez L., Perales H., Sanay R., Lozano M. A., Cabrera V. H. D., Anaya C. C. Red-billed tropicbird Phaethon arthereus occurrence patterns in the state of State of Veracruz, Gulf of México: posible causes and implications. Marine ornithology. 2014. 42:119-124.

[19] Maldonado G. A. L., González G. E. J., Cruz S. G. E. Una aproximación a la presentación del cambio climático en habitantes de dos cuencas del estado de Veracruz, México. Revista pueblos y fronteras digital. 2017. 12. 23: 149-174.

[20] Navarro D. C. L. Organización territorial e identificados de los pueblos de la laguna de Mandinga, Veracruz. Tesis de maestría, Centro de investigaciones en geografía y geomática "Ing. Jorge L. Tamayo", A.C., Centro público de investigación CONACYT. 2017. 92 p.

[21] Aldeco J., Cortés A. G., Jurado M. J. Adaptaciones culturales y económicas a cambios provocados por tala de mangle y deterioro pesquero en Mandinga, Veracruz. Sociedades rurales, producción y medio ambiente. 2015. 15.29: 137-157.

[22] Gonzales G. E. J., Maldonado G. A. L., Méndez A. L, M., Mesa O. S. L. Un estudio sobre vulnerabilidad y resiliencia social en poblaciones de alto riesgo a inundaciones en el estado de Veracruz. Revista Aidis de ingeniería y ciencias ambientales: Investigación, desarrollo y práctica. 2017. 11. 3:401-414.

[23] Maldonado G. A. L., González G. E. J., Cruz S. G. E. Una aproximación a la presentación del cambio climático en habitantes de dos cuencas del estado de Veracruz, México. Revista pueblos y fronteras digital. 2017. 12. 23: 149-174.

[24] Morales G. M. M. Presencia del insecticida thiamethoxam en aguas superficiales de los ríos Cotaxtla y Jamapa. Tesis de Maestría en Ciencias en Ingeniería Ambiental. Tecnológico Nacional de México/Instituto Tecnológico de Boca del Río. 2019. 72 p.

[25] Reyes V. C. Contaminación del agua por *Vibrio cholerae* y sus patrones de distribución en el sistema lagunar de Mandinga, Veracruz. Tesis de doctorado en agroecosistemas tropicales, Colegio

de Postgraduados-Campus Veracruz. 2013. 88 p.

[26] Domínguez M.M. E., Juárez C. S. F. Ocurrencia temporal de parásitos en *Macrobrachium acanthurus* de la cuenca baja del río Jamapa. Memoria del 2 encuentro de biometría y la V reunión de la región centroamericana y del caribe de la sociedad de biometría. Editores Juárez-Cerrillo S. F., Ojeda-Ramírez M. M., 2010: 59-63.

[27] Torres B. B., González L. G., Rustrían P. E y Houbron E. Enfoque de cuenca para la indentificación de fuentes de contaminación y evaluación de la calidad de un río, Veracruz, México. Revista Internacional Contaminación Ambiental. 2013. 29. 3:135-146.http://www.scielo.org.mx/scielo.php?script=sci_arttext&pid=S0188-49992013000300001&lng=es&tlng=es.

[28] Ramírez S. D. Estimación de la producción de sedimentos en cinco microcuencas del río Jamapa bajo seis condiciones de uso de suelo. Tesis profesional. Universidad Autónoma Chapingo. 2012. 93 p.

[29] Avendaño A. O., Salas M. D. Marín H. M., Salas de L. D. A., Monreal G. M. A., 2017. Annual hydrogical variation and hypoxic zone in a tropical coral rref sytem. Regional Studies in Marine Science. 2017. 9:145-155.

[30] Larson E. A., Gilliam D. S., López P. M., Walker B. K. Possible recovery of *Acropora palmata* (Scleractinia: Acroporidae) within the Veracruz reef system, Gulf of Mexico: a survey of 24 reefs to assess the benthic communities. Revista Biol. Tro. 2014. 62.3: 75-84.

[31] Celis H. O., Rosales H. L., Cundy A. B., Carranza E. A., 2017. Sedimentary heavy metal (loid) contamination in the Veracruz shelf, Gulf of Mexico: A baseline survey from a rapidly developing tropical coast. Marine pollution bulletin. 119(2): 204-213. https://doi.org/10.1016/j.marpolbul.2017.03.039.

[32] Montoya M. J., Alarcón R. E., Castañeda C. M. del R., Lango R. F. y Zamudio A. R. E., 2019. Heavy metals in muscle tissue of *Pterois volitans* from the Veracruz reef system national park, México. International journal of environmental research and public health. 2019. 16. 4611: doi:10.3390/ijerph16234611.

[33] Winfield I., Cházaro O. S., Horta P. G., Lozano A. M. A., Arenas F. V. Macrocrustáceos incrustantes en el Parque Nacional Sistema Arrecifal Veracruzano: biodiversidad, abundancia y distribución. Revista mexicana de biodiversidad, 81(Supl. oct). 2010. 165-175.http://www.scielo.org.mx/scielo.php?script=sci_arttext&pid=S1870-34532010000400011&lng=es&tlng=es.

[34] Arguelles J. J., Brenner J., Pérez E. H. Línea base para el monitoreo de los arrecifes del Sistema Arrecifal Veracruzano (PNSAV) a través de la metodología AGRRA (Atlantic and Gulf Rapid Reef Assessment). Universidad Veracruzana. The Nature Conservancy. Sea&Reef. Boca del Rio. 2019. 26 pp.

[35] Jiménez B. M. de L., Cruz R. S., Lozano A. M. A., Rodríguez Q. G., 2014. Problemática ambiental y socioeconómica del Parque Arrecifal Veracruzano. Investigación y Ciencia. 2014. 22. 60: 58-64. http://www.redalyc.org/articulo.oa?id=67431160007

[36] Celis H. O., Rosales H. L., Carranza E. A. Heavy metal enrichmet in surface sediments from the SW Gulf of Mexico. Environ Monit Asses. 2013. 185:8891-8907.

[37] Castañeda C. M. del R., Sosa V. C. A., Amaro E. I. A., Galaviz V. I., Lango R. F. Eutrophication in the lower

coastal basin of the Jamapa river in Veracruz, Mexico. International journal of research-Granthaalayah. 2017. 5. 12: 206-216. https://doi.org/10.29121/granthaalayah.v5.i12.2017.495

[38] García V. A. M., Montoya M. J., Chávez L. R., 2019. Aproximación histórica de la composición de especies de peces en arroyo Moreno, Veracruz, México. Biocyt Biológica, Ciencia y Tecnología. 2019. 12. 48: 895-908.DOI: 10.22201/fesi.20072082.2019.12.72323.

[39] Gonzales V. J. A., Hernández V. E., Rojas S. C., Del Valle M. J. Diagnosis of water circulation in an estuary: A case study of the Jamapa River and the Mandiga lagoons, Veracruz, Mexico. Ciencias marinas. 2019. 45.1: 1-16.

[40] Vázquez B. A., Villanueva F. S., Rosales H. L. Distribución y contaminación de metales en el Golfo de México. Diagnóstico ambiental del Golfo de México. Caso M., Pisanty I. Excurra E. 2004. 2:681-710.

[41] Ponce-V. G., Vásquez B. A., Díaz G. G., García R. C. Contaminantes orgánicos persistentes en núcleos sedimentarios de la Laguna el Yucateco, Tabasco en el sureste del Golfo de México. Hidrobiología. 2012. 22.2: 161-173.

[42] Rosales H. L., Kasper Z. J. J., Carranza E. A., Celis H. O., Geochemical composition of surface sediments near isla de sacrificios coral reef ecosystem, Veracruz, Mexico. Hidrobiologia. 2008. 18.2: 155-165.

[43] Celis H. O., Rosales H. L., Carranza E. A. Heavy metal enrichmet in surface sediments from the SW Gulf of Mexico. Environ Monit Asses. 2013. 185:8891-8907.

[44] Zamudio-Alemán, R. E., Castañeda-Chávez, M. D. R., Lango-Reynoso, F., Galaviz-Villa, I., Amaro-Espejo, I. A., & Romero-González, L. Metales

pesados en sedimento marino del Parque Nacional Sistema Arrecifal Veracruzano. Rev. Iberoam. Cienc. 2014.1.4: 159-168.

[45] Horta P. G., Cházaro O. S., Winfield I., Lozano A. M. A., Arenas F. V., Heavy metals in macroalgae from the Veracruz reef system, southern Gulf of Mexico. Revista bio ciencias. 2016. 3.4: 326-339.

[46] Castañeda C. M. del R., Sosa V. C. A., Amaro E. I. A., Galaviz V. I., Lango R. F. Eutrofhication in the lower coastal basin of the Jamapa river in Veracruz, Mexico. International journal of research-Granthaalayah. 2017. 5. 12: 206-216. https://doi.org/10.29121/granthaalayah.v5.i12.2017.495

[47] Botello A. V., Villanueva F. S., Rivera R. F., Velandia A. L., de la Lanza G. E. 2018. Analysis and tendencies of metals and POPs in a sediment core from the Alvarado Lagoon System (ALS), Veracruz, Mexico. Archives of environmental contamination and toxicology. 75:157-173. https://doi.org/10.1007/s00244-018-0516-z.

[48] Celis-Hernández O., Rosales H. L., Cundy A. B., Carranza E. A., Croudace L. W., Hernández H. H., 2018. Historical trace element accumulation in marine sediments from the Tamaulipas shelf, Gulf of Mexico: An assessment of natural vs anthropogenic inputs. Science of the total environment. 622-623:325-336. https://doi.org/10.1016/j.scitotenv.2017.11.228

[49] Ruiz-Fernández A. C., Sánchez-Cabeza J. A., Pérez-Bernal L. H., Gracia A. 2019. Spatial and temporal distribution of heavy metal concentrations and enrichment in the southern Gulf of Mexico. Science of the total environment. 651 (15): 3174-3186.

[50] Mapel-Hernández M. D., 2019. Distribución de metales pesados en sedimento del parque nacional sistema

arrecifal veracruzano; variaciones estacionales y biodisponibilidad. Tesis de maestría en ciencias en ingeniería ambiental. 85 p.

[51] Landeros S. C., Lango R. F., Castañeda C. M. del R.., Galaviz V. I., Nikolskii G. I., Palomares G. M., Reyes V. C., Minguez R. M. Assessment of water pollution in different aquatic systems: Aquifers, aquatic farms on the Jamapa river, and coastal lagoons of México. Journal of Agricultural Science. 2012. 4. 7: 186-196. URL: http://dx.doi. org/10.5539/jas.v4n7p186.

[52] Castañeda C. M del R., Lango R. F. y Navarrete R. G. Hexachlorocyclohexanes, cyclodiene, methoxychlor and heptachlor in sediment of the Alvarado, Lagoon system in Veracruz, México. Sustainability. 2018. 10. 76: doi:10.3390/su10010076.

[53] Castañeda C. M. del R., Sosa V. C. A., Amaro E. I. A., Galaviz V. I., Lango R. F. Eutrophication in the lower coastal basin of the Jamapa river in Veracruz, Mexico. International journal of research-Granthaalayah. 2017. 5. 12: 206-216. https://doi.org/10.29121/ granthaalayah.v5.i12.2017.495

[54] Navarrete R. G., Landeros S. C., Soto E. A., Castañeda C. M. del R., Lango R. F., Pérez V. A., Nikolskii G. I. Endosulfan: Its isomers and metabolites in commercially aquatic organisms from the Gulf of Mexico and the Caribbean. Journal of agricultural Science. 2016. 8. 1: 8-24. URL: http://dx.doi.org/10.5539/ jas.v8n1p8

[55] Castañeda C. M. del R., Lango R. F., García F. J. L., Reyes A. A. R. Bacteria that affects coral health with an emphasis on the Gulf of Mexico and the Caribben Sea. Lat. Am J. Aquat Res. 2018a. 46. 5: 880-889. DOI: 10.3856/ vol46-issue5-fulltext-2

[56] Rangel D. J., Arredondo G. M. C., Espejel I. ¿Estamos investigando la efectividad de las certificaciones ambientales para lograr la sustentabilidad acuícola?. Sociedad y ambiente. 2017. 15: 7-37.

[57] Alfaro G. K., Bazant F. O., Bouchot A. J., Buen día H. A., Gonzalez O. L., Trejo L. B., Ortiz L. L. Diagnóstico de la problemática ambiental en estero del Río Jamapa dentro de los municipios de Boca del Río, Alvarado y Medellín, Veracruz, México. Informe Técnico, Manejo Integrado de Zonas Costeras, Maestría en Ecología y Pesquerías, Instituto de Ciencias Marinas y Pesquerías, Universidad Veracruzana, México. 2014. 23 p.

Chapter 7

The Impact of Embankments on the Geomorphic and Ecological Evolution of the Deltaic Landscape of the Indo-Bangladesh Sundarbans

Subhamita Chaudhuri, Punarbasu Chaudhuri
and Raktima Ghosh

Abstract

The deltaic landscape of the Ganga-Brahmaputra delta has evolved through a complex interplay of geomorphic processes and tidal dynamics coupled with the anthropogenic modifications brought over in course of the reclamation of the islands since the late 18th century. The reclamation process was characterized by clearing lands for paddy farms and fish ponds by building a mesh of earthen embankments along creek banks to restrict saltwater intrusion. The length of the embankments in the Indian Sundarbans alone is 3638 km (World Bank, 2014) which altered the tidal inundation regimes, sediment accretion and geomorphic character of the deltaic inlets. The mean annual sedimentation rate (2.3 cm y^{-1}) in the central Ganga-Brahmaputra delta is over two times higher than sedimentation within the natural intertidal setting of the Sundarbans (Rogers et al., 2017). The tidal range has also increased inland due to polder construc¬tion, with high water levels within the polder zone increasing as much as 1.7 cm y^{-1} (Pethick and Orford, 2013). Embankments have impacted on the biodiversity and physiological adapta-tions of mangroves within the sphere of tidal ingression, habitat fragmentation and seedling establishment. The chapter attempts to reappraise the impact of dykes on the geomorphology of the deltaic landscape and on the functionalities of mangrove forests.

Keywords: land reclamation, embankment, tidal dynamics, geomorphology, mangroves, biodiversity

1. Introduction

Tidal energy influences deltaic components, regulates morphodynamic pro-cesses and facilitates onshore sedimentation worldwide [1–3]. Regular inundation of the intertidal terrain sustains the Ganga-Brahmaputra (G-B) deltaic plain. Since the pre-colonial era, the unified G-B delta region in both India and Bangladesh, witnessed construction of earthen embankments along the tidal channels, rivers, coastal stretches and also within the deltaic plain to prevent flooding, salinity intrusion and land erosion [4, 5]. Referred to by the Dutch term 'polder', there are low-lying floodplain tracts which are enclosed by the dykes and used for rice culti-vation and aquaculture in Bangladesh [2]. These partly-engineered structures not only restrain standard range of tidal flooding but also seize large areas of formerly intertidal landscape. Over the years, the entangled network of rivers fuelling life to the vast stretch of mangrove forests in both India and Bangladesh Sundarbans, has been reduced and the fresh water discharge has been deterred since the rivers got disconnected due to upstream interferences like large barrage and dam installations [5, 6]. In addition, more complex human interventions due to increased population, irrigation, economic activities in terms of flow of goods and requirement for main-taining larger scale infrastructure have magnified the fragilities of the delta over time. For past centuries, a number of embankments have been built in the south-western side (the Indian Sundarban) totaling to about 3638 km in length [6, 7]. In Bangladesh Sundarban, emerging, more than 129 polders have been constructed in the upstream areas encompassing 13,000 km^2 of land or, about 44 percent of the total area in deltaic Bangladesh [6, 8].

Several researches in past few years have documented substantial lowering of coastal tracts indicating greater risk of submergence by sea-level rise and amplified inundation by storm surges [2, 7, 9]. This elevation loss is attributed to the sedi-mentation within embanked channels conjoined with sediment compaction within intertidal platform and removal of forest biomass [2, 9]. Middlekoop *et al.* [10] found an identical condition in their study on Rhine delta where the sediment trap-ping efficiency of the floodplains is low and largely governed by rivers and engi-neering works [10]. Again, Hoa *et al.* [11] suggest that the engineering structures in the delta raise the flow velocities in the rivers and canals, increase bank erosion, and cause the water to be deeper in the rivers and canals. This induces flooding in the non-protected areas of the delta and invites the risk of catastrophic failure of the dykes in the protected areas [11]. In this backdrop, this chapter aims to build an understanding on the influences of embankments on the major morphodynamic processes and ecological robustness of the deltaic landscape by appraising and synthesizing a large amount of published data. Artificial barriers the embankments mark abrupt changes in the deltaic landscape and truncate the habitat connectiv-ity. Thus, the specific objectives of the chapter are to (1) document impacts of the embankments and polders on sediment accretion dynamics within different hydro-geomorphic set up of the G-B delta, (2) ascertain the imperatives of seasonal behavior on sediment accretion-erosion processes, (3) appraise the ramifications of embankments on the functionalities of the ecosystem and (4) discuss wider man-agement implications.

2. Study area

Created by the confluence of the Ganges, Meghna and Brahmaputra rivers and their myriad distributaries, the Sundarbans constitutes the southern end of the Ganga- Brahmaputra delta in Bangladesh and West Bengal (India). The geographi-cally undivided Sundarban tract (**Figure 1**) stretches approximately 260 km. west–east along the Bay of Bengal from the Hugli River estuary at the western segment in India to the Meghna River at the east in Bangladesh. Situated at the littoral fringe of the world's largest deltaic landscape and shrouded by magnificent

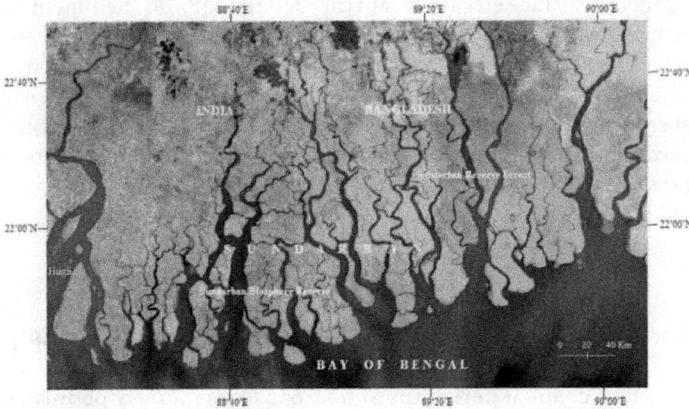

Figure 1.
The integrated indo-Bangladesh Sundarbans. Image details: True color image, (Landsat TM, 1992).

thicket of mangroves, the Sundarbans has a total area of roughly 10,200 km^2, about 60 percent of which is in Bangladesh and 40 percent in India [6]. This intertidal region receives saline influx from the Bay of Bengal twice a day, whilst is also bathed with freshwater flow from the Ganga-Brahmaputra-Meghna River System [12]. Constituted by total 102 islands, the Sundarban Biosphere Reserve (SBR) in India is bordered by 54 inhabited islands to the north with a repugnantly contrasting human-modified landscape and the rest are within the mangrove Reserve Forest area. In Bangladesh the mangroves are sheltered in the Sundarban Reserve Forest which is starkly surrounded by polders, shrimp farms and settlement to the north.

Constructed primarily by fluvial processes, the delta is now maintained almost exclusively by tidal actions and contributed by acute water-surface gradients along the interconnecting tidal channels. The major fluvial systems has either disjoined or migrated to the east and eliminated the most direct fluvial input at the Indian section of the delta due to the Farakka barrage or the Ganges Treaty which was operationalized between India and Bangladesh in 1975. Inevitably, it decreased the perennial flow by diverting a maximum 1133 cumec of water through the feeder canal through Bhagirathi-Hugli river system of West Bengal (India) [6]. Due to this upstream diversion, the distributaries of the Bhagirathi-Hugli have either been disconnected by siltation or dried. Presently freshwater in the rivers contributing to Sundarban is negligible or mostly absent except during monsoon months [6]. In eastern section (Bangladesh) Gorai-Modhumoti-Passur river systems still contribute substantially unlike the Mathabhanga-Kapotaksha-Sibsa river system which once was significant but deteriorated subsequently.

Coupled with this severe shortage in fresh water discharge from upstream, especially at the Bengal section, incessant embanking at the downstream water courses has notably affected the salinity regimes in the rivers and creeks of the lower section of the delta including Sundarbans. The lower-gradient fluvio-tidal section in the south-east (Bangladesh) is advancing into the sea with comparatively steady-state channels contrary to the fluvially discarded tidal section in the south-west (India) which is accreting vertically but also declining irreversibly in certain sections [6, 13]. However, periodic flooding of the land surface during the tidal cycle coupled with enormous sediment delivery during the monsoon (wet season) promotes sediment accumulation and heterogeneous surface elevation gain through time [9, 14]. Sediment transport is landward and to the south-west in the Bengal section (India) [13]. Indeed, the flood defensive structures have terminated the pathways of sediment conveyance and inland sediment transport for much of the lower delta region, especially at the south-west section. Thriving at the southern end of the G-B delta and constituting about 47 percent of the deltaic coastline, the Sundarbans as a complex socio-ecological system with i) extended fluvio-deltaic plains, ii) agriculture and shrimp farming and iii) tide-dominated morphodynamic processes, is the focus area of the chapter.

3. A geo-historical account of embankments in the Sundarbans

At the dawn of colonial period the Sundarbans was sparsely populated teeming with profuse flora and fauna as pointed out by Hunter in an essay published in his book 'A Statistical Account of Bengal' [15]. The East India Company presented an image of the tracts that were not under the government's lease and thus, they conferred tenure and activated their revenue-yielding process through the conversion of inaccessible 'wastelands' into paddy lands as early as 1770s [16, 17]. At this time, they also noticed the dilapidated condition of the hitherto built 'public' embankments or mud bunds which were supposed to be maintained by *zamindars* (feudal landlords) as per the reports from resident Collectors and local offices. To secure the returns from the agricultural fields against the tidal inundation, the erection of mud embankments hastened in tandem with land reclamation by the *zamindars* and simultaneously, a complication arose between the British Government and *zamindars* with respect to embankment maintenance responsibilities. Even as the granting of tenure for reclaimed lands proceeded, private landlords encroached to the forestland in order to extend their landholdings beyond the leased plots by setting up 'private embankments' [18]. More than two decades later, migrants were appointed for clearing the land and constructing embankments. During this time, the British Government used to provide subsidy to the *zamindars* for maintaining embankments, but often the work was neglected. In 1793 Permanent Settlement Act was implemented to regulate property rights over land and it also included, for the first time, the provisions about embankment maintenance [18]. But the responsibility for supervision and maintenance of the embankments was finally handed over to the Embankment Committee in 1803 [18]. Despite the decentralization of embankment management, private participation in maintenance work remained inadequate and subsequently, the maintenance of embankments was recognized as public works in the Bengal Embankment Act of 1873 [18, 19]. The 1857–67 Public Expenditure data on embankment maintenance for the districts of Bengal, reported that embankments in the Sundarbans required high repair expenditures on an average per mile relative to other districts because of the fragilities and hazardous nature of the sites [20].

According to Hunter [15] a major part of reclamation work included in keeping out the lateral flow of saline water and thus, bringing the marsh lands under rice cultivation. In availing this method, all the inlets from the channels were embanked, and smaller channels called *poyans* were excavated around their ends. This embanking was usually done in November, after the river water goes down. When the tide is low, the channels were opened, and the water from the inside drains off; when it is high, the channels were closed [21]. These embankments were major means of communication. Eventually, the islands were joined by filling up intervening water courses and raised permanently above the high water levels. During the

19th century, the process of reclamation was speeded up by building embankments which ultimately, retarded the island formation as the silt accumulation was possible only inside the embanked channels instead of the islands. O' Malley's [22] findings stated that the human habitation was eased by the construction of embankments which kept the salt water out from the stretches of the plains. After the abolition of *zamindari* system in West Bengal (India), Irrigation and Waterways Department (IWD, Govt. of West Bengal) took over the responsibility of managing embankment constructions jointly with the Village Panchayats (village councils) [18]. On the other hand, many islands in Bangladesh witnessed rapid poldering as a part of 1960s and 1970s programs to increase available land for shrimp farming and agriculture [23, 24]. Approximately, 5000 km of polder embankments were built by hand which generated 9000 km^2 of new farmland, but also eliminated the semi-diurnal exchange of water and sediment between the tidal channels and tidal platform [25, 26]. Now these areas lie 1.0–1.5 m below mean high water level due to sediment deprivation coupled with auto-compaction of sediments within the land [2].

4. Morphodynamic processes of the delta

In almost all shallow and active tide-dominated platforms like the southern part of G-B delta, tidal inlets or creeks play a crucial role in governing local morphodynamic behavior and maintaining equilibrium at the landscape level. The equilibrium is achieved by the dynamic interplays among the landscape components which are materials, processes and forms. When the river influx and tidal forcing, averaged over a seasonal cycle are sufficiently steady, tidal channels can evolve towards near-stable morphology. In high discharge phases during wet seasons sediment export is promoted by high transport capacity of rivers which reduces tidal deformation. In contrast, stunted river discharge as a sequel of dry seasons, advocates flooding and facilitates sediment imports only from the estuarine mouth [14]. Hence, the tidal motion balances the delta morphology at the landscape scale as the sediment import during low flow balances sediment export during maximum flow within a seasonal cycle [14, 27]. In fact, the tidal distributaries or creeks, exhibit lesser migration rates and also, are not easily flooded by the river because of opposing non-linear interactivity between river discharge and the tide [28]. This results in more uniform and balanced distribution of discharge across the channels. Historical satellite imagery from the Sundarban mangrove forest depicts that the tidal-channel network in the region has been quasi stable since the 1970s with <2 percent net change in the length of waterways and meager changes in channel widths. Indeed, morphological equilibrium of the tidal inlets is not only dependent on the freshwater discharge, but also sediment transport coupled with grain size characteristics, tidal prism and tidal asymmetry i.e. high energy short duration flood tides, and low energy long duration ebb tides [29].

The bi-directional flow associated with tidal rise and fall, generating huge volume of water inland with amplitude reaching more than 4–7 m during spring tides, is called tidal prism. As the tidal oscillations conduct sediment from the shelf across the delta plain, up to 120 km towards hinterland the tidal prism retains substantial suspended sediment concentrations [30]. Suspended sediment amount may be estimated by the widely applied form:

$$Q_s = aQ^b \qquad (1)$$

where, Q_s represents suspended sediment discharge (kg s^{-1}) and Q indicates water discharge (m^3 s^{-1}). With the propagation of tidal wave up channel non-linear frictional distortions and relative depths of channel beds induce differences in magnitude and duration between flood and ebb tidal waves, also known as tidal asymmetry [31, 32]. The tidal approach of within G-B delta of south-western section (India) depicts pronounced flood dominance and rising tide turns the estuarine channels into sediment sink. Sediment convergence takes place at the turbidity maximum zone (**Figure 2**) where sediments are trapped by outrushing freshwater discharge and onshore tidal deluge. The sediments typically precipitate as velocities decline at the end of the transport path along the upstream reaches of the tidal channels or on the intertidal mangrove platform and small creeks and the process is further sustained by relatively low flow velocity of the falling tide [30, 35]. This tidal pumping [31] is imperative in facilitating vertical accretion in the plains which receive little or no sediment influx from upstream barring Hugli (India) and Baleshwari river (Bangladesh). Sediment deposition at the Sundarbans mangrove platform is measured up to 96x10^6 tons per year which is 10 percent of the annual sediment content of the Ganga-Brahmaputra-Meghna river system [6].

A unit-scale analysis [2] at 48 stations of deltaic Bangladesh, reports that the annual sediment content of the Ganga-Brahmaputra-Meghna river system (~1.1 Gt yr.$^{-1}$) is competent to aggrade the entire deltaic system at rates of >0.5 cm yr.$^{-1}$, provided there is an effective riparian energy enabling dispersal of sediment to various reaches of the delta and also, substantial tidal exchanges carrying the sediment inland supplementing to vertical accretion of the delta. Through their seasonal investigations Rogers *et al.* [31], Auerbach *et al.* [2] and Hale *et al.* [24] inferred steady rates of deposition that are sustained by the large magnitude conveyance

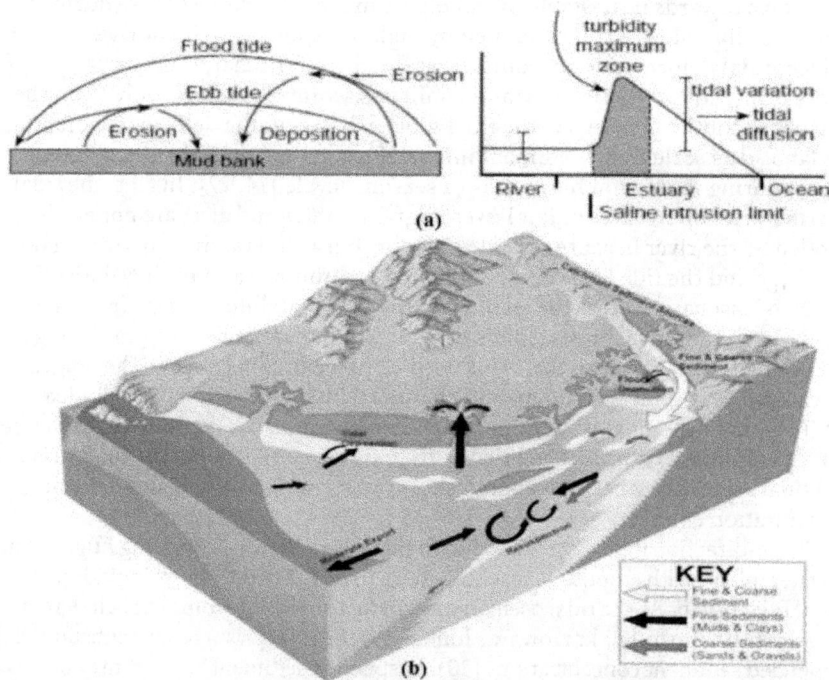

Figure 2.
(a) Sediment transport pattern within the tide-dominated estuarine landscape, after ozCoasts (https:// ozcoasts.org.au/) [33], (b) key processes generating turbidity maximum zone, after Alongi [34].

of sediment through the tidal channels or estuaries and ultimately supplied by monsoonal discharge of the main-stem rivers. Clearly, regional suspended sediment concentration begins to increase in August, coincident with a decrease in local salinity, indicating the arrival of the sediment-laden, freshwater plume of the combined Ganga-Brahmaputra-Meghna rivers (**Table 1**). Hence, a strong seasonal signal is apparent in this dynamic setting: surface elevation change is positive during monsoon season (July–August) and often negative in dry seasons. These patterns are further supported by onshore sediment transport in a tidal cycle.

As implied by the studies of Hale *et al.* [24], the net sediment transport in a tidal cycle is typically 106–107 kg, with magnitude varying largely with tidal phase and spring tides generating 1.5 to 3 times greater net transport than during neap tides [24]. Together mean sediment discharge and net sediment transport patterns reflect an overall flood-oriented asymmetry and net onshore transport of sediment by the tidal pumping. Such tidally supported sedimentation yields mean accretion rates of ~1 cm yr.$^{-1}$, with local observations often reaching 3–5 cm yr.$^{-1}$, which indicate favorable sediment delivery to the unified Sundarbans [2, 24]. Bomer *et al.* [37] utilized an array of surface elevation tables, sediment traps, and groundwater piezometers to provide longitudinal trends of sedimentation and elevation dynamics with respect to local platform elevation and associated hydro-period. They compared two hydro-geomorphic settings of the Sundarbans mangrove forest in Bangladesh: higher elevation stream-bank and lower elevation interior. Seasonal measurements over a time span of five years reported elevation gain in all settings, with highest rates reported from elevated stream-bank zones.

On other hand, Syavitski *et al.* [38] estimated values of variables within the delta systems worldwide to compile an index of vulnerability indicating the degree of natural and anthropogenic co-action. Relative changes between the local land-surface elevation and water levels (ΔRSL) are defined by the rates of sediment accretion (A), eustatic sea-level rise (ΔE), compaction (C), and tectonic subsidence (M), as:

$$\ddot{A}RSL = A - (\Delta E + C + M) \tag{2}$$

Based on the model of Syvitski *et al.* (2009), the G-B delta is among the 33 deltas worldwide with risk to a 50 percent increase in severe flooding over the next century as a result of relative sea-level rise. Also, in the past decade, 85 percent of the deltas experienced severe flooding, resulting in the temporary submergence of 260,000 km^2 [38]. Their study specifically constrained the variables, such as sediment aggradation, compaction and subsidence which contribute to the relative sea-level rise. In the case of the naturally flooded and forested Sundarbans, the mean ΔRSL deterministically approaches ~0.0 ± 0.3 cm yr.$^{-1}$ (that is, no net

Province	Surface area (km^2)	Storage (10^6 t a^{-1})	Percent of total sediment discharge of the Ganga and the Brahmaputra*
Bangladesh	4800	62.4	6.2
India	3000	13–32	1.3–3.2
Total	8000	77.4–96.4	8–10

*Total Ganga–Brahmaputra sediment discharge is estimated as 10^9 t a^{-1} by Milliman and Syvitski [36].

Table 1.
Sediment budget of the Sundarban forests, after Rogers et al. [35].

change) because abundant sediment supply and regular tidal inundation compensate for local variation in compaction and subsidence [38]. This net balance reflects the historical pattern of dynamic stability in this landscape where sediment accretion has remained in equilibrium with relative sea-level rise (RSL) rise. In contrast, we calculate a ΔRSL of ~1.3 cm yr.$^{-1}$ for the sediment starved polder, plus ~20 cm of lowering from wood extraction. These values account for 85 ± 35 cm of elevation loss in the 50 years since embankment construction [2].

5. Modifications within the embanked landscape

Reclamation of the landscape (**Figure 3**) with hard, consolidated embankment causes wave and current deflection from the base of the wall causing basal scouring and subsequent subsidence and overtopping. As the cross-sectional area is reduced due to exclusion of intertidal part by reclamation, sedimentation takes place on the channel bed and thereby, raises its elevation [39]. On the flip side, the interior part of islands remains sediment starved and thus subsides due to compaction and de-watering [40, 41]. This develops a reverse gradient with lower central position. Once flooded, water enters into the island and is unable to drain back to the channels (**Figure 4b**).

Formerly, widespread sedimentation could be possible in the active stretches as the tidal distributaries tended to migrate or witness landward transgression across the plain for maintaining their position within coastal energy gradient [38]. But unplanned construction of these flood defensive walls have hindered the channel migration process and reduced the capacity of small channels to navigate into the larger one. Estuarine transgression, however, is a two-stage process in which lateral erosion of the seaward margins of salt marshes and upper mudflats is balanced by vertical accretion on their surfaces so that their landward margins frontier forward. But the embankments are standing between these two processes, permitting

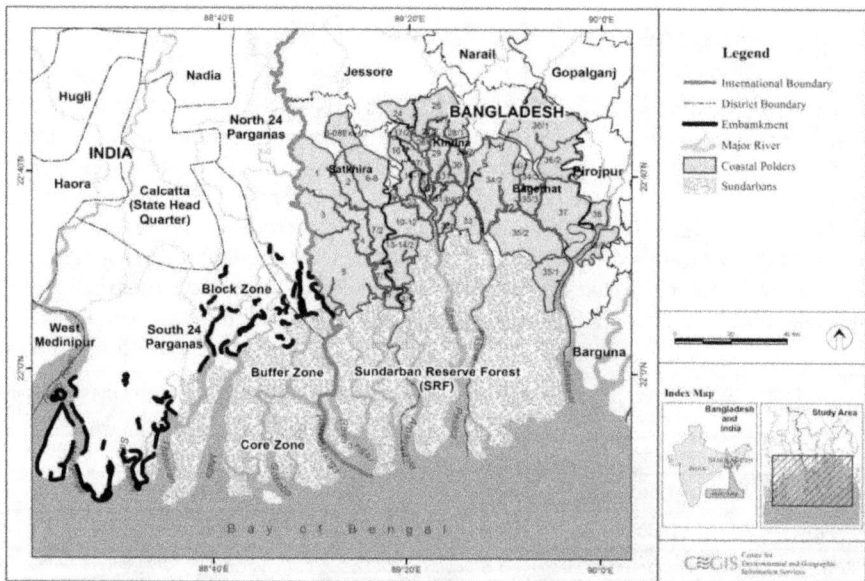

Figure 3.
Embankments and polders in the indo-Bangladesh Sundarbans [6].

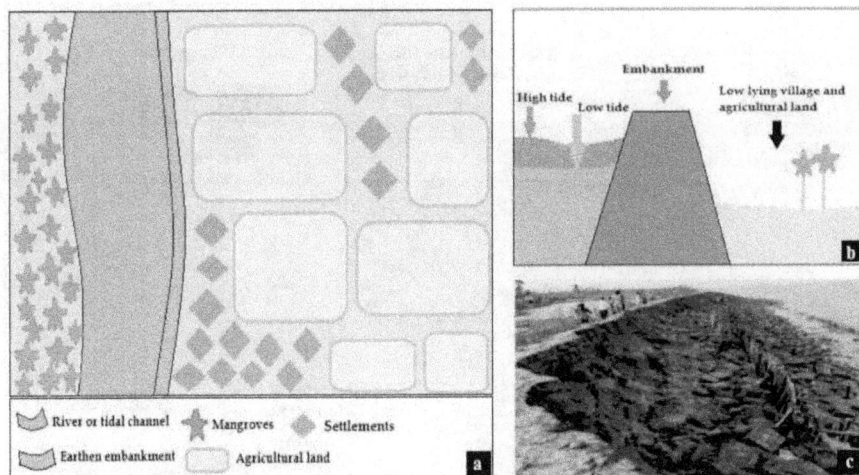

Figure 4.
(a) Typical south-western deltaic landscape with embankment, (b) cross-section of the delta plain: River bed lies above the land behind embankment, (c) post-cyclone ('Amphan', 2020) damages of earthen embankment in Indian Sundarban.

the erosion to take place to seaward preventing the accretion to landward [41]. Moreover, a number of interior creeks and estuaries are exposed to unprecedented siltation, especially at the western part of the lower delta stretch, as a consequence of reduction in tidal spill [42].

Four types of embankments are commonly built at the south-western section of the delta: i) 2 m high earthen wall bordering small tidal channels, ii) 2 m high earthen wall with brick pitching on island margins, iii) 3 m high wall with brick pitching on wave exposed coastal tracts, iv) 3.67 m high wall with boulder pitching on eroding stretches [6]. On other hand the tidal creeks of Sundarbans are classed as meso-macro tidal having amplitude ranging from 2 to 5.5 m. The incoming tidal waves during spring tides and wet months often achieve the height of more than 3 m (often 4 m) to overtop the barrier and thereby, enter into the agricultural lands [40]. Clearly, this positive feedback loop discards the embankments as a protective wall from saline water. Further, deep foundation of embankments impairs the dynamic interconnection between ground water table and river by retarding influent and effluent seepage [43]. The embankment on the bank is affected by the hydrostatic pressure causing uplift and dislodgement of such structures. The scouring of the bank along the concave side of meander and heavy weight of overlying structure often leads to collapse of the banks.

Beginning in the late 1960s and continuing to the early 1980s, ~5000 km² of the low-lying tidal delta plain of Bangladesh section was embanked and converted to densely inhabited, agricultural islands i.e., polders. From this section of Bangladesh Wilson *et al.* estimated the closure of >1000 km of primary creeks due to direct blocking by embankments and sluice gates [13]. This precludes natural exchange of water and sediment that defines the delta plain [44]. Without the regular delivery of sediment to the land surface from tidal overbank flooding over the last 50 years, significant loss in elevation (1–1.5 m) relative to mean high-tide levels has occurred, culminating in enhanced flood risk in the event of embankment failure [45]. Moreover, Polder construction reduced local tidal prism by remaining channels, lowered local current velocities and favored enhanced sediment deposition. Most infilling takes place along one or more channels which have been beheaded from

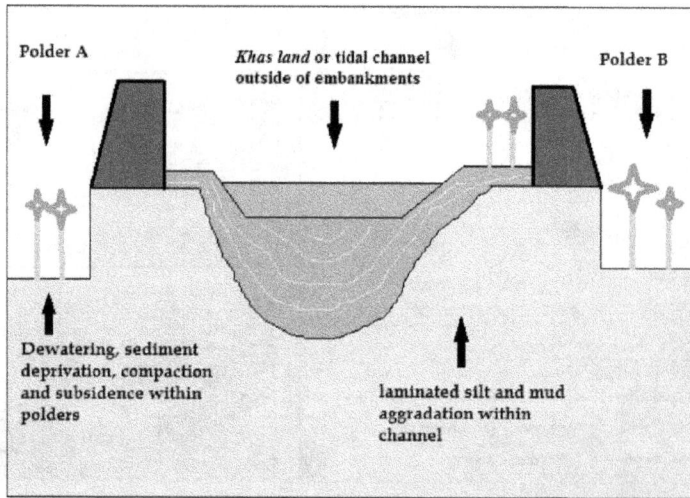

Figure 5.
Standard polder system of the south-eastern section (Bangladesh) of the delta. Siltation within tidal channels result in formation of khas land, *but increased waterlogging within depressed polders; modified after Wilson and Goodbred [13].*

Year	Poldered Area			Natural Area		
	1973	2003	2013	1973	2003	2013
Length of tidal channels outside of polders (km)	1891	783	782	1964	1987	1981
Length of tidal channels obstructed by polders (km)	0	1108	1108	0	0	0
Length of tidal channels outside of polders with >50% obstruction [*khas land*] (km)	0	355	420	0	0	0
Change in drainage network		−59%			1%	
Conduit channels converted to *khas land*		45%	54%		0%	

Table 2.
Summary of results from GIS analysis (1973–2013), after Wilson et al. *[13].*

the tidal network, reducing their local discharge and permitting sediment deposition. Subsequently, new lands are constructed on the infilled, shallower tidal channels surrounding the compacted, sediment-scarce polders. These are called *khas land* (**Figure 5**) [13]. The northern section of Sundarban Reserve Forest (SRF, Bangladesh) constitutes of the polders and the *khas land* only (**Table 2**).

6. Discussion

6.1 Geomorphic ramifications of dykes

The massive shrinkage in inter-tidal areas by the artificial barriers affected natural ability of the system to respond to and dissipate flood energy. In many parts of the delta, the land behind the barriers is now lower than the inter-tidal areas or channel beds: an apparent sign that the system dynamics has been disrupted (**Figure 4**). Construction of marginal levees increases channel depth and accentuates flood dominance which further, aggravate sedimentation at the channel bed as the channel tries to restore its equilibrium [46]. The major waterways lose their original widths. Consequently, the 'conduit' channel beds procure substantial elevation by infilling processes as opposed to the land behind the embankments. In the state of morphological equilibrium, length of a resonant macro-tidal estuary like the Hugli (Indian section), tends to equal a quarter of the tidal wavelength (L) [46, 47], i.e. 0.25 L. As explained by Bandyopadhyay (1997), the tidal wavelength (L) is determined by the following equation:

$$L = T\sqrt{gD} \qquad (3)$$

Where, T is tidal period (constant in a given locality), g is gravitational acceleration (a constant) and D is mean depth of the estuary (the only variable). Reclamation by restricting the area of tidal spill through marginal embankments increases the mean depth and leaves the estuary out of equilibrium to which it tries to return by active in-channel sedimentation and bank erosion to decrease mean depth [5]. With its length 0.17 (i.e. less than a quarter) of its tidal wavelength, this is the situation of the Hugli estuary [5].

Post-storm (category-I cyclone 'Aila', 2009) investigation of Auerbach *et al.* [2] at five sites of breached embankments in south-eastern section (Bangladesh) of the delta, imparted that four of the five breaching has been occurred at the mouths of former tidal channels which are now blocked by embankments. The storm surge height was only 0.5 m above the usual spring tide level. All the breached sites have witnessed ~50–200 m of bank erosion in the decade before the storm and the beach tidal channels have scoured into fine silt at all the spots. During 2 years before the repair, much of the polder was submerged during each high tide, whilst the adjacent Sundarban, at ~100 cm higher, was inundated only in spring tides. This reflects the sediment scarcity within the polders and historical loss of surface height (**Figure 5**). Auerbach *et al.* [2] reported that the embanked islands in southwest Bangladesh have lost 1.0–1.5 m of elevation, whereas the neighboring Sundarban mangrove forest has remained comparatively stable. Sediment starvation has impacted on tidal inundation pattern by inflating local tidal prism i.e. volume of water moving on and off the polder within the polder. In this case, the substantial loss of elevation has severely exacerbated the effects of tidal inundation by increasing the tidal prism accounting for the exchange of ~62 × 10⁶ m³ of water through the breaches during each tidal cycle.

On other hand, there has been substantial channel infilling by interlaminated silt and mud in association with reduced tidal prism and current velocities outside polder walls. The combination of these geomorphic responses has led to many channel beds becoming shallower than polder elevations, which exacerbates water-logging in polders [2, 13]. Wilson *et al.* [13] have estimated the blockage of >1000 km of primary creeks in Bangladesh section directly by embankments. Additionally, the infilling of tidal channels outside the polders often disconnect the channels from the tidal network thereby reducing local discharge. Through the field-based experiments they accounted for 60 percent decrease in total length of channels across ~3000 km² poldered area which corresponds to a loss of nearly two thirds of the region's waterways over the past 40 years. Indeed, the infilled water systems of Bangladesh section are altered from navigable tidal channels that proffered public fishing grounds and aquatic habitat, to well-defined land plots that are, inevitably, reclaimed and typically used for either paddy cultivation or shrimp farming [48, 49].

6.2 Impacts of dykes on ecological evolution

Mangrove swamps, with interspersed tidal channels flanked by the estuary, serve as a depocentre for fine sediment. It is important to note that, interaction between fresh water and saline tidal water at the mineral-rich turbidity maximum zone leads to flocculation or aggregation of clay or silt particles, followed by the settlement of macro-flocs [26, 50]. Diurnal and semi-diurnal tidal deluge deforms significantly inside the mesh of mangrove trees and intertwining roots. As the sediment-laden tidal water enters to the platform, the flow speed is slackened by the dense distribution of aerial roots and further, turbulence is generated by flow around the trees and roots. The turbulence induces eddy formation leading to collision of particles and development of flocs. Here, settling of flocculated sediments takes a shorter time (<30 minutes) during the flood-ebb transition. Settling is also enabled by sticking of microbial mucus and percolation of water through animal burrows on the swamp floor [50]. Hence, sediment trapping on the floor plays a dominant role in building new land and further, sheltering mangrove ecosystem.

Reversing tidal flows complement exchange of materials, such as nutrients, dissolved oxygen, mangrove litter etc. between mangrove areas and coastal waters [36]. In addition, viviparous germination of mangroves in association with unique propagule dispersal and subsequent, seedling establishment processes are supported by the gentle slope of bottom substrate, loose bottom sediments and water inundation into the swamp area and channels during the tidal period. Mangrove propagules have an obligate dispersal phase for several weeks before the radicle extends for root development. The dispersal depends on their buoyancy, longevity and action of tides and currents. After the period of obligate dispersal the healthy propagules anchor at the soft, muddy slope of channel banks and swamps. Mangrove propagules are not only the key to mangrove propagation and regeneration, but also are significant storage potentials to atmospheric carbon [51]. Moreover, species-level zonation and distribution of mangroves are attributable to the dispersal and seedling anchorage patterns. For instance, freshly fallen gray mangrove propagules, such as *Avicennia,* are floated on the flood tide and strand near high water of the ebb; subsequently the pericarp is shed and the newly released seedling does not refloat on the next tide. These attributes describe the tendency of *Avicennia* to find higher densities of seedlings in the upper mangrove zone than in the lower zone [52]. The brick slopes at the forest-fringed island margins hinder the seedling anchorage at the channel banks; thereby the propagules often become unhealthy after refloating several times. Moreover, embankment construction along the tidal channels often leads to felling of mangroves, especially at the southwestern section.

The mangrove environment should be understood as a component of the total ecosystem that comprises the river basin, river, and estuarine and coastal waters, forming an ecohydrology system that should be considered holistically [17]. Physical processes within the mangrove swamps of the Sundarbans, especially, processes of water movement, are imperative in maintaining the functional characteristics of the mangrove ecosystem and building new land. Water circulation and the dispersion of material in mangrove swamps control both the aquatic and terrestrial biodiversity and nourish mangrove colonies. Frequency of tidal inundation, volume of tidal prism, duration of flooding at long time scales and salinity gradients determine species-level structural and phenological variability of mangrove forests [36]. But the brick dykes along the water bodies have fragmented the network of tidal channels and mangrove areas in the lower stretch of the delta, especially at the south-western section. As the embankments disjoin forest-river alliance in the buffer and transition areas of the Sundarban Biosphere Reserve (SBR, India), it inevitably, influence the tidal inundation pattern at the core area forest swamps.

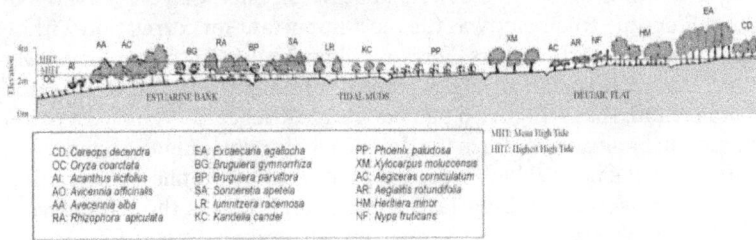

Figure 6.
Representative vegetation profile along a mud bank without embankment [41].

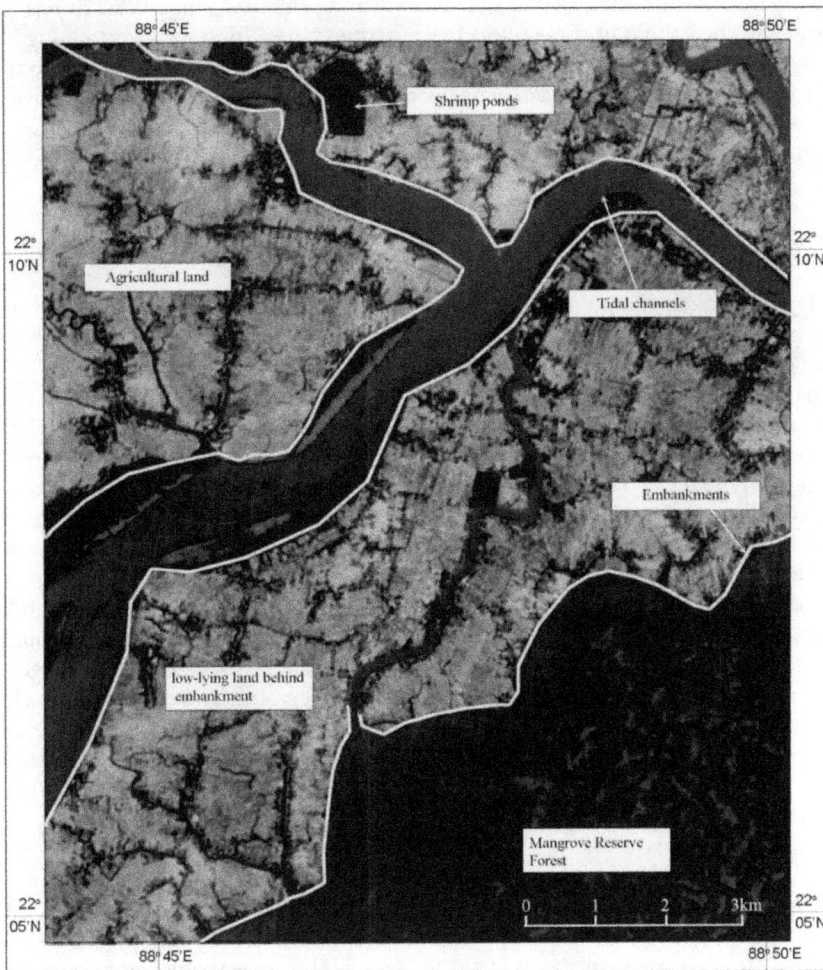

Figure 7.
Embankment structures, surrounding the islands, hinder free dispersal of mangrove propagules and thereby, affect mangrove regeneration in the Indian Sundarban (IRS 1D, LISS-III, 2002).

As mentioned before, inflated tidal prism within the enclosed tidal channel further reduces sediment entrainment towards the forest swamps, instead deposit it on the channel bed giving it a higher elevation. Hence, the swamp elevation gets lowered by sediment deprivation over time which, in turn, affects the zonation pattern of mangroves. Looy *et al.* [53] studied the effect of disruption of alluvial forests from natural river flooding on their vascular plant diversity in the river Meuse floodplain in Belgium. Flooding frequency was the most important for correlation within community composition of the forests. Forests still under influence of the river were significantly richer in species diversity pattern from the stream bank and across in the deltaic Sundarbans. **Figure 6** depicts the ideal vegetation zonation pattern from the stream bank and across in the deltaic Sundarbans. Along-channel artificial armor interrupts the normal hydro-geomorphic set up, disconnects their association to the channels and thereby, affects habitat conditions of the mangroves.

Excavation of burrows or 'bioturbation' by fiddler crabs (genus *Uca*) is an important component in mangrove ecosystem functioning as they influence sediment distribution, quality and composition and also aid in aeration and enrichment of the soil organic matter. Moreover, it influences the functions other soil associated organisms. But, the construction of brick and cemented embankments has threatened the habitat of these tillers [54]. Frequent breaching and overtopping of the dykes lead to salt water ingression and flooding in the low-lying agriculture and aquaculture tracts within the islands affecting the floral community which thrive on fresh water. **Figure 7** depicts the embankment enclosures opposite to the reserve forest region in Indian Sundarban.

7. Management implications

The embanked and poldered landscape is incompetent to withstand the current pace of sea-level rise and storm surges. Even, implementation of management plan is seemingly complicated at the present form of social and political alignments in both the countries. However, there prevails severe lack of researches about environmentally and socially suitable management aspects. Wilson *et al.* [13] asserted for a potential restoration method of many polders by providing sufficient sediment input from the infilled water bodies. Dredging of *khas land* and infilled tidal channels could restore the waterways as well as fulfill the sediment demands of low-lying polders. In some parts of Bangladesh this restoration process has been implemented through the local approach of Tidal River Management; but the output has been mixed with social and engineering challenges. Nevertheless, long term sustainability of the delta plain with respect to embankment and polder tract management could be feasible through integration of both local knowledge and scientific knowledge.

The role of mangroves in protecting coasts against storms and coastal erosion has been widely acknowledged. Sea level rise and tidal hydraulics entanglements often result in erosion of sea-facing islands and estuary margins and thereby curtail the land area progressively. Indeed, conservation and restoration of mangrove forests are crucial in order to protect the island as well as prevent the surge of sediment input inland. At this juncture, planting mangroves on the artificially built sediment terraces alternated by brick blocks on the embankment slopes might occupy some amount of sediment as well as protect the banks. However, Aamir and Sharma [55] have experimented with the novel 'Porcupine system' which regulates flow energy at a definite reach of river, prevents scouring at the bottom river bank and induces sedimentation. A Porcupine is a unit with six bars which are joined by iron nuts and bolts giving it a tetrahedral frame. In real sense, the frame acts as a substitute of a tree in effectuating turbulence generation and resisting sediment entrainment by flow of water (**Figure** 8). The technique, however, might be applicable in poldered tracts by linking the area with the surrounding tidal channel through a narrow conduit. Once a part of the excessive sediment-laden water is passed to the depressed polder area the porcupine frames on the depressed area could retard the flow energy and facilitate sediment accumulation. Indeed, the height and spacing between the Porcupine frames need to be relative to the amount of incoming water within a tidal cycle.

Already established earthen walls pressurize the soft, compressible riverine soils and the collapsing tendency temporarily brings disastrous impacts on the livelihoods of the region. Geotextile as a state-of-the-art geosynthetic-reinforced structure is considered to be one of the useful and cost-effective environmental applications, especially for embankment management [56, 57]. Geotextiles are flexible and permeable continuous sheets of woven, non-woven or knitted fibers or yarns. Geogrids which are uniformly placed array of apertures between their longitudinal and transverse elements, permit direct contact between soil particles on either side of the sheet. Geocells are relatively thick, three-dimensional networks constructed from strips of polymeric sheet. The strips are conjoined to form

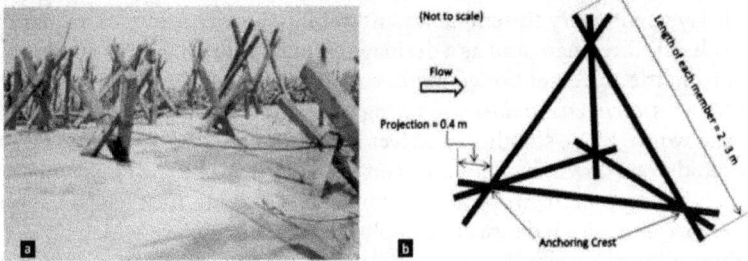

Figure 8.
(a) Sediment deposition around porcupine units in river Brahmaputra, (b) three-dimensional sketch of a porcupine unit, after Aamir and Sharma [55].

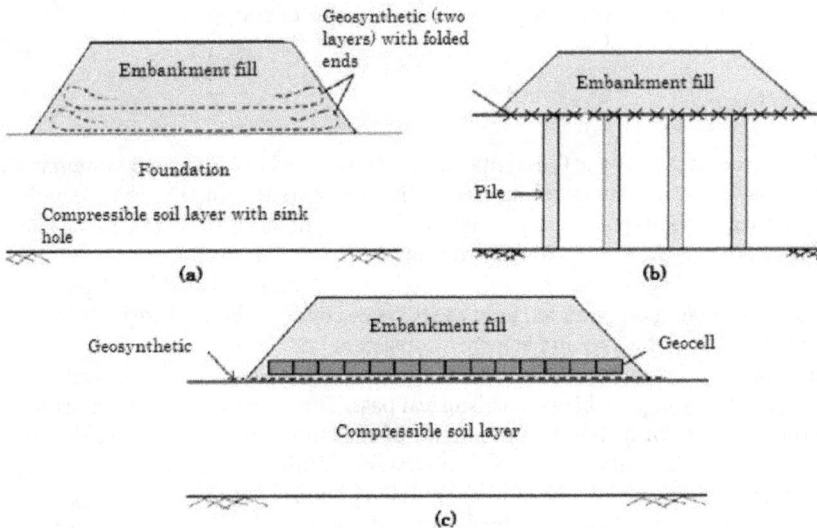

Figure 9.
Different types of geosynthetic methods to control stability and settlement of embankments. (a) Two-layered geotextiles with folded ends, (b) embankment with vertical piles or geogrids and (c) embankment with geocells [56].

Figure 10.
Naturally grown clump of vetiver grass at Kuakata, Bangladesh. Image courtesy: Islam [59].

interconnected cells which are filled with soil and sometimes concrete (**Figure 9**). Geotextile layers intensify the embankment stability by virtue of two primary functions: tensile reinforcement and as a drainage element reducing pore pressures [58].

Apart from the aforementioned methods, plantation of long-rooted vetiver grass (**Figure 10**) (*Vetiveria zizanioides*) is getting exposure in management applications in India and world. Slope stability and river bank protection capability of vetiver grass by rapid draw down, has already been acknowledged by the environmental scientists. Inhabiting in the tropical and subtropical regions, it develops a simple vegetative barrier of rigid, dense and deeply rooted clump grass, which slows runoff and retains sediment on site [59]. Islam [60] studied the performance of *binna* or vetiver grass on 18 coastal polders over 87 kilometers of earthen coastal embankment of Bangladesh during the period from September 2000 to October 2001. His studies inferred to manifold guidelines for applications of vetiver grass coupled with its successful propagation and better performances. This cost-effective bio-engineering method may be applied extensively in the embankment tracts of the Sundarbans in order to provide it an ecological mode of support.

8. Conclusion

Estuarine stretch of the G-B delta plain is considered to be a broad system with nature-wetland subsystem and human-social sub-system being interconnected to one another. Embankment as a product of the human-social processes plays a key role in governing the steady-state of the entire system. The ever escalating population has left this fragile plain densely settled. Furthermore, as livelihood generation and increasing demand for food are two key concerns behind agriculture, the inhabitants of this region perform a consistent task of protecting their lands from being swallowed by saline water. Embankments, in fact, made cultivation in apparent saline soil possible since historical past. The system is, however, disturbed and the functions of different components of the landscape are modified due to the stark separation between the two subsystems. While embankment breaching is seemingly destructive to the land use and livelihoods of the inhabitants, the long-term impacts of these walls by far more dubious to the deltaic physical and ecological processes. This chapter documents here, i) changing tidal inundation pattern and morphodynamic characteristics within the tidal channels or primary creeks due to the artificial structures, ii) shifting of sediment deposition and channel infilling as contrasted to previously intertidal deltaic plain and iii) gradual alteration to mangrove functionalities and habitat with respect to seedling establishment, swamp flooding and fragmentation. Most of the suspended sediment that is imported landward on flood tides is instead deposited within channels, resulting in the infilling and closure of >600 km of intertidal channels and emergence of ~90 km^2 of new land in the southwest delta [13]. Current researches as representative of mean annual sedimentation in the central fluvial-tidal G-B delta indicates that the cumulative vertical accretion rate (2.3 cm y^{-1}) is over two times higher than sedimentation within the natural intertidal setting of the Sundarbans [61]. Explicitly, the chapter attempts to reassess the previous studies and searches for feasible and sustainable routes to delimit the degree of nature-society conflicts in this embanked landscape.

Author details

Subhamita Chaudhuri[1*], Punarbasu Chaudhuri[2] and Raktima Ghosh[3]

1 Department of Geography, West Bengal State University, Barasat, India

2 Department of Environmental Science, University of Calcutta, Kolkata, India

3 Department of Geography, The Bhawanipur Education Society College, Kolkata, India

*Address all correspondence to: subhamita.chaudhuri@gmail.com

IntechOpen

References

[1] Davis Jr, R. A., & Hayes, M. O. (1984). What is a wave-dominated coast? Marine geology, 60(1-4):313-329.

[2] Auerbach, L. W., Goodbred Jr, S. L., Mondal, D. R., Wilson, C. A., Ahmed,K. R., Roy, K. & Ackerly, B. A. (2015). Flood risk of natural and embanked landscapes on the Ganges–Brahmaputra tidal delta plain. Nature Climate Change,5(2): 153-157.

[3] Cahoon, Donald R., & R. Eugene Turner (1980). Accretion and canal impacts in a rapidly subsiding wetland II. Feldspar marker horizon technique.*Estuaries*12, no. 4:260-268.

[4] Ishtiaque, A., Sangwan, N. & Yu, D. J. (2017). Robust-yet-fragile nature of partlyengineered social-ecological systems: a case study of coastal Bangladesh. *Ecology and Society*, 22(3).

[5] Bandyopadhyay, S.(1997). Coastal erosion and its management in Sagar Island, South 24 Parganas, West Bengal. Indian Journal of Earth Sciences 24, no. 3-4:51-69.

[6] Nishat, B. (2019). Landscape narrative of the Sundarban: Towards collaborative management by Bangladesh and India, The World Bank Report (No. 133378):1-207.

[7] Hazra, S., Ghosh, T., Dasgupta, R., & Sen, G. (2002). Sea level and associated changes in the Sundarbans. Science and Culture,68(9/12):309-321.

[8] EGIS, I. (2001). Environmental and social management plan for Khulna Jessore drainage rehabilitation project (Hari river system). Dhaka: Ministry of Water Resource. 65.

[9] Brown, S., and R. J. Nicholls (2015). Subsidence and human influences in mega deltas: the case of the Ganges–Brahmaputra–Meghna. Science of the Total Environment 527 : 362-374.

[10] Middelkoop, Hans, Gilles Erkens, and Marcel van der Perk (2010). The Rhine delta—a record of sediment trapping over time scales from millennia to decades. Journal of soils and sediments 10 (4): 628-639.

[11] Le, T. V. H., Nguyen, H. N., Wolanski, E., Tran, T. C., & Haruyama, S. (2007). The combined impact on the flooding in Vietnam's Mekong River delta of local man-made structures, sea level rise, and dams upstream in the river catchment.Estuarine, Coastal and Shelf Science, 71(1-2): 110-116.

[12] Jalais, A. (2014). Forest of tigers: people, politics and environment in the Sundarbans. Routledge. 251.

[13] Wilson, C., Goodbred, S., Small, C.,Gilligan, J., Sams, S., Mallick, B., & Hale,R. (2017). Widespread infilling of tidalchannels and navigable waterways in the human-modified tidal deltaplain of southwestBangladesh. *Elementa-Science of the Anthropocene*, 5(78):12.

[14] Bain, R. L., Hale, R. P., & Goodbred, S. L. (2019). Flow Reorganization in an Anthropogenically Modified Tidal Channel Network: An Example From the Southwestern Ganges-Brahmaputra-Meghna Delta. Journal of Geophysical Research: Earth Surface, 124(8):2141-2159.

[15] Hunter, W. W. (1877). A statistical account of Bengal (Vol. 20). Trübner & Company:383.

[16] Richards, J. F., & Flint, E. P. (1990). Long-term transformations in the Sundarbans wetlands forests of Bengal. *Agriculture and Human Values*, 7(2):17-33.

[17] Bhattacharyya, P (2009). Determinants of yields in shrimp culture: scientific vs. traditional farming systems in West Bengal. The IUP Journal of Agricultural Economics VI (1): 31-46

[18] Harrison, H. L. (1875). The Bengal Embankment Manual. The Bengal Secretariat Press.

[19] Roy, T. (2010). 'The Law of Storms': European and Indigenous Responses to Natural Disasters in Colonial India, c. 1800-1850. *Australian Economic History Review*,50(1): 6-22.

[20] Sarkhel, P. (2012).Examining Private Participation in Embankment Maintenance in the Indian Sundarbans. SANDEE. 45.

[21] Bhattacharyya, S. (2011). A history of the social ecology of Sundarbans the colonial period. 317.

[22] O'Mally, L. S.S. (1914). Rivers of Bengal, Bengal District Gazetteers. The Bengal Secretariat Book Depot: 1-31.

[23] Ghosh, A., Schmidt, S., Fickert, T. & Nüsser, M. (2015). The Indian Sundarban mangrove forests: history, utilization, conservation strategies and local perception.*Diversity*, 7(2): 149-169.

[24] Hale, R., Bain, R., Goodbred Jr, S. & Best, J. (2019). Observations and scaling of tidal mass transport across the lower Ganges-Brahmaputra delta plain: implications for delta management and sustainability. Earth Surface Dynamics, 7(1):231-245.

[25] Islam, M. R. (2006). 18 Managing Diverse Land Uses in Coastal Bangladesh: Institutional Approaches.*Environment and livelihoods in tropical coastal zones*, 237.

[26] Nowreen, S., Jalal, M. R., & Shah Alam Khan, M. (2014). Historical analysis of rationalizing South West coastal polders of Bangladesh. *Water Policy*, 16(2): 264-279.

[27] Hoitink, A. J. F., Wang, Z. B., Vermeulen, B., Huismans, Y., & Kästner, K. (2017). Tidal controls on river delta morphology. *Nature geoscience*, 10(9), 637-645.

[28] Hoitink, A. J. F., & David A. J. (2016) Tidal river dynamics: Implications for deltas." *Reviews of Geophysics* 54,(1): 240-272.

[29] Gao, S., & Collins, M. B. (1994). Analysis of grain size trends, for defining sediment transport pathways in marine environments. Journal of Coastal Research. 70-78.

[30] Barua, D. K., Kuehl, S. A., Miller, R. L., & Moore, W. S. (1994). Suspended sediment distribution and residual transport in the coastal ocean off the Ganges-Brahmaputra river mouth. *Marine Geology*, 120(1-2):41-61.

[31] Postma, H. (1967). Sediment transport and sedimentation in the estuarine environment. *American Association of Advanced Sciences*, 83: 158-179.

[32] Dronkers, J. Tide-induced residual transport of fine sediment. *Physics of shallow estuaries and bays* 16 (1986): 228-244.

[33] Australian Online Coastal Information. Estuarine biophysical models. Available from: https://ozcoasts. org.au/ [Accessed on June 25, 2020].

[34] Alongi D. The Energetics of Mangrove Forests. (2009). Germany: Springer Science & Business Media.

[35] Rogers, K. G., Goodbred Jr, S. L. & Mondal, D. R. (2013). Monsoon sedimentation on the 'abandoned'tide-influenced Ganges–Brahmaputra delta plain. Estuarine, Coastal and Shelf Science, 131, 297-309.

[36] Milliman, J.D., Syvitski, J.P.M. (1992). Geomorphic/tectonic control of sediment Discharge to the 938 ocean: The importance of small mountainous rivers. *Journal of Geology*, 100(5): 525-544.

[37] Bomer, E. J., Wilson, C. A., Hale, R. P., Hossain, A. N. M. & Rahman, F. A. (2020). Surface elevation and sedimentation dynamics in the Ganges-Brahmaputra tidal delta plain, Bangladesh: Evidence for mangrove adaptation to human-induced tidal amplification.Catena,187, 104312:12.

[38] Syvitski, J. P. (2003). Supply and flux of sediment along hydrological pathways: research for the 21st century.*Global and Planetary Change*, 39(1-2), 1-11.

[39] Allison, M. and Kepple, E.(2001). Modern sediment supply to the lower delta plain of the Ganges-Brahmaputra River in Bangladesh,Geo-Mar. Lett., 21: 66-74.

[40] Maiti, R, Das T, K and Maji (2014). A. Embankment Breaching and Related Hazards–A Study through System Approach at Indian Sundarbans. 2-15.

[41] Paul Ashis (2002). Coastal Geomorphology and Environment. Acb publications.582.

[42] Bandyopadhyay, S.(2019). Sundarban: A Review of Evolution and Geomorphology. World Bank Group. Washington, D.C.,36.

[43] Higgins, S, Overeem, I, Rogers, K & Evan Kalina, E. (2018). River linking in India: Downstream impacts on water discharge and suspended sediment transport to deltas. Elem Sci Anth 6 (1).

[44] Pethick, J & Julian, D. Orford J.D (2013). Rapid rise in effective sea-level in southwest Bangladesh: its causes and contemporary rates. *Global and Planetary Change* 111 (2013): 237-245.

[45] Wallace Auerbach, L., S. L. Goodbred, D. R. Mondal,... & S. L. Nooner. (2013). In the Balance: Natural v. Embanked Landscapes in the Ganges-Brahmaputra Tidal Delta Plain. *AGUFM* 2013 : EP31A-0845.

[46] Pethick, J. (1994). Estuaries and wetlands: function and form. In Wetland management: Proceedings of the international conference organized by Institution of Civil Engineers and held in London on 2-3 June. Thomas Telford Publishing. 75-87.

[47] Wright, L. D., James M. Coleman, & Bruce G. Thom.(1973) Processes of channel development in a high-tide-range environment: Cambridge Gulf-Ord River Delta, Western Australia. The Journal of Geology, 81, (1):15-41.

[48] Goodbred Jr., S. L. & Kuehl, S. A. (1994). Holocene and modern sedimentbudgets for the Ganges-Brahmaputra river system: Evidencefor highstand dispersal to flood-plain, shelf, and deep-sea depocenters, Geology, 27: 559-562.

[49] Barkat, A., Zaman, S. & Raihan, S. (2000). Distribution and Retention of Khas Land in Bangladesh. 2000a. http:// www. hdrcbd. com.

[50] Chaudhuri, P., Chaudhuri, S.& Ghosh, R. (2019). The Role of Mangroves in Coastal and Estuarine Sedimentary Accretion in Southeast Asia. In *Sedimentary Processes-Examples from Asia, Turkey and Nigeria*. Intech Open.

[51] Mazda, Y. (2014). Outline of the Physical Processes Within Mangrove Systems. 63.

[52] Clarke, P. J. & Myerscough, P. J. (1993). The intertidal distribution of the grey Mangrove (Avicenna marina) in southeastern Australia: the effects of physical conditions, interspecific competition, and predation on

propagule establishment and survival. Australian Journal of Ecology,18(3): 307-315.

[53] Van Looy, Kris, Olivier Honnay & Martin Hermy (2003). The effects of river embankment and forest fragmentation on the plant species richness and composition of floodplain forests in]the Meuse valley, Belgium. Belgian Journal of Botany. 97-108.

[54] Sen, S, & Homechaudhuri. S (2015). Spatial distribution and population structure of fiddler crabs in an Indian Sundarban mangrove. *Scientia Marina* 79 (1): 79-88.

[55] Aamir, M., & Sharma, N. (2015). Riverbank protection with Porcupine systems: development of rational design methodology. *ISH Journal of Hydraulic Engineering*, 21(3): 317-332.

[56] Pinto, M. I. M. (2003). Applications of geosynthetics for soil reinforcement. Proceedings of the Institution of Civil Engineers-Ground Improvement, 7(2):61-72.

[57] Wu, H., Yao, C., Li, C., Miao, M., Zhong, Y., Lu, Y. & Liu, T. (2020). Review of application and innovation of geotextiles in geotechnical engineering. Materials, 13(7), 1774.

[58] Christopher, B. R., Holtz, R. D. & Berg, R. R. (2001). Geosynthetic reinforced embankments on soft foundations. In *Soft Ground Technology.* 206-236.

[59] Islam, M. S. (2010). Performance of Vetiver Grass in Protecting Embankments on the Bangladesh Coast against Cyclonic Tidal Surge.8.

[60] Islam, M. N. (2003). Use of vetiver in controlling water-borne erosion with particular reference to Bangladesh coastal region. In Proc. 3rd Int. Conf. on Vetiver (ICV3), Guangzhou, China, October. 358-367.

[61] Rogers, K G., and Overeem. I. (2017). Doomed to drown? Sediment dynamics in the human-controlled floodplains of the active Bengal Delta. *Elem Sci Ant* 5:15.

Chapter 8

Danube Delta: Water Management on the Sulina Channel in the Frame of Environmental Sustainability

Igor Cretescu, Zsofia Kovacs, Liliana Lazar, Adrian Burada,
Madalina Sbarcea, Liliana Teodorof, Dan Padure
and Gabriela Soreanu

Abstract

The Danube Delta is the newest land formed by both transporting sediments brought by Danube River, which flows into the Black Sea and by traversing an inland region where water spreads and deposits sediments. Diurnal tidal action is low (only 8–9 cm), therefore the sediments would wash out into the water body faster than the river deposits it. However, a seasonal fluctuation of water level of 20 cm was observed in the Black Sea, contributing to alluvial landscape evolution in the Danube Delta. The Danube Delta is a very low flat plain, lying 0.52 m above Mean Black Sea Level with a general gradient of 0.006 m/km and only 20% of the delta area is below zero level. The main control on deposition, which is a combination of river, wind-generated waves, and tidal processes, depends on the strength of each one. The other two factors that play a major role are landscape position and the grain size distribution of the source sediment entering the delta from the river. The Danube Delta is a natural protected area in the South-Eastern part of Romania, declared a Biosphere Reserve through the UNESCO "Man and Biosphere" Programme. Water is a determining factor for all the human settlements in the Biosphere Reserve, the whole Danube Delta being structured by the three branches of the Danube (Chilia, Sulina and Sfantu Gheorghe (Saint George)). Our case study is focused on the Sulina branch, also named Sulina Channel, which offers the shortest distance between the Black Sea (trough Sulina Port) and Tulcea (the most important city of the Danube Delta from economic, social and cultural points of view) for both fluvial and marine ships. The improvement of water resources management is the main topic of this chapter, in terms of water quality indicators, which will be presented in twenty-nine monitoring points, starting since a few years ago and updated to nowadays. During the study period, significant exceedances of the limit value were detected in case of nitrate-N (3.9–4.6 mg/L) at the confluence (CEATAL 2) with the Saint George branch and in the Sulina Channel after the Wastewaters Treatment Plant (WWTP) discharge area, as well as near two settlements, namely Gorgova and Maliuc. The higher concentrations of Nitrogen-based nutrients were caused by the leakage from the old sewage systems (where these exist) and the diffuse loads.

Keywords: Danube Delta Biosphere Reserve; water quality indicators, surface water monitoring, Sulina channel between Tulcea and Sulina port, water resources management

1. Introduction

Water quality is a highly important issue that should concern all of us, taking into consideration that our health is directly dependent on the water sources. At the core of the Water Framework Directive (2000/60/EC) [1] is an integrated approach for sustainable water management in river basin district.

After the Volga River which is the largest river in Europe, the Danube River is second, with a basin surface of 801,463 km^2 covering more than 10% of the territory that belongs to nineteen countries [2]. The main course of the Danube River passes through ten countries (Germany, Austria, Slovakia, Hungary, Croatia, Serbia, Romania, Bulgaria, the Republic of Moldova, Ukraine) and four capitals (Vienna, Bratislava, Budapest, Belgrade) (**Figure 1**).

Due to its geological and geographic conditions, the Danube River Basin is divided into three main parts: The Upper Danube, the Middle Danube and the Lower Danube [2]. The Lower Danube „risk" is generated by the nutrient pollution and hazardous substances (including persistent organic compounds such as pesticides and petroleum products) and is in large part due to hydro morphological alterations [2].

The Danube River is the collector of all discharges from upstream countries, affecting the quality of the Danube Delta waters and the Black Sea coast. The EU Marine Strategy Framework Directive aimed at achieving or maintaining a Good Environmental Status (GEns) by 2020 in the territorial waters of the EU Member States [4]. The water quality in the Danube Delta is the result of complex processes having the genesis in the whole river basin, but local factors lead, inside the Delta, to specific differentiation for the Danube branches, delta lakes and in general for each ecosystem. The entire deltaic ecosystem complex of the Danube has been

Figure 1.
The main course of the Danube River (adapted from [3]).

declared a Biosphere Reserve and a UNESCO site since 1991, but, nevertheless, the works of arranging the main in order to reduce the length and increase the flows necessary for navigation, as well as digging various secondary on the entire surface of the delta had a negative impact on the global ecosystem of the delta, as they changed the natural environments, disrupting the reproduction of fish, intensifying the erosion of the banks and the deposition of alluvium [5]. Other works carried out especially during the communist period aimed at draining wetlands and transforming them into agricultural land (over 100,000 hectares), intended for crops, forest plantations or fish farming. As a result of these changes, with the increase in pollution and eutrophication of Danube waters and the intensive exploitation of fishing in the absence of regulations, the fish stocks have visibly decreased [3, 5].

The Danube Delta is the newest land formed, starting since more 12000 years ago (a relatively new delta) by the sediments that are transported by the Danube River before discharged into the Black Sea. Therefore, the Danube Delta has a size and respectively a shape, controlled by the balance between the watershed processes that supplied the sediment, and the receiving basin processes that redistribute, isolate, and export these sediments [6]. Water is a determining factor for all human settlements of the Biosphere Reserve of Danube Delta and therefore its quality plays a key role for the development of the local communities.

The course of the Danube has a dynamic character, which although satisfactory at a certain stage, can be influenced by human activities determining an evolution towards an unsatisfactory quality. Based on previous studies, the following main classes of pollutants were identified:

- Nutrients based on nitrogen (mineral i.e. nitrates, nitrites, ammonium and organic i.e. amino acids, peptides, proteins, urea, etc.), and based on phosphorous (mineral: orthophosphates and organic: organic phosphates (as phospholipids and nucleotide phosphates), phosphatides, etc.) which lead to the increasing of water eutrophication potential;

- Petroleum substances, which form large surface films on the waters, as a result of intensification of port activities and river traffic;

- Specific pollutants with persistent character and high toxicity: heavy metals (copper, zinc, iron, manganese, etc.), organoclorurate pesticides (i.e. terbuthylazine, metolachlor, acetochlor, atrazine, etc.) and benzimidazole fungicide (i.e. carbendazim, etc.), used for the protection of agricultural crops.

The main pressure on the surface water bodies, and not only, is exerted by human activities like discharging untreated or insufficiently treated wastewater into natural emissaries, a practice that must be stopped, in order to protect the water resources.

The water resources represent the hydrological potential formed by the surface waters and groundwaters in the natural regime or under hydrotechnical arrangements, which ensures the supply for multipurpose water uses (drinking water preparation, irrigation of cultivated plants, use as process/industrial water, use in fish farming (aquaculture), or for leisure/sport/tourism and respectively as a means for river transportation). Human health and the environmental protection and wastewater treatment are the main challenges for a healthy environment, both in urban and rural areas and, especially in protected areas of the Danube Delta.

The uncontrolled discharge of the wastewater endangers the population health and the environmental quality, and these could be tackled by improving the monitoring system of surface water quality.

The water supply sources in the Danube Delta settlements are mostly represented by surface water, but also, in isolated cases, by groundwater. The Danube with its arms and the adjacent and the lakes of the Razim and Babadag Complexes represent the surface water sources, while the underground sources are represented by some random drilling of small and medium depth [7]. According to the data presented in the Environmental Quality Report related to the monitored area [8], only 60% of the rural population consumes drinking water of high/ medium depth or spring water, which falls within the parameters stipulated by regulations in force; 10% consume water directly from the Danube; 20% consume well or shallow water that does not fall within the regulated values, and the remaining 10% of the rural population consumes treated water that is not compliant with the terms of potability from a microbiological point of view.

The main anthropogenic activities (e.g. agricultural and fish farms), including tourism and fish poaching, frequently involve some discharges (in terms of detergents, domestic waste, agricultural fertilizer, animal manure, and oil products) in the water of Sulina Channel, which lead to the enrichment of its content with dissolved nutrients. Such nutrients are especially Nitrogen and Phosphorous based, promoting the growth of algae and other aquatic plants, which take oxygen from the water, causing the death of fish and bringing again an additional contribution to the water pollution and finally the eutrophication process occurs. Eutrophication is known as a major environmental issue for the management of water resources, affecting the full exploitation potential.

In order to avoid the installation of the eutrophication process, this should be stopped in its early stage of development, by specific measures starting with the monitoring of both nutrients concentration and algae development in water (usually with optical measurements for the colored pigments from the algae structure (e.g. Chlorophyll, Phycocyanin, etc.)).

Based on the above mentioned aspects, the first objective of the present study is to reveal the evolution over time of the water quality indicators especially in terms of nutrients (Nitrogen and Phosphorous forms) on one of the 3 branches through which the Danube is discharged into the Black Sea, trying to correlate the natural factors with the anthropogenic ones that have recently led to the destabilization of the ecological balance, having a negative impact both on the environment and on the conservation of the traditional heritage of the local communities and of the authentic elements specific to the protected areas of the Danube Delta.

Another objective of this chapter is to review, in the last part, the most important data involved in the improvement of the water management (focused on drinking water preparation and wastewater treatment) in the main human settlements located along the Sulina Channel, in order to be further correlated with the water quality, using of some dedicated software solutions, assuring the environmental sustainability and the primary water needs of the local inhabitants.

1.1 The deltaic morphology of the Danube River and their ecosystems before the discharge into the Black Sea

The Danube Delta is an environmental buffer between the Danube River and the Black Sea, filtering out the pollutants and enabling both adequate water quality conditions and natural habitats for fish in the delta and in the environmentally vulnerable shallow waters of the north-western Black Sea [2].

As it is known, a river Delta is a form of land created by the sediment deposition, as a result of the river water loaded with fluvial sediment, when this leaves the riverbed, before purging into other waters with slower or stagnant movement [9].

The Danube Delta was formed in two stages: a pre-deltaic one (geologically located in the Pleistocene, being marked by climate change) and a deltaic one (started 12000–15000 years ago), in which the territory is no longer submerged and on which we focused our attention in the present chapter, although probably 40% of the actual Delta has been built in the last 1000 years [10]. Therefore, the modern Danube Delta began to form after 4000 BC, in a Gulf of the Black Sea, as a result of rising water levels in the Black Sea and the development of a sand deposit that partially blocked the discharge of the Danube, so the initially formed delta advanced outside the estuary blocked by sediments, after 3500 BC, building several successive lobes: St. George I (3500–1600 BC), Sulina (1600–05 BC), St. George II (0 BC) and Chilia or Kilia (1600 CE - present). Thus, several internal lobes were formed in the lakes and lagoons that border the Danube Delta to the north (Chilia) and to the south (Dunavatz). Most of the alluvium in the delta resulted from soil erosion associated with degraded biomass in previous millennia in the Danube basin, thus causing the expansion of its surface in the form of lobes [10, 11].

The Danube Delta formation was dependent on many factors, including the marine wind waves from the Black Sea (up to 7 m high), and plays an important role in coastline defense and drinking water supply for local communities [12]. Thus, the Danube Delta was formed by both transporting sediments brought by water that flows into the sea and by traversing an inland region where water spreads and deposits sediments. Diurnal tidal action is low (only 8–9 cm); therefore, the sediments would wash out into the water body faster than the river deposits it. However, a seasonal fluctuation of water level of 20 cm was observed in the Black Sea, contributing to alluvial landscape evolution in the Danube Delta. The delta formation is a long process (in a permanent metamorphosis) in which the amount of sediment carried by water should be significant, and due to the decrease of water velocity (as a result of the flow section increasing), alluvium deposition is achieved which contributes to the building of a deltaic system [13].

The Danube Delta is a very low flat plain, lying 0.52 m above Mean Black Sea Level (MBSL) with a general gradient of 0.006 m/km, and therefore the hypsometry is limited to a very narrow range of values. The maximum difference in altitude is 15 m and is given by the highest point (−12.4 m) of the Letea dunes and the lowest lake bottom (−3 m) from the marine part of the delta. Compared to the Black Sea level, only 20.5% of the delta area is below 0 m. The rest (79.5%) is above 0 m, the most of which (54.6%) is in the range of 0–1 m above MBSL. If the 1/2 range (18.2%) and that of below 0 m are added to this range, more than 93% of the delta area is within the 3 m range of hypsometry [14].

Deltas are typically classified according to the main control on deposition, which is a combination of river, wind-generated waves, and tidal processes [15], depending on the strength of each [16]. The other two factors that play a major role are the landscape position and the grain size distribution of the source sediment entering the delta from the river [17]. In wave dominated deltas, wave-driven sediment transport controls the shape of the delta, and much of the sediment emanating from the river mouth is deflected along the coast line [15].

The relationship between waves and river deltas is quite variable and largely influenced by the deep water wave regimes of the receiving basin. With high wave energy near shore and a steeper slope offshore, waves will make river deltas smoother. Waves can also be responsible for carrying sediments away from the river delta, causing the delta to retreat [12]. For deltas that form further upriver in an estuary, there are complex yet quantifiable linkages between winds, tides, river discharge, and delta water levels [18, 19].

In the case of the Danube Delta, initially there was a triangular bay of limanic type that stretched over a distance of 180 km, in which low-amplitude tides do not significantly contribute to the process of sediment removal, and coastal currents and transport of sedimentary materials have led to the deposition of a significant amount of alluvium. The Danube flows into the Black Sea causing complex interactions between sediments carried by its huge flow rate and marine dispersal forces that create complex configurations with competing morphologies, i.e. influenced by the river versus wave and by deposition instead of erosion, which materializes in the formation of plain-type relief forms with low monotonous ridges or covered by transgressive dune fields, which are a common feature of deltaic lobes influenced by waves [20]. Climate and environmental changes played also an important role in the Danube Delta formation [21, 22].

The inflow rate of the Danube into the Delta is 6,350 m^3/s. The Danube Delta is located mostly in Dobrogea - Romania (82% - 3446 km^2) and partially in Ukraine (18% - 732 km^2) (**Figure 2**).

The Danube Delta covers a total area of 4,178 km^2, which makes it the second largest and best preserved of the European deltas [23].

As early as 1856, in the Danube Delta, a series of works began to arrange the navigable, which aimed mainly at correcting the meanders in order to shorten the distances between the main localities under the coordination of Sir Charles Augustus Hartley, Civil Engineer (1825–1915: the Father of the Danube). He was designated as designer and executor of these works by the European Commission of the Danube. In this sense we can mention 1862 as the beginning of the first correction works (cutting the bends and meanders of the watercourse, consolidating the banks and dredging the riverbed/bottom of the water) carried out on the Sulina Channel, which continued until 1902 and maintained until nowadays in order to maintain navigation on the channel. Therefore, the length of the Sulina Channel was reduced from 92 to 71 km, and its flow was two times increased, making it suitable for

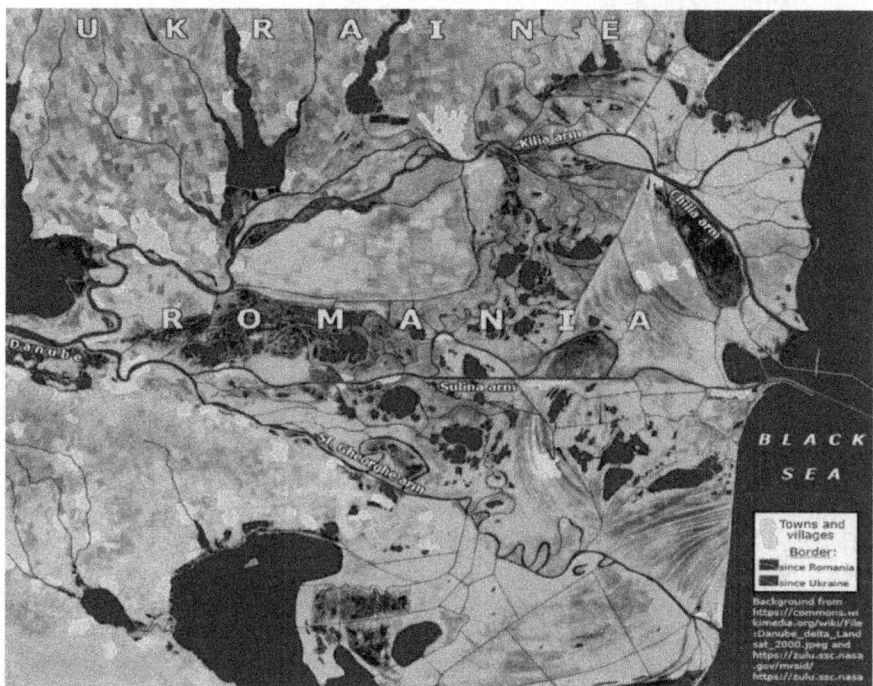

Figure 2.
Satellite view of the Danube Delta (adapted from [23]).

navigation with both large fluvial ships and respectively with maritime boats [24, 25].

The type of global ecosystem found in the Danube Delta (considered as a young region in continuous development), is the Pannonian steppe of Eastern Europe, with Mediterranean influences. It consists of 23 specific natural ecosystems, mostly aquatic due to the rejuvenation of wetlands, along with existence of terrestrial ones, between which a swampy strip is interposed, easily flooded by authentic flora and fauna provided with means of adaptation to the aquatic or terrestrial environment, depending on the season or on the hydrological regime.

The Danube Delta, located on the main migration routes of birds, is a unique place in Europe [2], due to the conditions offered by the development of an extremely diverse flora and fauna, with many rare species. Thus, the Danube Delta, through its nesting and hatching conditions, attracts birds from six major Eco-regions of the world, over 320 species of birds gathering here during the summer, of which about half are hatching species, and the rest are migratory. During the winter, there is an impressive population of over one million individual birds (swans, wild ducks, etc.) [26].

The flowing water ecosystem is characterized by a fairly well-oxygenated environment, rich in plankton, and numerous species of fish, such as carp, pike, perch, catfish and freshwater sturgeon, which is found in all arms of the Danube and a series of their channels in which the water circulation is visible.

Another ecosystem typical for the Danube Delta is that of stagnant water, which is found in many lakes, as well as in various ponds, streams and, being characterized by a rich floating flora and submerged rootless floating plants, which have a negative effect on aquatic bioproductivity.

Another widespread ecosystem in the Danube Delta is the one represented by swampy and floodable areas, characterized by reed plants (surrounding lakes and slowly invading the water surface), floating reed islands and vegetation represented by rush, alternating with other species, offering ideal land for reproduction and nesting for a very large and varied population of birds, some of which are very rare.

Shoreline ecosystems refer to the land on the riverbanks in the delta, mostly represented in the past by willow (*Salix alba, Salix fragilis, Salix purpurea, Salix petandra, Salix triandra*, etc.), that was more recently cut and replaced with Canadian poplars. The characteristic ecosystems of forests are of mixed oak type (*Quercus robur, Quercus pedunculiflora*) with various trees (*Fraxinus pallisae, Ulmus foliacea, Populus tremula*), shrubs (*Prunus spinosa, Crataegus monogyna, Rosa canina, Berberis vulgaris* etc.) and climbing plants reaching up to 25 m (*Vitis sylvestris, Hedera helix, Humulus lupulus, Periploca graeca*) that grow in sand dune areas.

2. Case study: water quality status in the Sulina Channel and perspectives for a better water management

Once the Danube reaches Pătlăgeanca village, in Romania, it forks at CEATAL-1 (Chilia) into two branches: to the North - the CHILIA arm and to the South - the TULCEA arm. At Sfantu Gheorghe Fork (CEATAL-2), the Tulcea arm is divided into two other arms: the SULINA Channel and the SFANTU GHEORGHE branch (**Figure 3**).

At the contact between freshwater and seawater, certain physico-chemical and biological processes take place, which led to the emergence of an ecosystem that is very different from the classical ones, called by specialists the" pre-delta": Musura Bay, north of Sulina and Sfantu Gheorghe Gulf are considered the most representative examples for this type of ecosystem.

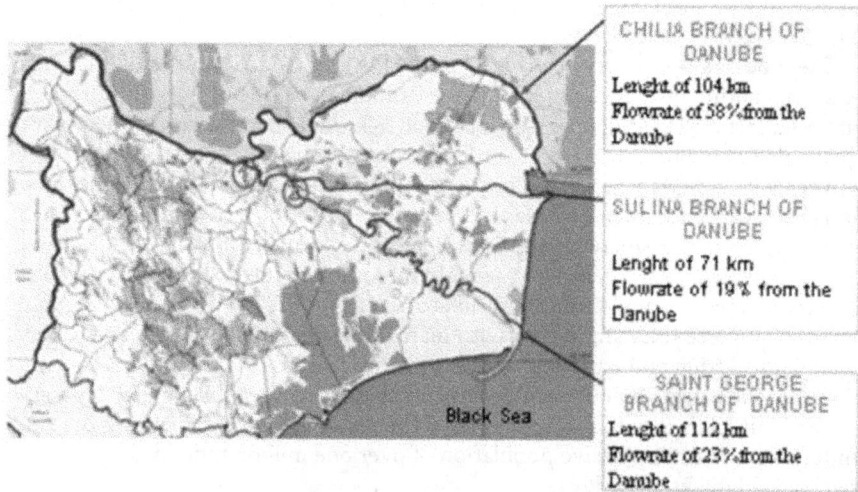

CHILIA BRANCH OF DANUBE

Lenght of 104 km
Flowrate of 58% from the Danube

SULINA BRANCH OF DANUBE

Lenght of 71 km
Flowrate of 19% from the Danube

SAINT GEORGE BRANCH OF DANUBE
Lenght of 112 km
Flowrate of 23% from the Danube

Black Sea

Figure 3.
The branches of the Danube on the territory of Romania (adapted from [23]). 1-CEATAL-Patlageanca;
2-CEATAL-Sfantu Gheorghe.

The dimensional characteristics of the Sulina Channel are: length of 71 km, 50 m maximum width, 18 m maximum depth and 7.32 m minimum depth. This Danube arm is regularized and channeled, being maintained for the maritime navigation of seagoing vessels with a draft of up to 7 m, under the management of the Lower Danube River Administration (AFDJ) based in Galati (the biggest city on the Danube on Romanian territory) [28]. It can be observed that the Sulina arm, which is also denominated as Sulina Channel, is the shortest and straightest arm of the Danube, flowing directly into the Black Sea near the town of Sulina (**Figure 4**), the easternmost settlement in the European Community.

The regularization of the Danube Delta and the appearance of the Sulina Channel meant the opening of the Romanian space to the Black sea, being considered an opportunity for economic development of Romania, based on the implementation of their own projects.

It can be considered that the transformation of one arm of the Danube crossing the Delta, into a navigable channel for heavy ships, in the same time with keeping the specific features of the unique delta in Europe, was a right decision for better water management on the international river, which is subject to the needs of the human community while preserving the biodiversity characteristic of delta-specific ecosystems.

2.1 Some actual and historical data on the water quality on the Sulina Channel

For the settlements located along the Sulina Channel (**Figure 4**), the water from the Danube River is the main source for drinking water preparation. Surface water can only be used as drinking water after its treatment in the drinking water treatment plants. The raw water captured from the Danube is subjected to the water purification procedure through a series of successive processes such as: decantation, flocculation, filtration and chlorination, followed by the temporary storage of drinking water and sending to the consumers, through the drinking water distribution networks.

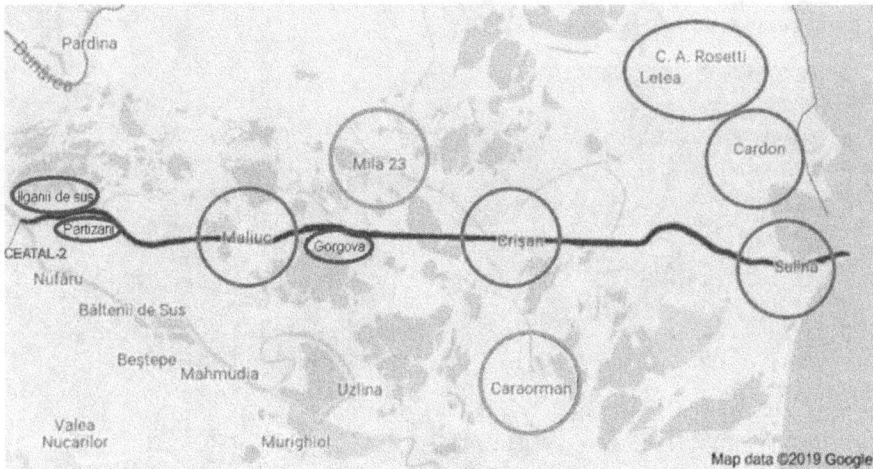

Figure 4.
The Sulina Channel with populated settlements (adapted from Google maps). Red circle-main settlements; blue circle-small settlements belonging to Maliuc; green circle-small settlements belonging to Crisan.

In order to assure the best water management in the frame of environmental sustainability, the domestic wastewater must be collected from the consumers through sewage networks and pumped to wastewater treatment plants, where the water quality indicators are adjusted to the prescription values to allow the discharge of the treated waters into the natural emissary in accordance with the Romanian Legislation [7], harmonized with European Legislation [1]. The treatment of the domestic wastewaters is usually assured by successive combination of the mechanical, physico-chemical and biological processes, which are designed to have a treatment capacity (m^3/h) in relation to the number of the city inhabitants (from which the wastewaters is collected), that should ensure the water quality before discharge into the natural emissary, also represented in our case by the Sulina Channel.

Maintaining a better quality of water in the Sulina Channel is closely linked to the ensuring of sustainability for the water resources management in the Danube Delta settlements, in accordance with the Integrated Strategy for Sustainable Development of the Danube Delta [8] and with the proper management of the Danube Delta Biosphere Reserve [27].

Even if the Sulina channel is formed in fact at the CEATAL 2 Sfantu Gheorghe by splitting the Tulcea Channel in 2 arms (Sfantu Gheorghe and Sulina), the last section of Tulcea Channel from Tulcea Harbor to CEATAL 2 (with 2 sampling points) was included in our monitored area, as well. In other words the Sulina Channel runs from Tulcea via Crisan and Sulina to be further discharged into the Black Sea, and therefore the most of the sampling points were only accessible by boat. These are numbered for the three different monitoring areas, which are presented in **Figure 5**. Area 1 is illustrated in **Figure 6**, located upstream and downstream, respectively, of Sulina city, up to the sea.

Area 2 (**Figure 7**) is located around Crisan, while area 3 (**Figure 8**) is located around Maliuc (Maliuc and Gorgova are not shown on the map in **Figure 8**).

Some of the obtained results carried out according to the recommended methodology [28] during the summer monitoring campaign (July–August 2019/2020) are presented as the average values in **Figure 9**.

The Sulina channel was considered to have an entrance (namely input) at sampling point S8 from **Figure 8** (CEATAL 2- Saint George/Sfantu Gheorghe), and

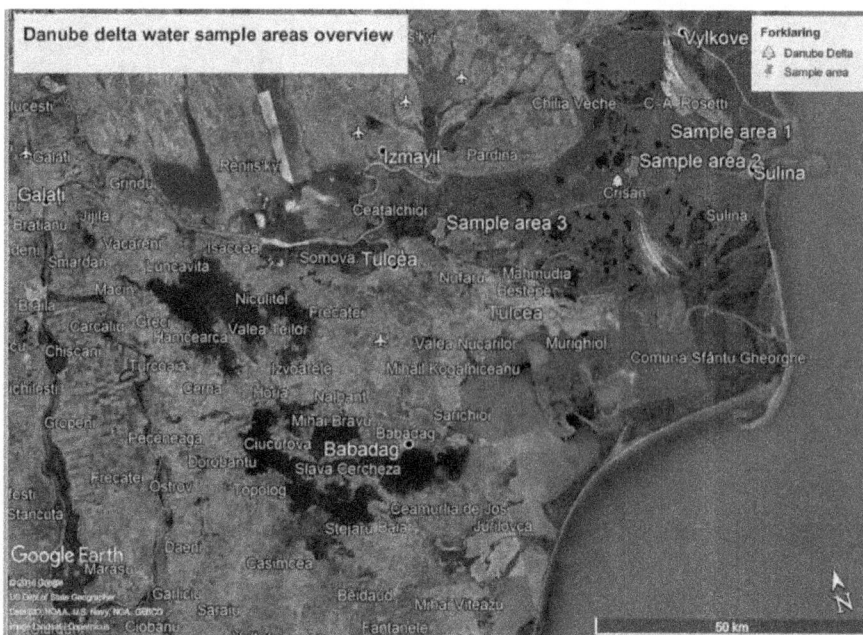

Figure 5.
Overview map of the sampling areas.

Figure 6.
Sulina – Sampling area 1. (Sample 1 – Sulina after the old light tower on the Sulina side; Sample 2 – Sulina after the old light tower opposite of Sample 1; Sample 3 – Side channel at the outskirts of Sulina by the large floating crane; Sample 4 – Sulina after Sample 3 by horse pasture; Sample 16 – Sulina drinking water treatment plant; Sample 17 – Sulina port; Sample 18 – Sulina WWTP after the old light tower; Sample 19 – Sulina WWTP discharge tube; Sample 20 – After the discharge of BUSURCA channel; Sample 21 – After Sulina weather station; Sample 22 – BUSURCA Channel near the beach; Sample 23 – Black sea at the beach; Sample 24 – Black sea at the end of Sulina channel; Sample 25 – Sulina city - artificial channel (Busurca); Sample 26 – Sulina city – artificial channel at the intersection with interior channel).

Figure 7.
Crisan – Sampling area 2. Sample 5 – Crisan camping Main street no. 122; sample 6 – Crisan intersection with old Danube Vis-à-Vis of camping (north); sample 7 – Crisan intersection with CARAOMAN Channel south by camping; sample 12 – Crisan intersection with CEAMURLIA Channel, Main street no. 612; sample 13 – Crisan WWTP; sample 14 – Crisan WWTP docks; sample 15 – Intersection old Danube between Crisan and Sulina.

Figure 8.
Maliuc-Gorgova-Tulcea – Sampling area 3. Sample 8 – Sulina channel before intersecting Tulcea channel & St. George channel water level station; sample 9 – St. George channel; sample 10 – Tulcea channel; sample 11 – Tulcea harbor; sample 27 – Gorgova; sample 28 – Maliuc; sample 29 – Partizani.

Nitrogen and Phosphorus species [mg N-P/L]

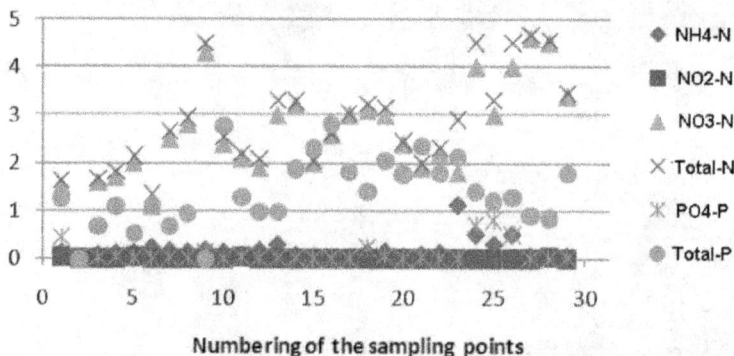

Numbering of the sampling points

a/Nutrients concentration based on nitrogen and phosphorus, respectively

COD [mg O₂/L]

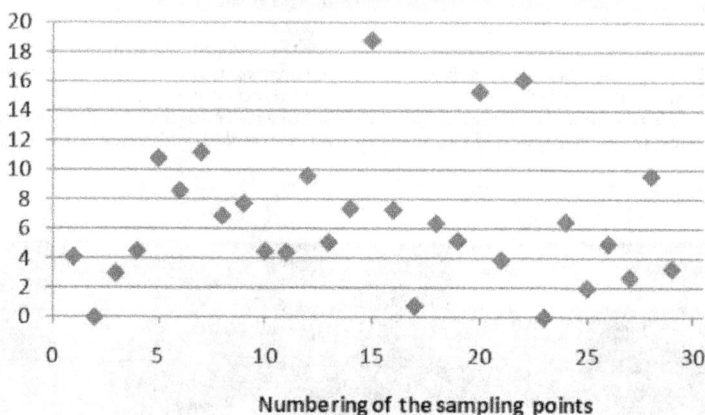

Numbering of the sampling points

b/Organic load content expressed as chemical oxygen demand (COD)

Figure 9.
The water quality data in the sampling points 1–29 ((a) nutrients concentration based on nitrogen and phosphorus, respectively; (b) organic load content expressed as COD).

respectively an exit (namely output) after Sulina city S21 from **Figure 6**. In case of nutrient loading, high concentration values for nitrate-N were measured at point S9 in the St. George Canal (4.3 mg /L), at point S13 and S14 after the Wastewaters Treatment Plant (WWTP) discharge point in the Sulina Channel (3.1 mg/L and 3.3 mg/L) and at point S27 and S28 near Gorgova and Maliuc (4.4 mg/L and 4.6 mg/L). Also around Sulina City the measured values exceed the limit values [29].

The following historical data related to the nutrient concentrations based on nitrogen and phosphorus are presented in **Figure 10(a–d)**. The points from the graph represent the year average values calculated based on trimestrial

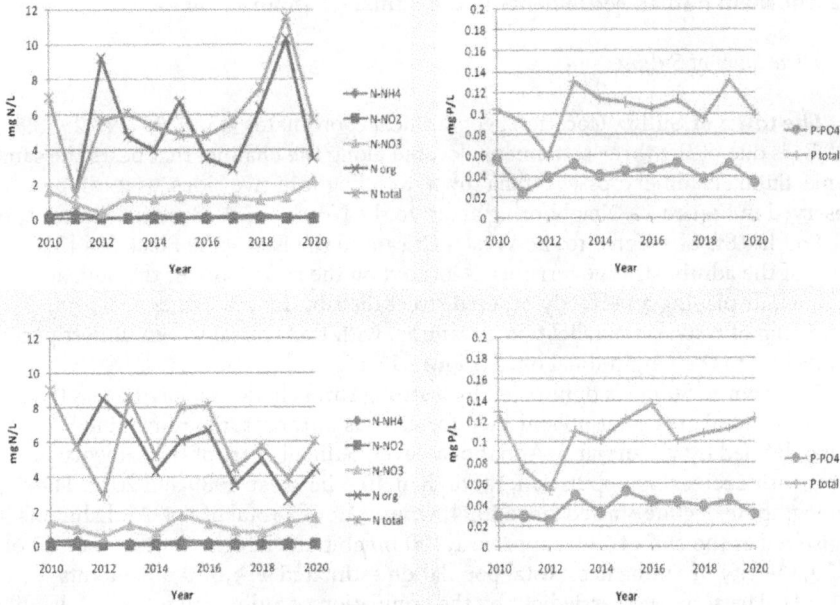

Figure 10.
Historical evolution of nutrient (N, P) concentration at the input (a, b) and output (c, d) of Sulina channel, neglecting the most polluted places from Sulina City. (a) N-based nutrients - Input Channel; (b) P-based nutrients - Input Channel; (c) N-based nutrients - output channel; (d) P-based nutrients - output channel.

measurements (usually 4–6 measurements carried out at different months of the year, from spring to autumn).

As a general statement, the input is usually more polluted as visually observed (e.g. due the anthropogenic activities in the Tulcea municipality) than the output (after Sulina city, before discharge into the Black Sea). This aspect can be explained on the basis of self-treatment phenomenon and on the basis of the dilution with clean water resulted from the adjacent channels, where the vegetation acts as a self-treatment system. However, some exceptions appear due to some anthropogenic activities around the Sulina city as well, as can be seen in **Figure 10**.

This study tried to point out some particular aspects of the environmental management in the monitored area, by scanning it with a higher resolution (many sampling points correlated with the visual observation from the field) in comparison with the dedicated sampling points of the water management plan, which included in our case CEATAL 2 - Sf. Gheorghe (as the channel input) and Sulina-discharge in the Black Sea (as the channel output), without intermediary sampling points.

According to **Figure 4**, the human communities are concentrated in the following main settlements located along the Sulina Channel; Sulina city, Crisan commune and Maliuc commune. For each of them, the management of the water resources will be further presented, in order to improve it in the context of environmental sustainability, according to the EU recommendations on the improvement of river basin management [30]. As it was mentioned earlier, the water management includes several components. Among these we will further focus on the drinking water preparation and wastewater treatment in the main human settlements located along the Sulina Channel, providing the most relevant data requested for further using dedicated software solutions in order to satisfy the water needs of the local inhabitants, in the frame of environmental sustainability.

2.2 The main human settlements in the Sulina Channel

2.2.1 The town of Sulina

The town of Sulina (location geographical coordinates: 45° 9′ 34" N - 29° 39′ 10" E) is one of the three settlements located along the channel that bears the same name. Sulina channel crosses Sulina town, dividing it in two sides, as it can be observed in **Figure 11**, Neighboring areas: to the North – C.A. Rosetti commune; to the South – Sf. Gheorghe; to the West – Crisan, to the East – the Black Sea [31]. Most of the administrative territory is located on the right bank of the Sulina channel, displaying a perfectly ordered street distribution, in orthogonal grid, consisting of 6 streets parallel to the Danube, with transversal streets almost perpendicular to the longitudinal ones (**Figure 12**).

The town of Sulina is dominated by a strong rural character, having less than 50% of its total area built up and used for various purposes, the remaining 50% being defined by vacant land. At national level, Sulina is part of the category of cities with decreasing population, more than 40% between 1989 and 2012. The demographic decline started since 1994, when, out of a total of 5,432 inhabitants registered at the end of 1993, approx. 1,400 inhabitants left the city. At the end of 2020, the city of Sulina has a total population estimated at 4,000 inhabitants [31, 34]. The activities carried out by the population of Sulina fall into the following categories: fishing and economic activities specific to the fishing profile, dismantling and specific manufacturing economic activities, naval transport, public transport (naval and terrestrial), agriculture, telecommunications, trade, tourism. The

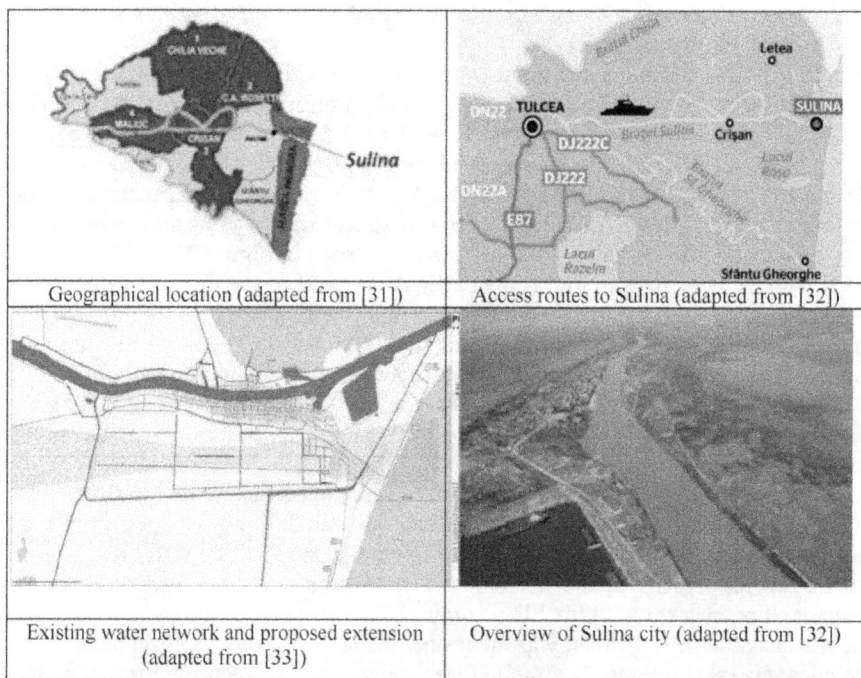

| Geographical location (adapted from [31]) | Access routes to Sulina (adapted from [32]) |
| Existing water network and proposed extension (adapted from [33]) | Overview of Sulina city (adapted from [32]) |

Figure 11.
Illustrated information about Sulina City. Geographical location (adapted from [31]). Access routes to Sulina (adapted from [32]). Existing water network and proposed extension (adapted from [33]). Overview of Sulina city (adapted from [32]).

Figure 12.
Aerial view images of Sulina City [31, 34].

administrative territory of Sulina is part of the Danube Delta Biosphere Reserve (RBDD), established in 1993 by the Romanian Law No.82.

The altitude is the lowest in Romania, on average 0.31 m, and the climate is temperate - continental; the year maximum absolute temperature was 37.5°C (August 20, 1946) and the absolute year minimum temperature was −25.6°C (February 9, 1929). The total administrative area of the city is about 32956 ha.

Sulina has a special status as a city, being considered a port both on the Danube and at the Black Sea. It has port constructions along the city seafront, which allow the mooring of maritime ships and of those that serve the local traffic of goods and passengers. This settlement is connected to the rest of the country only by water: the Sulina channel to Tulcea, the Black Sea to Constanta and the channels that cross the Danube Delta, to the surrounding localities, with the amendment that sometimes you can drive on the dam between Sulina and Sfantu Gheorghe, using off-road vehicles (tractor, sport utility vehicle, or even motorcycle in certain periods) (**Figure 11**).

However, port activity also comes with negative impacts, especially in terms of environmental pollution, actively participating in the phenomenon of global warming and the degradation of water quality, flora and fauna.

About 95% of the administrative area of Sulina is covered by water (ponds). The hydrological network consists mainly of the Sulina Channel (fragment of the maritime Danube), Musura bay, Roșu, Roșuleț, Lumina, Vătafu, Rotund lakes and numerous gorges and canals. Due to the low altitude, the groundwater is at shallow depths [35]. The water management in Sulina City is schematic illustrated in **Figures 13** and **15**.

The water management works on the territory of Sulina aim at: ensuring the navigation, ensuring the necessary water for various uses (especially for drinking water), protection against floods and land improvements. The town of Sulina is part of the group of localities for which investments were financed for the rehabilitation and expansion of the drinking water systems and the domestic wastewater sewage system, using European funding through the Large Infrastructure Operational

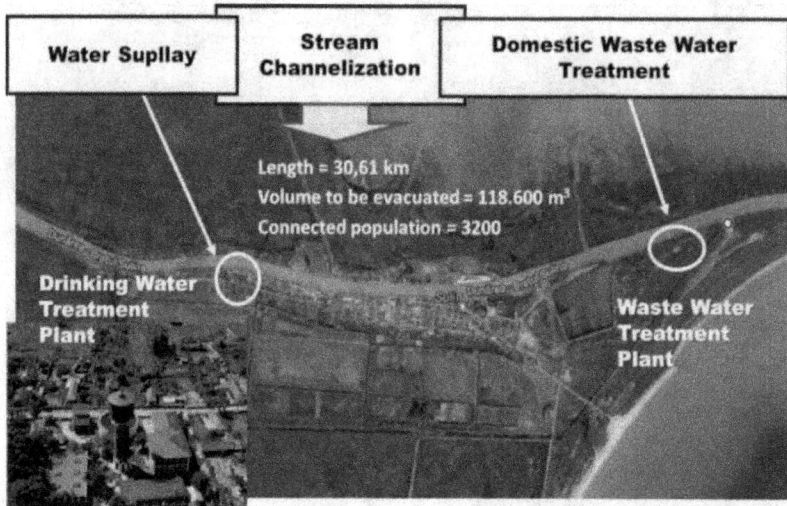

Figure 13.
The water management in Sulina City.

Program (POIM) and Sectoral Operational Programme Environment (POS Mediu 2007–2013, 156/N/23.08.2007) [35].

2.2.1.1 The water management in Sulina City (drinking water treatment plant)

The construction of the drinking water treatment plant of the Sulina city (**Figure 14**) began in 1886 and was completed in 1905, with funds provided by the Dutch Royal House, taking into consideration that the Sulina City was a very important transportation link on international waters, hosting at that time the European Commission of the Danube.

The access of unauthorized persons is restricted inside the water treatment plant, but the water distribution tower can be admired by tourists, as the tallest building in Sulina, so a visible objective from a great distance.

The drinking water treatment plant, together with the Water Tower is still operational today (after modernization based on the European project) and is an important part of the local industrial cultural heritage.

Figure 14.
Drinking water treatment plant with the distribution tower in Sulina City.

Figure 15.
The most important water sampling points in Sulina City which request a continuous water quality monitoring system (with online data transmission [30]) in order to assure the environmental sustainability.

In the city of Sulina, the water supply source is the Sulina channel, from which the water is captured to be decanted, filtered and treated with chlorine for disinfection in order to meet the quality parameters before distribution to the population. The water plant provides a water flow rate of 64 L/s, covering the current water needs of the city [31, 35].

The drinking water distribution network (**Figure 11**) has a length of approximately 32 km and serves all consumers on the right bank of the Sulina Channel and some of those on the left bank. There is no adduction of treated water in the water supply system of Sulina. Drinking water produced in the drinking water treatment plant is distributed directly from the plant, through a water distribution tower (presented in **Figure 14**). In 2018, the drinking water distribution network included: 1156 home connections; 37 connections to flats in apartment buildings; 90 connections to small business companies; 20 connections to public institutions; 970 is the number of homes that benefit from drinking water in their indoor space [35].

2.2.1.2 The water management in Sulina City (wastewater treatment plant)

Prior to the implementation of the Sectoral Operational Program (SOP), in Sulina there was already a sewage network for domestic wastewater discharge and a meteoric water network which collected the water fallen on the quay surface, which was also used for the discharge of domestic wastewater in the Danube. The total length of the rainwater network utilized as well as a domestic wastewater network was 4.1 km and facilitated the discharge of untreated wastewater for a number of 1394 inhabitants (1197 inhabitants in apartment buildings and 197 inhabitants in houses). Since 2019, the sewage network has a total length of 20.3 km and includes a treatment plant that was extended and modernized in the previous programming period of EU funding through a European project, implemented by the Aquaserv SA water company. The water management infrastructure also includes flood protection dams [35].

2.2.2 The commune of Crisan

Crisan commune is located in the center of the Danube Delta and is crossed by the Sulina Channel on the east–west direction. It comprises three villages: Crisan, Caraorman and Mila 23 (**Figure 16**). The neighbors are: the communes of Chilia Veche and C.A. Rosetti (North), Sfantu Gheorghe commune (South), Murighiol and Maliuc communes (West) and Sulina city (East). The administrative area of the commune is of 38,333.85 ha, and the total urban area is of 389,587 ha. The total population of the commune is of 1237 inhabitants [36].

The settlement of Crisan is the capital of the commune with the same name. It is a linear village, developed on both banks of the Sulina Channel, the houses of the inhabitants lining up along the Danube for a distance of over 5 km. It is considered a starting point for the visitor's trips, both to the North (to Mila 23, Matita, Letea) and to the South (to Caraorman, Litcov, Roşu - Roşuleţ). Mila 23 village develops on the right bank of the Old Danube, being crossed by multiples that discharge the excess water into the main channel and from here into the Old Danube. The village of Caraorman is located between the Sulina Channel and Sfantu Gheorghe branch of Danube on the Caraorman maritime ridge.

2.2.2.1 The water management in Crisan commune

Two of the three component settlements of Crisan commune - Crisan and Mila 23 - have partially solved the drinking water supply issues, and *for Caraorman, the drinking water supply system is under construction*. Crisan and Mila 23 villages have a centralized water supply system, which consists of: water intake (capture), adduction pipe, water treatment and pumping station, water storage tanks, water distribution networks [7].

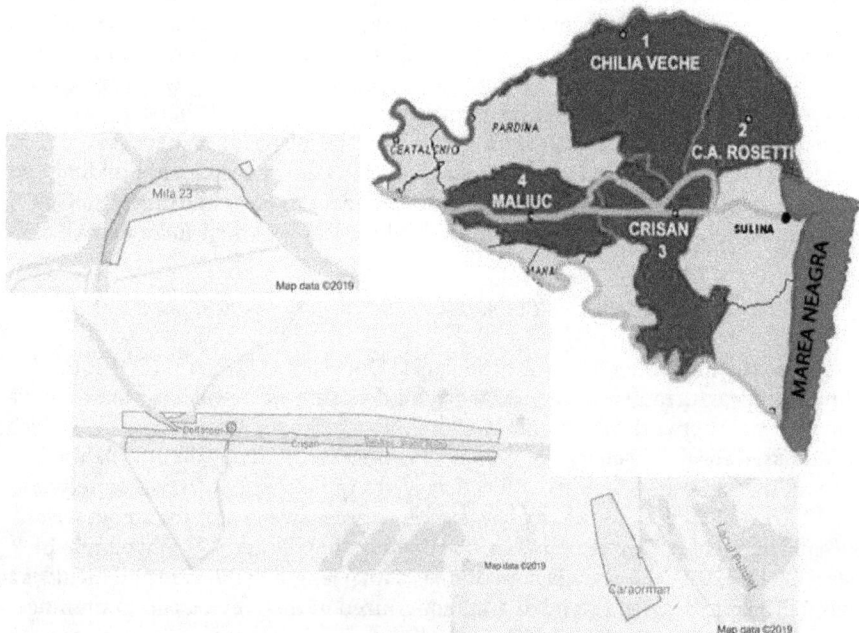

Figure 16.
Settlements in Crisan commune (adapted from Google maps).

No settlement belonging to the Crisan commune actually has the sewage network system as operational; Crisan and Mila 23 being in the process of completion/commissioning, and Caraorman is in the project phase. The settlements of Crisan and Mila 23 have a sewage system and a treatment plant, built, but not functional, due to the non-existence of connections between them (started to be partially implemented last year) and the lack of authorization for the electricity transformation station which should bring the necessary power supply for the wastewater treatment plant (kept in conservation).

Caraorman does not have a sewage system and a treatment plant yet. Domestic wastewater from homes and socio-cultural objectives is discharged into drainable basins, many of them using dry toilets.

Ensuring the evacuation of domestic wastewater in the commune of Crisan is one of the major, acute and difficult to solve problems [7, 35].

2.2.3 The commune of Maliuc

Maliuc commune is located in the eastern part of Tulcea County, in the first western third of the Danube Delta, being crossed by the Sulina Channel on the east - west direction (**Figure 17**). It comprises five villages: Maliuc, Gorgova, Vulturu, Partizani and Ilganii de Sus (**Figure 18**). Its neighbors are Pardina commune (North), Nufăru, Mahmudia, Murighiol and Beştepe communes (South), Tulcea town (West) and Crisan commune (East). The administrative area of the commune is of 25962.02 ha, and the total urban area of 315.62 ha. The total population of the commune is of 1060 inhabitants [37].

Maliuc is the capital of the commune, located about 25 km from Tulcea and situated on the left bank of the Sulina Channel. Vulturu settlement is located on the right bank of the Sulina Channel at approx. 3.44 km from the commune capital. Gorgova is the village located on the right bank of the Sulina Channel at approx. 4.22 km from the commune capital. The village of Ilganii de sus is located on the left bank of the Sulina Channel at approx. 13.63 km from the commune capital. The village of Partizani is located on the right bank of the Sulina Channel at approx. 13.51 km from the commune capital.

Figure 17.
Maliuc commune (adapted from Google maps).

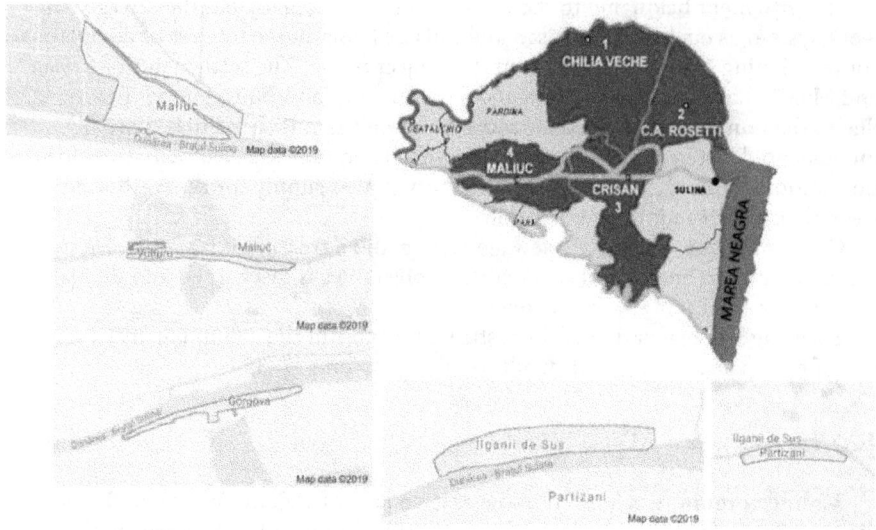

Figure 18.
Settlements in Maliuc commune (processed on Google maps support).

2.2.3.1 The water management in Maliuc commune

Of the five settlements belonging to Maliuc commune, three - Maliuc, Partizani and Gorgova - are provided with centralized water supply systems [7, 27, 37]. The villages of Partizani and Gorgova have modern water purification stations, which consist of: water intake (capture) from the Sulina Channel, adduction pipe, module type of water treatment and pumping station, water storage tanks, water distribution networks [7, 37].

A large part of the population in Maliuc commune is supplied with water from underground sources, the total length of the supply network being of 17 km, the rest of the inhabitants getting water from their own sources (wells with a depth of 2–6 m). The village of Maliuc is provided with sewage network, mostly in a damaged state, which takes wastewater from consumers and discharges it into the Danube. No settlement in the Maliuc commune has a domestic wastewater treatment plant. It is necessary to refurbish the existing sewage networks and to set up new domestic sewage networks, pumping stations and sewage treatment plants for each settlement. The discharge of wastewater into the recipient of the area must be done with the provision of sanitary water protection areas in the Sulina Channel [38].

3. Discussions

The Danube Delta is a natural protected area in the South-Eastern part of Romania, declared a Biosphere Reserve through the UNESCO "Man and Biosphere" Programme, where the water is a determining factor for all the human settlements, including these located on the Sulina Channel. However, the eutrophication process based on the increasing of nutrients concentration started in a few locations due to the increasing of the anthropogenic activities. The improvement of water resources management especially related to the drinking water preparation and wastewaters treatment, based on the water quality indicators, which were screened during the summer of 2019/2020 in a twenty-nine monitoring points distributed along the

Sulina Channel, and compared with historical data obtained since a few years ago, is the main topic of this study.

Nutrient pressures in the Danube catchment come from both point and diffuse sources, reflecting the different drivers [39]. The results of the water quality study on the Sulina Channel (at 29 sampling points) show that the values of nitrate-N concentration are outstanding after the WWTP discharge point of Sulina and near the settlements of Gorgova and Maliuc, which could be caused by the leakage of the aging drain pipes. It should be emphasized that the inhabitants of Maliuc obtain drinking water from shallow, 2–6 m deep drilled wells, due to the risk of exceptionally high nitrate concentrations in surface waters, and therefore a special attention should be paid to the drinking water quality monitoring. Residents living along the Sulina Channel that engage in fishing and agricultural activities, bring an additional significant impact on the nutrient load of the surface waters. These findings show that the major point source is the wastewater (insufficiently treated and untreated) and the main diffuse source is the run-off of agricultural fertilizers [39]. The ICPDR modeling studies [39] suggest that in the Danube catchment 86% of nitrogen emissions and 71% of phosphorus emissions now come from diffuse sources.

In order to have an overview on the water quality status in different locations along the Danube River, including the monitored area of this study (highlighted in Bold), the average concentrations for the selected indicators (nutrients and organic load (COD)) during (2013–2018), are presented in **Table 1** [40, 41].

These data belong to the historical data and represent the 6 years average (calculated based on the annual average between 2013 and 2018); some of these were determined by the coauthors of this chapter [28]. It is observed that the water quality has a better quality in the Middle Danube basin in comparison with the Lower Danube basin, where the Sulina Channel is included with 2 sampling points

Name of the sampling point	COD dichromate (mg/L)	NH$_4$-N (mg/L)	NO$_2$-N (mg/L)	NO$_3$-N (mg/L)	Total N (mg/L)	Ortho-phosphate-P (mg/L)	Total P (mg/L)
Vámosszabadi (Hungary-Slovakia border)	7.97	0.05	0.01	1.74	1.25	0.04	0.09
Esztergom	8.71	0.05	0.02	1.77	2.38	0.05	0.12
Upstream of Budapest	11.5	0.06	0.01	1.93	2.61	0.02	0.08
Downstream of Budapest	11.95	0.06	0.01	1.92	2.06	0.02	0.09
Dunaföldvár	11.39	0.06	0.01	2.01	2.54	0.04	0.10
Paks	11.26	0.05	0.01	1.65	2.18	0.05	0.11
Hercegszántó (Hungary-Croatia border)	11.03	0.05	0.01	1.72	2.23	0.05	0.11
CEATAL 1 Chilia	14.50	0.12	0.02	1.55	5.21	0.05	0.12
CEATAL 2 Sfantu Gheorghe	**14.00**	**0.13**	**0.02**	**1.20**	**5.69**	**0.05**	**0.11**
Sulina	**13.89**	**0.15**	**0.02**	**1.09**	**6.67**	**0.04**	**0.11**

The highlighted values in Bold are related to the studied area i.e. Sulina Channel.

Table 1.
The average concentrations for the main water quality indicators (nutrients and organic load (COD)) during (2013–2018) in different sampling points distributed along the Danube (from Slovakia to Danube Delta).

(CEATAL 2- Sfantu Gheorghe and Sulina- at the Black Sea). This could be explained taking into consideration the pollution generated by the anthropogenic activities, especially brought by the tributary rivers from the Romanian territory, which are more polluted than the Danube River. In order to support this information we will present a very recent study [41], including the map of the Lower Danube basin with 15 sampling points (related to both Danube River and respectively in the tributary rivers at the discharged points), which is presented in **Figure 19**.

This study presents a complex methodology for assessing water quality based on WQI (Water Quality Index), applied at 15 monitoring points from the Lower Danube and Romanian tributaries for a period of 10 years. The water quality has improved in time at the most sampling points, but some of the Romanian tributary rivers are more polluted than the Danube and still require efforts to improve wastewater treatment from urban agglomerations [41].

Based on the same database [41], in order to have an overview on the water quality discharged into the Black Sea by Danube Delta trough each branch (Chilia, Sulina, Sfantu Gheorghe), including the monitored area of this study (highlighted in Bold), the average concentrations for the selected indicators (nutrients and organic load (COD)) during (2013–2018) are presented in **Table 2**.

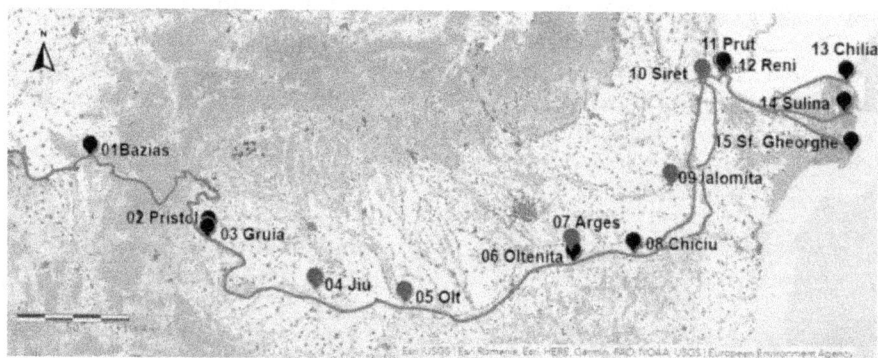

Figure 19.
CORINE land cover map of the monitored area in the lower Danube with representation of the sampling points (01–15): 01-Bazias, 02-Pristol, 03-Gruia; 0.4-Jiu; 05-Olt; 06-Oltenita; 07-Arges; 08-Chicium; 09-Ialomita, 10-Siret; 11-Prut; 12-Reni; 13-Chilia (at the Black Sea); 14-Sulina (at the Black Sea); 15-Sfantu Gheorghe (at the Black Sea); Danube River—Black points, tributary rivers—Blue points [41, 42].

Name of the sampling point	COD dichromate (mg/L)	NH$_4$-N (mg/L)	NO$_2$-N (mg/L)	NO$_3$-N (mg/L)	Ortho-phosphate-P (mg/L)	Total P (mg/L)
Chilia (at the Black Sea) [41]	14.78	0.10	0.02	1.10	0.04	0.10
Sulina(at the Black Sea)	**13.89**	**0.15**	**0.02**	**1.09**	**0.04**	**0.11**
Sfantu Gheorghe (at the Black Sea) [41]	14.41	0.10	0.02	1.09	0.04	0.09

Table 2.
The average concentrations for the main water quality indicators (nutrients and organic load (COD)) during (2013–2018) corresponding to discharge into the Black Sea for each Danube branch, including Sulina Channel (original data, highlighted in bold).

A very good correlation of ±0.01 mg/L was found between our results and these reported by [41], for all water quality indicators, except the value for NH4-N which was higher with 0.05 mg/L and for Total-P which was higher with 0.01 mg/L.

In order to have another overview on the international context related to the water quality status in different similar locations from other European Deltas, a comparison will be focused on the Volga River, which runs through Russia, having a length of 3530 km, cradling more than half of the important Russian cities in the Volga basin, including the capital of Moscow [43], as it is presented in **Figure 20**. The Volga River is discharged into the Caspian Sea, through a huge Delta (10,400 km^2) with a classic triangular shape, formed by many channels, from which Nikitinsky Channel was selected to be presented in comparison with Sulina Channel.

Figure 20.
The Volga River basin, with Nikitinsky Channel, located in the western part of the Volga Delta.

Name of the sampling point	COD dichromate (mg/L)	NH4-N (mg/L)	NO2-N (mg/L)	NO3-N (mg/L)	Ortho-phosphate-P (mg/L)	Total P (mg/L)
Sulina Channel (at the Black Sea)	13.89	0.15	0.02	1.09	0.04	0.11
Nikitinsky Channel	29.63	—	—	—	—	—

Table 3.
Comparative water quality in Sulina and Nikitinsky Channel.

Name of the sampling point	NH₄-N (mg/L)		NO₂-N (mg/L)		NO₃-N (mg/L)		Total N (mg/L)		Ortho phosphate-P (mg/L)		Total P (mg/L)	
	Min	Max	Min	Mean	Min	Max	Min	Max	Min	Max	Min	Max
Volga Delta at the Caspian Sea [46, 47]	—	—	—	—	0.007	0.140	0.147	3.640	0.004	0.025	0.066	0.103
[48]	0.390*	**	0.020*	0.04***	—	0.950	—	—	—	0.090	—	—
[49]	—	0.350	—	0.047	0.209	—	—	—	—	—	—	0.444

*Maximum Permissible Concentration (MPC) - Characteristics of the surface and sea water quality in Russia.
**The occasional maximum concentrations ranged between 3.9–4.4 mg/L, while the annual mean concentration is lower than 0.39 mg/L.
***The annual mean concentration is around 0.04 mg/L, while the occasional maximum concentrations ranged between 0.2–0.3 mg/L.

Table 4.
Nutrients concentration in the Volga at discharge in the Caspian Sea.

This was a difficult task because the sources of information for water quality was varied (generally not systematic and not comprehensive) and the measurements were insufficient both temporally and spatially to characterize the nutrients concentrations, and therefore the measurements were focused especially on pollution generated by the dissolved metals (heavy metal ions) and organics (in terms of oil products, phenols, etc.) in water, due to the sampling efforts, expenses, and low levels reported elsewhere in the recent literature. However, some information related to the organic content (in term of COD) in Nikitinsky Channel was relatively recently reported in [44], as it is presented in **Table 3**.

In comparison with the Sulina Channel the organic pollution in Nikitinsky Channel is more than twice, but is hard to draw a general conclusion, taking into account that the characterization of water column concentrations at any single time does not provide much information other than a screening, since concentrations vary rapidly with currents. It was therefore decided to place more effort on characterizing sediments, which could provide a more historical perspective about the pollution. In this respect more data related to the priority pollutants in sediments were comparatively reported in different zones of the Volga and Danube delta in comparison with the Rhine delta [45]. However, even in this paper no data about the nutrients pollution, but as a general conclusion of the three deltas that Winkels [45] investigated – the Volga, Rhine and Danube, the Volga was the cleanest, followed by Danube.

The Hydrometeorology Service in the soviet time performed a routine monitoring of the water and sediments in the part of the Caspian Sea, including the Volga Delta area. The coverage was quite comprehensive, consisting of shore-normal transects around the coast. There are some concerns about the reliability of some of these data and about methodology, particularly with ammonia, phenols, and heavy metals, tacking into consideration that the above mentioned pollutants are quite difficult to be measured in the water. At present, the Hydrometeorology Service are not providing the same level of measurement, and quality-controlled data are sparse for the past decade. These data were carefully collected from the literature and harmonized in order to have the same units and to be easily compared with EU standards. Some representative data are centralized in **Table 4**, even if there are some uncertainties in these values.

As a general overview the concentrations of nutrients in Volga and in Danube (Sulina Channel) are in the same range, but their annual mean concentrations in Volga are 2–3 times higher than in Sulina Channel, with occasional very high values. However, no evidence of widespread eutrophication in the Caspian Sea, though some deltaic and lagoonal areas are slightly eutrophied.

4. Conclusions

The flow of the Danube with an increased contribution of alluvium in the Black Sea gave birth, in time, to the Danube Delta, which is a plain in formation, advancing by 40 square meters in the sea each year. The Danube Delta was formed in two major stages, of which a pre-deltaic stage (prehistoric corresponding to the glacial period) and a deltaic stage, which includes several phases: the gulf phase (at the beginning of the postglacial era, about 10,000 years ago) and the lagoon phase (fluvial - lacustrine, 9,000 years ago when a maritime cord was formed between the strips of land advancing into the seawater due to the currents in the Black Sea, which brought alluvium from coastal erosion and those from the Dniester, Dnieper and Bug discharge areas), followed by other successive phases of the formation of secondary deltas, which in turn have undergone changes due to sea level rise and alluvial blockage of the Sulina Channel. From the relief point of view, the Danube Delta consists of ridges and islands, swamps, ponds, streams, countless small and large lakes, being the first in Europe as a protected area with swampy territories and wetlands, which is home to unique ecosystems hosting hundreds of flora and fauna species.

In order to avoid health risks for the population in the Danube Delta, it is necessary to assure a continuous monitoring of the drinking water quality by regional operators who ensure the water supply, as well as the monitoring of water quality from the wells. Potential sources of surface water pollution, like direct or uncontrolled discharge of untreated wastewater, must be eliminated, as well as for soil and groundwater pollution, through the occurrence of leaks in sewage networks. In settlements where there is no water supply system, it is necessary to build and expand such a system to provide the population with the necessary drinking water.

From the settlements located along the Sulina channel, only Maliuc, Crisan and Sulina have sewage systems and domestic wastewater treatment plants, but not completely in function, and therefore there are still many issues related to the wastewater management in terms of domestic wastewater canalization, treatment and discharge.

Existing treatment plants involve only mechanical stage treatment, using grates, desanders, septic tanks, grease separators and decanting spaces.

All settlements in the monitored area must have centralized sewage systems that collect wastewater and domestic water throughout the entire locality, as well as treatment plants. Increasing the degree of wastewater treatment, refurbishment and improvement of the treatment process require the implementation of the following measures:

- controlled discharge of domestic wastewater in emissaries;

- improving and making the treatment process of wastewater discharged by economic agents more efficient;

- rehabilitation and extension of sewage networks and old wastewater treatment plants;

- establishment of new sewage networks and treatment plants in all settlements;

- proper treatment of the sludge from wastewater.

In order to respect the Danube Delta Biosphere Reserve, a permanent concern is required from the local authorities, from the Romanian state authorities for: new sources of financing and new investors; accountability of the responsible actors.

The entire deltaic ecosystem of the Danube has been declared a Biosphere Reserve and a UNESCO site since 1991, but, nevertheless, destruction and pollution continue unhindered. The lack of efficient protection mechanisms and of appropriate fines make the upstream pollution of the Danube, private concessions of water surfaces, illegal constructions, races with high speed boats that disturb the birds, poaching without limit, chaotic burning of reed surfaces to remain the main and serious still unsolved problems of the Danube Delta in order to ensure environmental sustainability.

In order to improve the water management in terms of water quality, the implementation of an automated water quality monitoring system on a passengers ship, could provide an early warning message to the water authorities and stakeholders responsible with the water management on the Sulina Channel.

Acknowledgements

This work was cofinanced by a grant of the Romanian National Authority for Research and Innovation, CCCDI - UEFISCDI, project number: 107/2019, Cod: COFUND-WW2017-WATER HARMONY, within PNCDI III and EU project (Horizon 2020): "Closing the Water Cycle Gap – Sustainable Management of Water Resources".

Author details

Igor Cretescu[1*], Zsofia Kovacs[2], Liliana Lazar[1], Adrian Burada[3], Madalina Sbarcea[3], Liliana Teodorof[3], Dan Padure[4] and Gabriela Soreanu[1]

1 "Cristofor Simionescu" Faculty of Chemical Engineering and Environmental Protection, "Gheorghe Asachi" Technical University of Iasi, Iasi, Romania

2 Institute of Environmental Engineering, Pannon University, Veszprem, Hungary

3 "Danube Delta" National Institute for Research and Development, Tulcea, Romania

4 Faculty of Hydrotechnics, Geodesy and Environmental Engineering, "Gheorghe Asachi" Technical University of Iasi, Iasi, Romania

*Address all correspondence to: icre@tuiasi.ro; icre1@yahoo.co.uk

IntechOpen

References

[1] Water Framework Directive (WFD); On 23 October 2000, the "Directive 2000/60/EC of the European Parliament and of the Council establishing a frame work for the Community action in the fie ld of water policy" EU Water Framework Directive (WFD), Available from: https://eur-lex.europa.eu/Legal-content/EN/TXT/?uri=CELEX%3A32000L0060

[2] Drainage Basin of Black Sea, Chapter 5, Available from: https://unece.org/fileadmin/DAM/env/water/blanks/assessment/black.pdf

[3] Dunarea. Available at: https://ro.wikipedia.org/wiki/Dunarea (accessed: 2020-11-16).

[4] Tim O'H., Andrew F., Georgi D., Stale K. and Laurence M., (2014): Achieving good environmental status in the Black Sea: Scale mismatches in environmental management, Ecology and Society 19(3):54 http://dx.doi.org/10.5751/ES-06707-190354

[5] Danube Delta Biosphere Reserve. Tulcea, Romania. Available from: http://www.ddbra.ro/en (accessed: 2020-11-16)

[6] Blum MD., Tornqvist TE., Fluvial responses to climate and sea-level change: A review and look forward. Sedimentology. 2000; 47:2-48. DOI: 10.1046/j.1365-3091.2000.00008.x.

[7] Environmental Protection Agency. Annual report on the state of environmental factors in Tulcea County – 2019, (in Romanian), Available from: http://apmtl-old.anpm.ro/docfiles/view/117378 (accessed: 2020-11-16)

[8] Dimache T., Burlacu M., Sârbu L., Popa I., Bufnilă L., Țîbîrnac M., Doba A., Nistorescu M. Environmental Report-Integrated Strategy for Sustainable Development of the Danube Delta, Romanian Government (Ministry of Public Works, Development and Administration), 2016 (in Romanian). Available from (accessed: 2020-11-16): https://mlpda.ro/userfiles/delta_dunarii/2.raport_de_mediu_SIDDDD_rev06.pdf

[9] Elliot T., Deltas. in: Reading H. G, Editor. Sedimentary Environments and Facies. Oxford: Blackwell Scientific Publications, 1986; p. 113-154.

[10] Giosan L., Donnelly JP., Constantinescu S., Filip F., Ovejanu I., Vespremeanu-Stroe A., Vespremeanu E., Duller GAT., Young Danube delta documents stable Black Sea level since the middle Holocene: Morphodynamic, paleogeographic, and archaeological implications. Geology, 2006; 34:9:757-760. DOI: 10.1130/G22587.1

[11] Giosan L. Early anthropogenic transformation of the Danube-Black Sea system, Scientific Reports 2, 2012. DOI: 10.1038/srep00582

[12] Anthony EJ., Wave influence in the construction, shaping and destruction of river deltas, a review, Marine Geology, 2015; 361:53-78. DOI: 10.1016/j.margeo.2014.12.004.

[13] Pasternack GB., TFD modeling, Available from: http://pasternack.ucdavis.edu/research/projects/tidal-freshwater-deltas/tfd-modeling/ (accessed: 2020-11-16)

[14] Gastescu P., The Danube Delta Biosphere Reserve, Geography, Biodiversity, Protection, Management, in Water Resources and Wetlands, Editors: Gâştescu P., Lewis Jr. W., Brețcan P., Conference Proceedings, 14-16 September 2012, Tulcea – Romania, ISBN: 978-606-605-038-8, Available at: 066.pdf (limnology.ro) (accessed: 2020-11-16)

[15] Galloway WE., Process framework for describing the morphologic and stratigraphic evolution of deltaic

depositional systems, in: Brousard ML., Editor. Deltas, Models for Exploration, Houston, Texas: Houston Geological Society, 1975, 87-98.

[16] Perillo GME., Geomorphology and sedimentology of estuaries. New York: Elsevier Science B.V., 1995, 470.

[17] Orton GJ., Reading HG., Variability of deltaic processes in terms of sediment supply, with particular emphasis on grain size, Sedimentology, 40(3), 1993, 475–512, Available from: https://doi.org/10.1111/j.1365-3091.1993.tb01347.x

[18] Pasternack GB., TFD hydrometeorology, Available from (accessed: 2020-11-16): http://pasternack.ucdavis.edu/research/projects/tidal-freshwater-deltas/tfdhydrometeorology

[19] Pasternack GB., Hinnov LA., Hydrometeorological controls on water level in a vegetated Chesapeake Bay tidal freshwater delta, Estuarine, Coastal and Shelf Science, 58(2), 2003, 367–387, DOI:10.1016/s0272-7714(03)00106-9.

[20] Vespremeanu-Stroe A., Preoteasa L., Zainescu F., Rotaru S., Croitoru L., Timar-Gabor A., Formation of Danube delta beach ridge plains and signatures in morphology. Quaternary International, 415, 2016, 268-285. DOI: 10.1016/j.quaint.2015.12.060

[21] Renaud F., Kuenzer C., Climate and Environmental Change in River Delta Globally: Expected Impacts, Resilience and Adaptation, Chapter 2, in: Renaud F., Kuenzer C., Editors, The Mekong Delta System – Interdisciplinary Analyses of a River Delta. Dordrecth, Springer, ISBN 978-94-007-3961-1, 2012, 7-46. DOI: 10.1007/978-94-007-3962-8.

[22] Giosan L., Syvitski J., Constantinescu St., Day J., Climate Change: Protect the world's Deltas, Nature, 516, 31–33 (04 December 2014). DOI: 10.1038/516031a

[23] Delta Dunării (Danube Delta). Available from: https://ro.wikipedia.org/wiki/Delta_Dun%C4%83rii (accessed: 2020-11-16).

[24] Ghidul Deltei. Available from: http://www.ghiduldelteidunarii.ro (accessed: 2020-11-16)

[25] Oamenii Deltei. Available from: http://www.galoameniideltei.ro/ (accessed: 2020-11-16)

[26] Török L. (Ed.), deltas and wetlands (book of abstracts: Conference: The 22nd international symposium "deltas and wetlands" 2013), pp. 226, Tulcea

[27] Tulcea County Council. Report on the environmental impact assessment study, Integrated waste management system in Tulcea county, 2017 (in Romanian). Available: http://www.smid judetultulcea.ro/index.php?p=proiect&t=despreproiect(access: 2020-11-16)

[28] Teodorof L., Despina C,. Burada A., Seceleanu–Odor D., Anuți I., Methods for monitoring physico-chemical indicators in the aquatic ecosystems of the Danube Delta, in Tudor M. Editor, « Ghid metodologic de monitorizare a factorilor hidromorfologici, chimici și biologici pentru apele de suprafață din Rezervația Biosferei Delta Dunării », Printing house « Editura Centrul de Informare Tehnologică Delta Dunării », Tulcea, Romania. Available:https://www.researchgate.net/publication/294694361_ (accessed 2020-12-02).

[29] Ordin nr.161 din 16/02/2006 the Norms regarding the classification of surface water quality in order to establish the ecological status of water bodies, Available from: http://Legislatie.just.ro/Public/DetaliiDocumentAfis/72574

[30] Cretescu I., Kovács Z., Cimpeanu S. M., Monitoring of Surface Water Status in the Lower Danube Basin, in River

Basin Management, Bucur D. Editor 2016, InTech, Zagreb, Croatia, ISBN: 978-953-51-2604-1, 205-223, Available from: Monitoring of Surface Water Status in the Lower Danube Basin | IntechOpen

[31] Sulina, Available at: https://ro. wikipedia.org/wiki/Sulina (accessed: 2020-11-16)

[32] Tulcea, Available at: https://ro. wikipedia.org/wiki/Tulcea (accessed: 2020-11-16)

[33] Tulcea City Hall. Romania. Available from: https://www.primaria tulcea.ro (accessed: 2020-11-16)

[34] Sulina, City Hall, Romania. Available from: https://www.primaria-sulina.ro (accessed: 2020-11-16)

[35] Sulina, City Hall, Romania, Sulina Local Development Strategy, 2018. Available from: https://www.primaria-sulina.ro/plan_integrat_dezvoltare.html (accessed: 2020-11-16)

[36] Crisan City Hall, Romania. Available from: https://www.primariac risan.ro/?p=consiliul_local (accessed: 2020-11-16)

[37] Maliuc City Hall, Romania. Available from: https://www.comuna-maliuc.ro/?p=consiliul_local (accessed: 2020-11-16)

[38] Badea Gh., Badea DG., Raport de Mediu Maliuc, 2015, Available from: http://www.ddbra.ro/documente/ad min/2015/RM_PUG_MALIUC_ 26.09.2018.pdf

[39] Danube River Basin District Management Plan, Update 2015, published ICPDR, Available from: https://www.icpdr.org/flowpaper/app/ #page=1

[40] River Basin Management Plan 2015, Hungary - Vízgyűjtő Gazdálkodási Terv https://www.vizugy.hu/index.php?mod ule=vizstrat&programelemid=149

[41] Frîncu R. M., Long-Term Trends in Water Quality Indices in the Lower Danube and Tributaries in Romania (1996–2017), International Journal of Environmental Research and Public Health, 18(4), 2021, 1665, available from: https://doi.org/10.3390/ijerph18041665

[42] European Environment Agency, Corine Land Cover Europe; European Environment Agency: Copenhagen, Denmark, 2018, Available from (accessed on 14 December 2020): https://hub.arcgis.com/datasets/129e 81fc75ec4426aa25a02943cebaf0?geome try=8.431%2C41.611%2C38.006% 2C47.103

[43] Taylor O'Connor, Russia, Geograph y - GCU 114 (weebly.com)

[44] Strelkov S., Boronina L., Sorokin A., Kondrashin K., Petrov R., Assessment of the Ecological State of the Surface Waters of the Nikitinsky Fish Passage Channel during Dredging Work, E3S Web of Conferences 135, ITESE-2019, 01016, 2019, https://doi.org/10.1051/ e3sconf/201913501016

[45] Winkels H.J., Kroonenberg S.B., Lychagin M.Y., Marin G., Rusakov G.V, Kasimov N.S., Geochronology of priority pollutants in sedimentation zones of the Volga and Danube delta in comparison with the Rhine delta, Applied Geochemistry 13(5), 1998, 581-591, DOI: 10.1016/S0883-2927(98) 00002-X

[46] Kosarev, A. N., Yablonskaya E. A., The Caspian Sea. The Hague: SPB Academic Pub., 1994.

[47] Dumont H. J., the Caspian Lake: History, biota, structure, and function, Limnol. Oceanogr., 43(1), 1998, 44-52,

Available from: https://doi.org/10.4319/
lo.1998.43.1.0044

[48] Bukharitsin P. I., Luneva, Z. D.,
Water Quality Characteristics of the
Lower Volga Reaches and the Northern
Caspian Sea, Water Resources (Vodnye
Resursy), 21(4), 1994, 410-416
(445-451), available from: 29334_doc.
pdf (oieau.org)

[49] Fashchevsky B., Human Impact on
Rivers and Fish in the Ponto-Caspian
Basin, Proceedings of the second
international symposium on the
management of large rivers for fisheries,
Volume I, in Phnom Penh, Cambodia,
11-14 February 2003, Edited by Robin L.
Welcomme and T. Petr, available from:
Proceedings of the second international
symposium on the management of large
rivers for fisheries: Volume I (fao.org)

Chapter 9

Anthropogenic Activities as a Source of Stress on Species Diversity in the Mekong River Delta

Charles Nyanga, Beatrice Njeri Obegi and Loi To Thi Bich

Abstract

Deltas are landforms, which come into existence when sediment carried by river or stream empties its load into another water body with slow flow rates or stagnant water. Sometimes, a river may empty its sediment load on land, although this is uncommon. The world's deltas are amongst the most productive and in some cases more populated than even land. This chapter reviews the formation of deltas, the ecology and habitats of deltas as well as the biodiversity in coastal habitats and delta habitats. Additionally, the chapter looks at recent advances in deltas such as the loss of sediment and other stressors currently facing deltas with a focus on anthropogenic activities in the Mekong River Delta (MRD) that is amongst the most resource rich deltas in the world. The Mekong River Delta (MRD) is currently known to be in peril due to anthropogenic activities such as dam construction for hydropower and irrigation, overfishing, agricultural production amongst many others. Additionally, demographical trends like population increase have also been scrutinized to see the impacts on the MRD. The results of the review process have shown that at least 85% of the deltas in the world are subsiding and losing their fertility to the sea. Finally, the chapter has endeavored to come up with suggestions on how best to overcome some of these stressors resulting from the anthropogenic activities.

Keywords: biodiversity, sediment load, stressors, anthropogenic, demographic, subsidence

1. Introduction

A Delta as defined by [1] is "the wetland which forms as the rivers flowing towards another body of water empty the water and sediments they carry into the other bodies of water. The other bodies of water may be oceans, lakes or other rivers".

NASA [2], explains that as the river enters the lake, the flow slows down, sediments drop out, this leads to delta formation. This formation causes a prism of sediment that tapers outwards to the lake to be created. Continual build – out of the delta as time progresses leads to sediment formation that are inclined in the lake-ward direction (**Figure 1**).

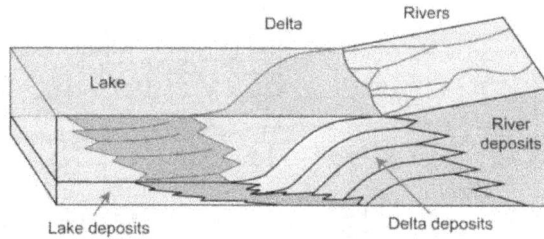

Figure 1.
The is figure which was produced by laboratory computer simulation by National Aeronautics and Space Administration (NASA) shows how a delta is formed as defined by [1]. Source: Adapted from [2].

1.1 Objectives and key scheme followed by this work

The major objective of this work is to: firstly, ascertain the impacts of the arch of anthropogenic activities of like hydropower construction works on the river deltas; secondly, to ascertain the impacts of other anthropogenic activities like agriculture, water abstraction and other industrial activities on biodiversity, thirdly, to ascertain the impacts of the actions by nature on the Mekong River Deltas (MRD).

1.2 Methods and approaches to assessment of anthropogenic activities impact on biodiversity

The methods utilized in this book chapter to achieve the above mentioned in 1.1 above are to review scientific articles, news reports and articles, other research works which are able to provide documented evidence of impacts of human activities on the Mekong River Delta.

Additionally, the methods followed are to review reports, news articles, to provide the levels of stress. These are then examined against the fifteen point stressor analysis framework developed by Scheltinga et al. [3]. The analysis carried out is based on findings by [4–12, 22–25].

1.3 Chapter structure

The chapter sets out by looking at (i) Mechanisms of Delta Formation, (ii) Types of Deltas – their formation, features and habitats, (iii) Deltas in catastrophe - the fifteen feature framework for analyzing the coastal stressors, (iv) The Mekong River Delta is discussed looking at location and then examining the delta against the fifteen point framework for stressor analysis, (v) The anthropogenic impacts on the Mekong is further discussed and supplementary materials are provided and (vi) Conclusions are stated.

2. Mechanisms of delta formation

Deltas are formed as the channel of the river flows across the earth's crust. When the river makes contact with the soil, its flow takes along with it sediments such as gravel, sand, silt and clay. Additionally, when a flowing river in a channel comes into contact with another water body, such a river loses some or most of its speed and tends to deposit the sediments it carries onto a flat area. This sediment which is deposited by a flowing river is termed "alluvium". The slowing speed of a flowing river coupled with the building up of alluvium causes the river to split up from its solitary channel as it gets closer to the mouth. As it flows on, if conditions are right, the river forms a deltaic lobe at its mouth. Additionally, as the deltaic lobe matures, it includes in its formation a distributary network – which is made up of a series of smaller channels which are less developed in depth termed distributaries [1, 13].

As the building-up of alluvium continues, completely new land is formed. This new land forms the mature delta. The delta tends to cause the river's mouth to extend right into the water body into which the river empties its alluvium laden water. River deltas are often divided into two components, namely: (i) the subaqueous, and (ii) the subaerial components [1, 13].

2.1 The subaqueous part of the river delta

The subaqueous component of a river delta is below the surface of water. It is the component which has the greatest slope and contains the silt which is finest. The most recently formed part of the subaqueous delta is referred to as the prodelta, and is most distant from the river mouth [1].

2.2 Subaerial part of the river delta

The subaerial component of a river delta is above the surface of water. The lower delta is that component of the subaerial delta component which is most influenced by waves and tides whilst the upper delta is that component of the subaerial delta most influenced by the river's flow [1].

3. Types of deltas

According to Seybod et al. [14], the terminology 'delta' is of Greek origin. It is believed by many people that the scholar who first coined the term delta nearly 2500 years ago was the ancient Greek historian Herodotus [15]. It comes from the Greek capital letter Δ. This coastal land feature became so called due to the fact that deltas are shaped like this Greek letter, Δ, delta. Accordingly, the delta can be said to be a sedimentary deposit brought to the coast by a flowing river channel with subaerial and subaqueous components. Therefore a river delta is formed by sediment laden river water that deposits its sediment at the edge of still water, an ocean or a lake. The structure of the river delta and sediment disposal processes all depend on the discharge levels, sediment amounts and magnitude of the tides. The sediment deposition characteristics depend on a complicated web of interactions amongst dynamic processes of climate, hydrologic characteristics, wave energy, tidal action amongst many other processes.

Galloway [16] as cited by Seybod et al. [14], provided a classification method on which [16], identifies and provides classes of deltas according to three main forces of river delta formation, namely: (i) river-dominated deltas; (ii) wave dominated deltas; (iii) tide dominated deltas (**Figure 2**).

3.1 River dominated deltas

3.1.1 Formation of River dominated deltas

Literature has not differentiated river-dominated deltas from fluvially-dominated deltas. The reason for this could be that all processes related to streams are referred to as fluvial. The word "fluvial" is obtained from the Latin word "fluvius = river" [18]. Additionally, [19], writes about fluvial systems as the systems in geomorphology which are dominated by rivers and streams.

1 Mississippi
2 Po
3 Danube
4 Ebro
5 Nile
6 Rhône
7 São Francisco
8 Senegal
9 Burdekin
10 Niger
11 Orinoco
12 Mekong
13 Copper
14 Ganges-Brahmaputra
15 Gulf of Papua
16 Mahakam

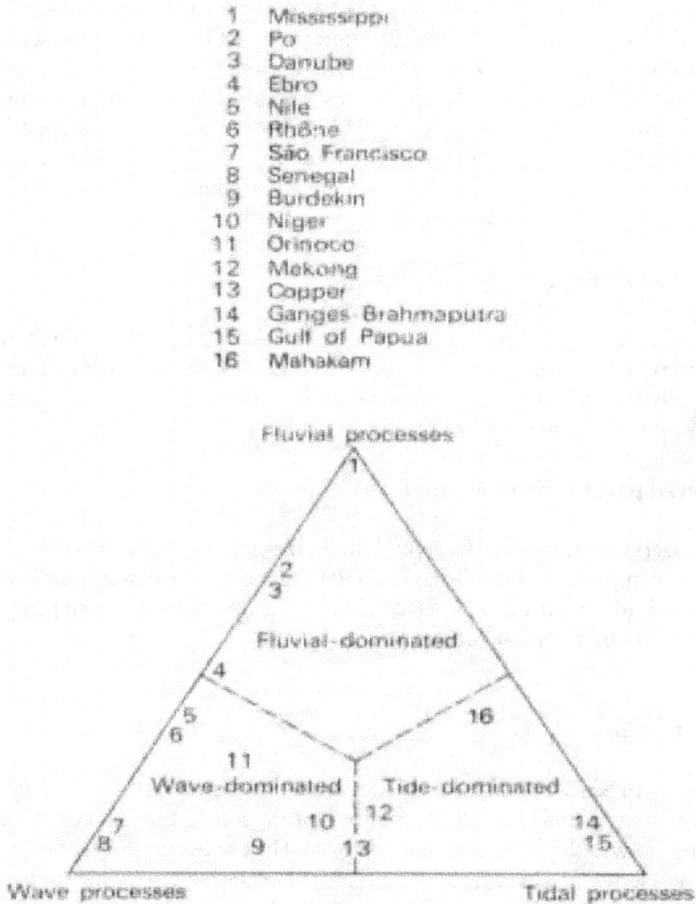

Figure 2.
This ternary is depicting the world's major deltas as well as how they are classified based on the processes involved with that particular delta. It shows that the Mekong River Delta (MRD) is a tide dominated delta and slowly changing to being a wave dominated delta as well. Source: Adapted from [17].

In the article by the Society of Economic Paleontologists and Mineralogists (SEPM), formerly called (Society of Sedimentary Geology) Stratigraphy Web [20] indicates that fluvially-dominated deltas are types of deltas which are mainly under the control of the difference in water density of the flowing water of the river and the standing water of the basin. Further, flow type differences are known to determine the sediments distribution and resulting sedimentary structures.

3.1.2 Features of river dominated deltas

Society of Economic Paleontologists and Mineralogists - SEPM [20], additionally explains that the sedimentary structures are: (i) homopycnal flow – which come into existence whenever the river water density is equal to the basin standing water density; (ii) hyperpycnal flow – which come into existence whenever the river water which flows into the basin has a higher density; (iii) hypopycnal flow – which comes into existence whenever the river water which flows into the basin has a lower density compared to the basin standing water density.

3.1.3 Habitats and ecology of river dominated deltas

Society of Economic Paleontologists and Mineralogists - SEPM [6], furthermore reports that fluvial dominated deltas with low tide and wave energy, after several investigations have shown three structures, namely: (i) inertia – dominated fluvial deltas – which are characterized by flow velocities which are high and large turbulence amounts. In this delta type, sediment deposition does not have a high lateral component; (ii) frictional – dominated fluvial deltas – are characterized by high frictional forces and shear stress at bed level. However, rapid slowing of flow arising from frictional forces and shear stress lead to sediment deposition with a wider lateral extent; (iii) buoyant – dominated fluvial deltas – are characterized by levee formation which result from the existence of a deeper river – basin region of interaction. Other deposit forms popular with buoyance – dominated deltas are distributary mouth bars, bar finger sands, distal bars as well as prodelta clays.

The aforementioned habitats support a variety of marine species. The productivity of plants gains as the tidal range widens. This happens due to pronounced flushing rates and the resulting renewal of nutrients. Grasses, sedges and herbs as well as freshwater wetland and floodplain vegetation occupy the areas above the influence of tides. The lower turbidity of the sea is able to offer support to the growth of phytoplankton.

3.2 Wave dominated deltas

3.2.1 Formation of wave dominated deltas

According to Nienhuis [21], wave dominated deltas are deltas in which the sea waves are the factors that play a dominant role in shaping the fluvial sediment. Examples of wave-dominated deltas are the Nile Delta in Egypt, the St George lobe of the Danube in Romania (Tulcea County) and partly in Ukraine (Odessa Oblast), the Rio Grijalva in Mexico amongst many others.

In another of his studies, Nienhuis [21], reported that the wave dominated deltas attain their shape due to the fact that the coarse-grained fluvial sediment flux which the river supplies to the river mouth becomes less than the largest quantity waves have capacity to transport away through the transportation of sediment along the shore along both flanks of the delta. This is in agreement with [22].

3.2.2 Features of wave dominated deltas

According to [23], the wave dominated deltas, also referred to as riverine estuaries, have the following identifying features:

i. They have variable habitats, mostly, brackish subtidal, intertidal and supratidal.

ii. They have narrow entrances which restricts marine flushing. This leads to very small proportion of water volume exchange at each tide.

iii. The river flow is very high. Whenever there is flooding, marine water is expelled and materials are flushed from the wave-dominated delta.

iv. Turbidity is naturally low. However, whenever there is high fluvial runoff, turbidity levels rise thus turbidity in wave-dominated deltas is dependent in catchment flows inwards.

v. The wave-dominated deltas usually are able to expel sediments into the coastal regions of the ocean.

vi. Wave-dominated deltas tend to have stable morphology due to the maturity of their evolution levels.

3.2.3 Habitats and ecology of wave dominated deltas

The wave-dominated deltas have habitats such as sandy beaches, intertidal flats of mud, marshes of salt and mangrove forests. These deltas are able to support euryhaline estuarine species as well as transient visitor species from the ocean environment whose presence in the delta depend on river flow states.

3.3 Tide dominated deltas

3.3.1 Formation of tide dominated deltas

Goodbred and Saito [24], explain that tide dominated deltas are quiet difficult to characterize. This is due to the major role that fluvial systems have put into action in pronouncing the deltas they are associated with as the rivers vary broadly in their discharge, sediment load, seasonal behavior as well as the sediment material grain size.

3.3.2 Features of tide dominated deltas

Ozcoasts [23], outlines the following as the key features of tide dominated deltas:

i. Tide dominated deltas support a wide range of both marine and brackish, subtidal, intertidal and supratidal estuarine habitats. Most of the delta area is covered by intertidal and supratidal regions whereas seagrass is precluded in certain areas due to turbidity.

ii. The marine flushing delta process is promoted due to the large entrance which tide-dominated deltas have.

iii. Since the flow of rivers is high in tide-dominated deltas, marine water may be expelled due to flooding and material is flooded from the delta by high flowing river.

iv. The turbulence which is induced by tides is strong in tide-dominated deltas which leads to high levels of turbidity.

v. In tide dominated deltas, terrigenous sediments and pollutants are susceptible to being trapped by adjacent environments such as intertidal flats, mangroves, salt marshes and salt flats.

vi. Trapping and processing of loads from land is encouraged by the tidal movements over environments of various features existing side-by-side.

vii. Tide-dominated deltas are stable in morphology since they are mature.

3.3.3 Habitats and ecology of tide dominated deltas

The habitats and ecology of tide dominated deltas are outlined by Ozcoasts [23], further explains that, typically, tide dominated deltas produce habitats such as channels, intertidal mudflats, mangroves, salt marshes, and salt flats.

4. Deltas in catastrophic situations

4.1 Loss of delta health - one of the recent advances in river deltas

According to InteGrate [25], one of the recent advances in river deltas which has been observed in the last decade is that there has been a decrease in the health of the major river deltas of our world. This reduction in the health earth's river deltas has come about due to several reasons. Some the reasons are: over-exploitation of the delta resources by humans, the introduction of pollutants, addition of excessive nutrients to the rivers from poor agricultural practices and industrial production as well as poorly managed river basins that feed the river deltas. All these have largely caused damage to the river deltas' environments which are very sensitive. Additionally, InteGrate [26] reports that one major anthropogenic activity which has upset the river deltas habitats is the reduced sediment loads in many deltas arising from dam construction. Global sea level rise has also resulted in widespread loss of delta based wetlands and other associated habitats like sand barriers along the shoreline.

4.2 Stressors of coastal habitats related to river deltas

4.2.1 Definition of coastal stressor

Coastal habitats and coastal community stressors are defined by Scheltinga et al. [3] as:" Physical, chemical, and biological components of the environment that, when changed by human or other activities, can result in degradation to the natural resources. "Furthermore, Scheltinga et al. [3], explain that stressors can be: (i) an element of the environment capable of transferring the impact of a pressure (for example: an anthropogenic activity) to other parts of the environment after it is changed from its natural state. Examples of such elements are nutrient concentrations which have been changed from the natural level of concentrations, habitat coverage which is less than the natural level, excess salt, amongst many others. There are several elements which are present in a healthy ecosystem. However, if the elements are different from the natural levels, they are taken to be stressors; (ii) an element of the environment that, whenever detected in an environment, might have the potential of causing shifts from natural levels. Such potential stressors are litter and pest species amongst many others.

A framework of fifteen elements on a stressor framework developed by Scheltinga et al. [3] is used here. Scheltinga et al. [3] provides a list of fifteen elements of the environment which have been included in a stressors' (physical, chemical and biological) indicators framework. These elements are:

   i. Aquatic sediments (altered from natural levels)

   ii. Bacteria/pathogens

   iii. Biota removal/disturbance

iv. Excess fresh water (hyposaline)

v. Excess salt (hypersaline)

vi. Fresh water flow regimes (altered from natural levels)

vii. Habitat removal/disturbance

viii. Hydrodynamics (altered from natural levels)

ix. Litter

x. Organic matter (altered from natural levels)

xi. Nutrients (altered from natural levels)

xii. Pests (plant, animal) species

xiii. pH (altered from natural levels)

xiv. Toxicants

xv. Water temperature (altered from natural levels)

5. The great Mekong River Delta

5.1 Geography and location

The Mekong River passes through a basin known as the Greater Mekong [27]. It is a region which holds riches which are irreplaceable. The riches ranges from rare wildlife, plant diversity, natural landscapes to communities with a variety of cultural heritages. The Greater Mekong covers an area of approximately 80.9 hectares which is has some habitats so diverse it is only second to the Amazon (**Table 1**) [27].

The Greater Mekong region is nick named the "rice bowl of Asia" and at the center of the region lies the Mekong River, a transboundary river in East Asia and Southeast Asia. The Mekong River runs through China, Myanmar, Laos, Thailand, Cambodia and Vietnam. It is number 12 in length on the world list and number 6 on

Type of Species	Estimated numbers of species
Mammals	> 430 species
Amphibians	> 800 species
Birds	~1200 species
Fish	> 1100 species
Plants	~20,000 species

Tiger, Soala, Asian elephants are endangered [27].
Mekong dolphin and Mekong giant catfish are endangered [27].
Source: Compiled by the authors based on information from [27].

Table 1.
This is depicting the species biodiversity in the Mekong.

the Asian list. It runs for an approximate length of 4909 km, drains a region covering 795,000 square kilometers and discharges 475 cubic kilometers of water per year [28]. Before the Mekong River spills its discharge of water into the China Sea in Vietnam, it forms an expanse of distributaries which together constitute a complex delta formation which is known as "The Nine Dragons". That is why the Mekong is sometimes referred to as the "River of the Nine Dragons" [29].

5.2 Biodiversity stressors in the Mekong River Delta

5.2.1 Evidence of altered aquatic sediment levels in the Mekong River Delta (MRD)

Alteration of aquatic sediment from natural levels in Mekong River has been reported in a study by [4]. Piman and Manish [4] report that the Mekong River Commission in 2013 produced sediment monitoring results (for the period before 2003 and after 2009) which showed that average sediment loads in the MRD reduced as follows: at Chiang Saen station, from 60 Million tons/year down to 10 Million tons/year (representing a reduction of 83%); at Pakse, from 120 Million tons/year down to 60 Million tons/year (representing a 50% reduction); at Kratie, the sediment reduction changed from 160 Million tons/year to 90 Million tons/year representing a reduction of 43%. Additionally, Piman and Manish [4] indicate that if all the hydropower stations proposed for the Lower Mekong Basin (LMB) were to be implemented the sediment load reaching the MRD region would reduce to 4% (a 96% reduction).

The implications of the foregoing for biodiversity are dire. Probably this might be indicating that the LMB is advancing towards a tipping point regarding the planetary boundaries and human opportunities for sustainably managing the future natural resources of the earth. Reductions in sediment loads might induce 12% - 27% reduction in primary productivity in the producers in the lower rungs of the food webs in the LMB aquatic ecosystems. Further, fish species such as Lithophils, Psammophils and Pelagophils which are dependent on sediments and nutrients loads might completely fail to adjust to the new nutrient and sediment regime which subsequently might lead to fisheries biodiversity depletion [4].

The table below, shows some of the fish species which depend on sediments for reproduction (**Table 2**).

The figure below (**Figure 3**) is adapted from Baran et al. [8] and fully credited to them. It is an expsotion of the composition of the fluvial sediments, both suspended sdiment and bedded sediment. It is clear that if there is a reduction in the sediment all the constituents will reduce accordingly and produce multi- dimensional stress effects on the biodiversity.

5.2.2 Evidence of biota removal, species, habitat and organic matter losses in the Mekong River Delta (MRD)

Allison et al. [5], report that in Lower Mekong River Basin (LMB), at the delta and ocean meeting region, there has been long term reduction in mangrove hectarage due to land-use conversion and utilization of forest products. In 1943 there were 306,000 hectares of mangroves in the delta. This reduced to 253,000 hectares by 1982. This represented a reduction of 17% in the mangrove forest area. However, after some replanting efforts by the Vietnamese government, in 2005 the hectarage stood at 270,000 hectares.

Additionally, the World Wide Fund for Nature (WWF) [6], resounds the dangers faced by Fiona in the MRD by noting that the MRD is home to species like

Species	Image	Habitats/Reproduction	Impacts of human activities
Probarbus jullieni		Important habitats for spawning are river rapids	
Pangasius macronema		Spawns in rapids at the beginning of the rainy season	
Boesemania macrolepis		Hard rock or pebble with silt or sand substrate, steep rocky sides descending in pools. Spawns in areas at a depth 20 meter with counter-current eddies.	
Probarbus labeamajor		Spawns in rapids	The large dams across the main streams in Stung Treng and Kratie provinces will remove the rapids which are important spawning habitats
Puatioplites proctozystron		Muddy river beds Spawns in slow moving water	If the sediment reduces it will stop spawning.
Channa gachua		River beds with silt or gravel. Spawns in shallow waters	

Source: Adapted and modified from [8].

Table 2.
Habitats and reproduction tendencies of some common fish species in the Mekong Delta.

tigers, giant catfish, self-cloning skink, fish with vampire fangs which are all now faced with a future which is uncertain in the face of the rapid development in MRD which is depriving them of their habitats which nature has provided in the form mangrove swamps and forests.

Other sources of information such as [7]; provide detailed accounts of how mangroves forests support fisheries by providing organic matter. These marvelous trees shed off about seven and half tons of leaf litter per acre per year. This litter fall is decomposed by bacteria and metabolized by fungi which release nutrients via the detrital food loop to organisms higher up in the food chains. Detritus is food for shrimp, mullet and many other organisms. Furthermore, Asokan [7], explains that mangroves support fisheries in two major ways, firstly, by providing well protected habitats for larvae and juveniles, secondly, by providing food for the fish from the leaf litter fall in the detrital food web.

The losses of mangrove forests have been producing a negative impact on the biota and the habitats.

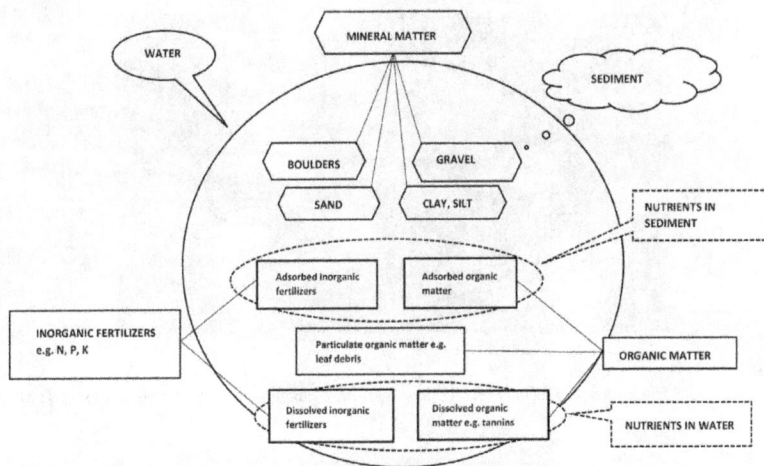

Figure 3.
Composition of river delta sediments. Source: Adapted from [8].

5.2.3 Evidence of nutrient alteration from the normal levels in the Mekong River Delta (MRD)

Alterations of nutrient transport from the natural levels has also been reported by Piman and Manish [4]. Furthermore, Piman and Manish [4] argue that if the full hydropower development which has proposed is implemented at all sites including Kratie, a total projected reduction of 47–53% in Nitrogen transport will be observed and Phosphorus will record a reduction of 57–62%. Additionally, Baran et al. [8] reports that dam developments and the advent of climate change both anthropogenic and natural will result in 53–59% reduction in the amounts of sediments the MRD receives and there will be a reduction of 47–84% in nutrient supply. Furthermore, Baran et al. [8], indicates that dam construction will cause a reduction of 30–38% in the net primary production. This will affect the food webs drastically.

5.2.4 Evidence of salt water intrusion (Hypersalinity) in the Mekong River Delta (MRD)

In their article, Allison et al. [5] report that reduced sediment load in MRD in some cases such as Sang Han distributary have led to salt water intrusion of up to 40 km into the Sang Han distributary channels.

5.2.5 Evidence of pollution from marine litter in the Mekong River Delta (MRD)

Plastics are amongst the most used and disposed off industrial products. The United Nations Environment Programme (UNEP) [12], reported that a colossal 300 million tonnes of plastics debris is produced by humans every year. Of this quantity, 2.7% (or 8 million tonnes) finds its way into the oceans. Additionally, the University of Hull [11] indicated that the Mekong River is amongst the most polluted rivers on Earth. The Mekong transports plastic litter estimated at 40,000 tonnes per year right into the world's oceans (**Figures 4–6**).

Furthermore, Plastics cause a lot of harm to both animals and plants. More than eight hundred marine and coastal species get affected by plastics through ingestion, entanglement, suffocation and many other dangerous ways. The figure below (**Figure 7**) shows how micro plastics affect micro organisms.

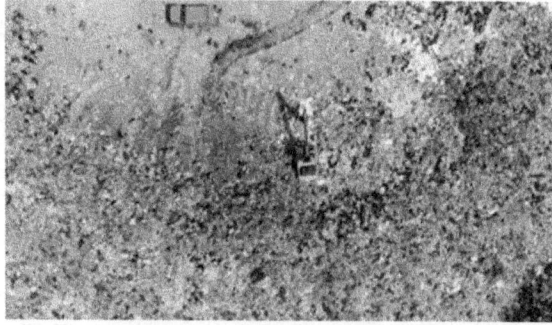

Figure 4.
Plastic litter in the Mekong River. Photo credit: United Nations environment programme (UNEP) [12]/ Adam Hodge.

Figure 5.
This figure shows the part of the great Mekong River with mixed litter. This is an opportunity to some sectors of the community while it poses great challenges to some sectors such as the institutions in charge of planning amongst many others. Photo credits: [30].

Figure 6.
This figure expresses the realities in the great Mekong posed by pollution emanating from huge populations in this region. This was sourced from the University of Hull's on-going projects. Photo credits: [11].

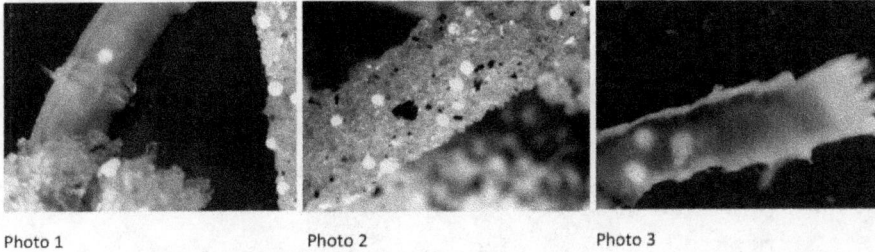

Photo 1   Photo 2   Photo 3

Figure 7.
This figure shows the danger of marine micro plastics to micro biodiversity. Photo 1: Fluorescent polyethylene microbeads incorporated into the tube and also ingested by the bamboo polychaete worm Clymenella torquata. *Photo 2: Fluorescent polyethylene microbeads incorporated into the tube of the bamboo polychaete worm C. torquata. Photo 3: Fluorescent polyethylene microbeads ingested by the bamboo polychaete worm C. torquata. Photo credits: Center for Coastal Resources Management, Virginia Institute of Marine Science, W&M'.*

5.2.6 Evidence of toxicants, PH and temperature variations in the Mekong River Delta (MRD)

Berg et al. [31], in their article report that in the Mekong River Delta, the Arsenic (As) concentrations in ground- water ranged from 1 microgram per liter to 1610 micrograms per liter (with an average of 217 micrograms per liter) in Cambodia; whilst in South Vietnam the Arsenic concentrations in groundwater ranged from 1 microgram per liter to 845 micrograms per liter (with an average of 39 micrograms per liter). In another research study, Shinkai et al. [9], carried out an assessment of Arsenic and other heavy metal in contamination of groundwater resources in the Mekong River Delta (MRD). Shinkai et al. [9] found that in Tien Giang Province and Dong Thap Province the total Arsenic (As) concentrations in groundwater resources which is utilized for domestic consumption ranged from 0.9 micrograms per liter to 321 micrograms per liter. This was well above the World Health Organization (WHO) guidelines of 10 micrograms per liter. Furthermore, Shinkai et al. [22] indicate that there was evidence of the presence of other heavy metals in groundwater. It was found that 91% and 27% of sampled shallow wells showed concentrations of Manganese (Mg) and Barium (Ba) which are higher than the World Health Organization (WHO) guidelines for drinking water.

The studies reviewed here show evidence of the presence of toxins. However, the levels of the impacts of these toxic elements on biodiversity such as capture fisheries, vegetation, aquatic life and others have not yet been clarified.

On the issue of temperature variation, the Mekong River Commission (MRC) [10] has clarified that there is little temperature variation in Mekong River Delta (MRD). The temperatures in Lower Mekong Basin (LMB) range from 32° C during the warmest months of March and April to 23° C. It can be seen that the temperature stress in the short term might not vary to levels which are dangerous. However, long term temperature variations due to climate change have not been clarified in the region yet.

5.2.7 Other sources of stress due to Anthropogenic actions

A traveler and researcher [32], the author of the famous book "The Last Days of the Great Mekong", expresses his observations in clear terms. Eyler [32] decided to come up with a book after traveling along the Mekong River and talking to the community members along this river. Eyler [32], provides one very important observation that due to massive dams which have been constructed by the Chinese on the Mekong before it leaves China on its flow route via Myanmar, Laos,

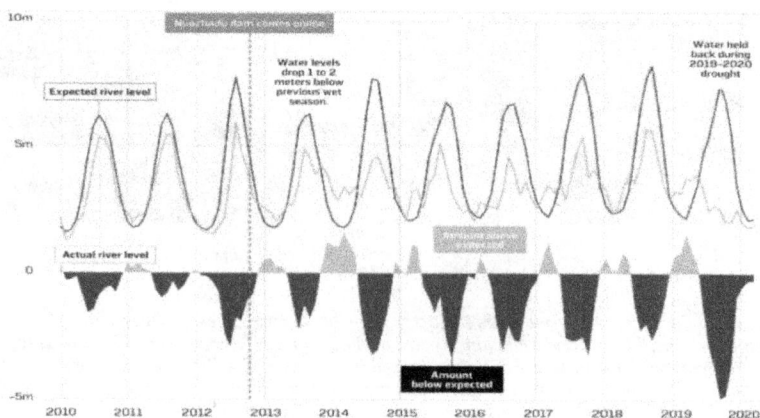

Figure 8.
This figure shows the impact of the Nuozhadu dam on the Mekong River. The dam came on the scene in 2013 and water levels' height in meters has been declining since then. The blue curve is the modeled river water height while the yellow curve is what is actually occurring. Source: Adapted from [32].

Thailand, Cambodia and finally into Vietnam [33, 34]; the Great Mekong which is the source of 20% of the world's freshwater fish catch is heavily dependent on the seasonal monsoon and flow of the river. However, Eyler [32] observes that it is heavily impaired by the Chinese dam construction (**Figure 8**).

6. Further discussions and supplementary materials

The deltas are currently facing many issues apart from the stresses imposed on biodiversity. Additional evidence of concerns is provided by Kazem [35], who indicates that there has been a growing concern owing to poor living conditions for the residents in deltas. Additionally, Safra de Campos [36] found that the Anthropocene has brought about marked changes which occur at scales which are different and speeds which are also varied from one delta region to another delta region.

6.1 Modern trends in deltas

Anthropogenic activities in river deltas and their basins, usually upstream, have dramatically affected delta regions to the level of changing them [37]. As alluded to in earlier sections of this chapter, [38] adds their voice to the devastating level of anthropogenic activities like anti-erosion agriculture, hydrological engineering works like dam construction have reduced river sediment delivery to several deltas in recent decades [39]. Here other key trends in recent decades in delta regions are discussed.

6.1.1 Trend number 1: Decreasing quality of life for populations in river deltas

Two groups of researchers using survey method and Focus Group Discussions (FGD's), namely [35, 40], both point out the environmental degradation that occurs in river deltas. This has subsequently affected the life styles of the populations in deltas. Szabo et al. [41] indicates that there is rapid onset of and creeping processes in deltas which bring about environmental hazards as well as lowering the quality of ecosystems services.

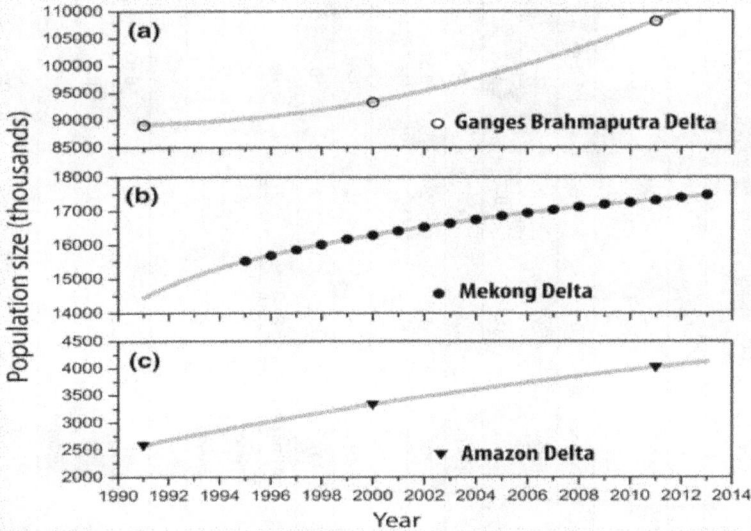

Figure 9.
This figure is depicting the population growths in the rivers deltas namely: Ganges Brahmaputra, Mekong and Amazon based on national statistics. Source: Adapted from [41].

6.1.2 Trend number 2: Increasing populations

Szabo et al. [41] also reports that population in some of the most important river deltas (namely: Ganges Brahmaputra, Mekong and Amazon, see **Figure 9** below) have been rapidly growing. The rapid increase in population has put a lot of stress on ecosystems services in these deltas as well.

6.1.3 Trend number 3: Massive construction activities and sediment losses

The world's river deltas as we know them today were built by long term deposition (aggradation or alluviation) of fertile river sediments over very long time periods [42]. It is due to this alluviation that the river deltas have been providing food-production areas and attracting huge populations [42]. Unfortunately, the river deltas are subsiding. The Mekong River Delta (MRD) is subsiding at the rate of 16 mm per annum [42]. However, the report from Syvitski [38] is showing similar trends but on a lower side (see **Table 3** below). The method Syvitski [38] utilized was to utilize high resolution data sets which were generated by National Aeronautics Space Administration (NASA)'s Shuttle Radar Topography Mission (SRTM). Additionally, the massive dam constructions (see **Table 4** below) has led to a lot of losses in the biodiversity habitats, loss of sediment and decreased levels of water along the Mekong River [32]. The results of studies by Syvitski [38] show that a small number of world river deltas are not under threat. However, Dang et al. [45] in studies of sediment budgets in river deltas using high frequency measurement further provided information on damming along the Mekong River as shown in **Table 4**.

6.1.4 Trend number 4: species threats

The anthropogenic activities have seen many species of fish being endangered, declining or decreasing. Some of the observed disturbances are those suffered by migratory fish species [27], see **Tables 5** and **6**.

Delta	No. Maps	Est. Area km² < 2 m ASL	Recent Area km² Storm Surge	Recent Area km² River Flood	Recent Area km² In situ Flooding	% Sediment Reduction	Floodplain or Delta Flow Diversion	% Distributary Channel Reduction	Subsurface Water, Oil & Gas Mining	Early 20th C Aggradation Rate mm/y	21st Century Aggradation Rate mm/y	Subsidence mm/y
Amazon, Brazil	6	1960*	0; LP	0	9340	0	No	0	0	0.4	0.4	?
Amur, Russia	—	1250	0; LP	0	0	0	No	0	0	2	1	0.5-2
Brahmani, India	6	640	1100	3380	1580	50	Yes	0	Major	2	1	0
Chao Phraya, Thai.	2	1780	800	4000	1600	85	Yes	30	Major	0.2	0	50-150
Colorado, Mexico	3	700	0; MP	0	0	100	Yes	0	Major	34	0	2-4
Congo§ DRC	—	460	0; LP	0	0	20	No	0	0	0.2	0.2	0?
Danube, Romania	4	3670	1050	2100	840	63	Yes	0	Minor	3	1	≈0
Fly, PNG	—	70*	0; MP	140	280	0	No	0	0	5	5	0.5
Ganges¶, Bangl.	9	6170*	10,500	52,800	42,300	30	Yes	37	Major	3	2	18
Godavari, India	6	170	660	220	1100	40	Yes	0	Major	7	2	≈4
Han, Korea	—	70	60	60	0	27	No	0	0	3	2	0
Indus, Pakistan	12	4750	3390	680	1700	80	Yes	80	Minor	8	1	1.3
Irrawaddy, Myan.	2	1100	15,000	7600	6100	30	No	20	Minor	2	1.4	6
Krishna, India	6	250	840	1160	740	94	Yes	0	Major	7	0.4	≈4
Limpopo, Moz.	—	150	120	200	0	30	No	0	0	7	5	0
Magdalena, Col.	14	790	1120	750	750	0	Yes	70	0	6	3	6.6
Mahakam, Borneo	—	300	0; LP	0	370	0	No	?	0	0.2	0.2	0.5
Mahanadi, India	6	150	1480	2060	1770	74	Yes	40	Moderate	2	0.3	0
Mekong, Vietnam	1	20,900	9800	36,750	17,100	12	No	0	Minor	0.5	0.4	>5
Mississippi, USA	15	7140	13,500	0	11,600	48	Yes	?	Major	2	0.3	5-25

Delta	No. Maps	Est. Area km² < 2 m ASL	Recent Area km² Storm Surge	Recent Area km² River Flood	Recent Area km² In situ Flooding	% Sediment Reduction	Floodplain or Delta Flow Diversion	% Distributary Channel Reduction	Subsurface Water, Oil & Gas Mining	Early 20th C Aggradation Rate mm/y	21st Century Aggradation Rate mm/y	Subsidence mm/y
Niger, Nigeria	9	350*	1700	2570	3400	50	No	30	Major	0.6	0.3	7.5
Nile, Egypt	15	9440	0; LP	0	0	98	Yes	75	Major	1.3	0	5
Orinoco, Venez.	10	1800*	0; MP	3560	3600	0	No	0	Unknown	1.3	1.3	0.8–3
Parana, Argentina	6	3600	0; LP	5190	2600	60	No	?	Unknown	2	0.5	3
Pearl¶, China	4	3720	1040	2600	520	67	Yes	0	Moderate	3	0.5	7.5
Po, Italy	20	630	0; LP	0	320	50	No	40	Major	3	0	4–60
Rhone, France	11	1140	0; LP	920	0	30	No	40	Minor	7	1	2–6
Sao Francisco, Bra.	—	80	0; LP	0	0	70	Yes	0	Minor	2	0.2	10
Tigris¶, Iraq	7	9700	1730	770	960	50	Yes	38	Major	4	2	5
Tone¶, Japan	—	410	220	0	160	30	Yes	§	Major	4	0	>10
Vistula, Poland	4	1490	0; LP	200	0	20	Yes	75	Unknown	1.1	0	0.3
Yangtze¶, China	8	7080	6700	3330	6670	70	Yes	0	Major	1.1	0	10
Yellow¶, China	11	3420	1430	0	0	90	Yes	80	Major	49	0	8

*Significant canopy cover renders these SRTM elevation estimates as conservative values.

¶Alternate names: Congo & Zaire; Ganges & Ganges-Brahmaputra; Pearl & Zhujiang; Tigris-Euphrates & Shatt al Arab; Tone & Edo; Yangtze & Changjiang; Yellow & Huanghe.

§The Tone R. has long had its flow path engineered, having once flowed into Tokyo Bay; the number of distributary channels has increased with engineering works.

Source: Adapted from [38].

Table 3.
This table is depicting the sediment losses and subsidence of the world's river deltas.

Country	Planned Dams	Proposed Dams	Status	Reference
China	11	2	11 completed	[43]
Laos	43	20	79 completed and planning to reach 100 by 2030	[43, 44]
Myanmar	7	0	No data available from reports	[45]
Thailand	7	0	No data available from reports	[45]
Cambodia	12	0	No data available from reports	[45]
Vietnam	1	0	No data available from reports	[45]
Total	74	22	90 completed	[45]

Source: Adapted and modified from [45].

Table 4.
The number of constructed dams and planned dams on the Mekong River. The data is obtained from various sources and reports as well news articles. Also.

English name	Latin name	IUCN list status	Population status
Goonch	*Bagarius Yarrelli*	near threatened	decreasing
Two head carp	*Bangana behri*	vulnerable	30–50% decrease
Boeseman croaker	*Boesemania microlepis*	near threatened	decreasing, local extirpations
Giant barb	*Catilocarpio siamensis*	critically endangered	80–90% decline
Striped river barb	*Mekongina crythrospila**	near threatened	decreasing
Giant Mekong Catfish	*Pangasianodon gigas**	critically endangered	>80% decline
Striped catfish	*Pangasianodon hypophthalmus*	endangered	~95% decline
Krempf's catfish	*Pangasius krempfi*	vulnerable	~30% decline
Giant pangasius	*Pangasius santiwngsei*	critically endangered	~99% decline
Jullien's barb	*Probarbus jullieni*	endangered	~50% decline
Thicklip barb	*Probarbus labeamajor**	endangered	~50% decline
Laotian shad	*Tenualosa thibaudeaui**	vulnerable	~30% decline
Giant sheatfish	*Wallago attu*	near threatened	decreasing

*Endemic to the Mekong basin.
Source: Adapted from [46] and other reports.

Table 5.
This table is showing the species threats as reported by on species biodiversity monitoring by the International Union for the Conservation of nature (IUCN) in MRC report.

Migratory guild	Potential range of habitat utilized	Typical characteristics*	Likely impact of mainstream dams on migrations.
Migratory main channel spawner guild	Floodplains to running river upstream	• Spawn in the mainstream, in tributaries and around floodplains	Very high
		• Adults and drifting larvae return to floodplains to feed.	
		• May migrate to deep pools in the mainstream during the dry season.	
		• Sensitive to damming	

Migratory guild	Potential range of habitat utilized	Typical characteristics*	Likely impact of mainstream dams on migrations.
Migratory main channel refuge seeker guild	Floodplains to slow river downstream	• Spawn in floodplains • Migrations between floodplains and mainstream deep pools in the dry sea son. • Sensitive to damming	Very high
Semi-anadromous guild	Estuary and lower slow river downstream	• Enters fresh/brackish waters to breed. • Enters freshwaters as larvae and Juveniles (bligate or opportunistic) • Impacted by river mouth dams that stop migration into the river.	High (for dams located in river mouths or lower potamon)
Catadromous guild	Marine to running river upstream	• Reproduction, early feeding and growth at sea. • Juvenile or sub-adult migration to freshwater habitats • Vulnerable to overexploitation and tend to disappear when river is dammed preventing longitudinal upstream migration. • May respond favorably to fish passage facilities.	Very high

Source: Adapted from [46].

Table 6.
This table is showing the impacts on the migratory pattern of fishery guilds caused by dam construction.

S/N	Resource	Resource contents	Provider
1	<GLOBAL_ANALYSIS_FORECAST_PHY_001_024> (model, 0.083degree x 0.083degree, from 2019 to 2101-01 to Present)	Salinity, Sea Surface Height (SSH)	Copernicus Marine Service
2	<GLOBAL_REANALYSIS_PHY_001_030> (model, 0.083degree x 0.083degree, from 1993 to 2001-01 to 2019-2112-31)	Salinity, SSH	Copernicus Marine Service
3	<GLOBAL_MULTIYEAR_BGC_001_033> (model, 0.083degree x 0.083degree, from 1998 to 2001-01 to 2019-2112-31)	Turbidity, Transparency	Copernicus Marine Service
4	https://help.marine.copernicus.eu/en/articles/5070873-what-are-the-marine-variables-available-to-monitor-the-ocean	All ocean variable monitoring variables	Copernicus Marine Service
5	https://www.mrcmekong.org/about/mekong-basin/geography/	All information about the Mekong River Region	Mekong River Commission (MRC)

Source: Compiled by the Authors for this book chapter.

Table 7.
This table is a collection of various additional resources which readers may refer to.

6.1.5 Trend number 5 – Climate change impacts

In the their article, Safra de Campos et al. [36] indicates that river deltas being in low-lying coastal areas are at risk from both natural climate change impacts and anthropogenic impacts. Examples of this risk are, namely, submergence of the sea front settlements, increased flooding of coastal land, salt water intrusion and changes in the frequency of cyclones.

7. Further resources for readers

The readers who wish to examine and study the issues pertaining to stressors, especially the marine stressors are invited to explore the following resources as indicated in the table below. The Mekong River Commission also has some useful resources available for users (**Table 7**).

8. Conclusions

In conclusion, the anthropogenic activities which have been seen to impact negatively on the shifts of stressor elements from normal levels to levels capable of inducing stress in biodiversity are the dam projects, agricultural practices, industrial operations amongst many other. However, the damming developments have been identified as the most ones prone to inducing stress due to the changes in the sediment loads. This change in sediment loading together with sea-level rise due to climate change have been causing the coastal water to change color from brownish hue to ocean blue in most deltas. This is an indication that the nutrient rich sediment carrying river water is intruded by ocean water with little nutrients impacting negatively on the productivity of coastal habitats.

Therefore, it is advisable to carryout dam development and other developmental activities in a precautionary manner. During planning, developers must take care of all issues pertaining to the sustainability of the projects. Similarly, during implementation care should be taken to ensure all bodied and professionals in the area of marine resources administration and research are consulted.

Acknowledgements

First and foremost, the authors of this chapter wishes to thank all the great writers who have been cited in here. In addition, the authors wish to express their gratitude to the Author Services Manager, Ms. Romina Rovan, for all the support rendered to make this chapter a success. Furthermore, many thanks go to the members of the Deep Ocean Stewardship Initiative (DOSI) from the Center for Coastal Resources Management, Virginia, Institute of Marine Science (W and M) for agreeing to send in the pictures showing the ingestion of micro plastics by small organisms, and of course the authors do not forget to thank the Editor(s) of the book.

Conflict of interest

The authors declare that there is no conflict of interest.

Note of thanks

The lead author wishes to express his thanks to Mr. Joackim Mambwe, from Zambia College of Agriculture - Mpika for helping in modifying Figure 1 from Baran and his team of researchers. Additionally, a hand of gratitude is extended to Mr. Mataa Muimui, a student at Zambia College of Agriculture - Mpika for making his device, the Lenovo Ideapad, available during the periods when the main desk top computer was down to make this work to be submitted on time.

Nomenclature

Aggradation	Also referred to as alluviation. This is the increase of the land elevation due to deposition of sediment in a river delta.
Alluvium	A deposit of clay, silt and sand left by flowing floodwater in a river or delta, typically producing fertile soil
Biodiversity	This term is derived from "biological diversity", it refers to the variety of life on Earth at all its levels, starting from genes to ecosystems.
Biota	Plant and animal life in general
Channel	A water way
Ecology	Ecology is the study of the relationships between living organisms and the environment in which they exist.
Estuary	This is a partially enclosed, coastal water body in which fresh water coming from rivers and streams mixes with the salt water from the ocean.
Fluvial	Of or associated with rivers and streams.
Fluvial process	Processes predominantly associated with rivers or streams.
Geomorphology	The scientific study of the origin and evolution of topographic and bathymetric features on the Earth's surface created by physical, chemical or biological processes (or a combination of these processes)operating at or near the Earth's surface
Habitat	A habitat is a place where an organism makes its home. The 3 components of a habitat are: shelter, water, food, and space.
Hydrodynamics	A branch of Physics that deals with motion of liquids and forces acting on bodies immersed in liquids.
Mangrove	This is a shrub or small tree that grows in coastal saline or brackish water. Mangroves can also grow in fresh water.
Mud flats	Also known as tidal flats,; are coastal wetlands that form in intertidal areas where sediments have been deposited by a tide or river.
River delta	A river delta is a land form which is created by sediment deposition by a river as the flow leaves the river mouth and enters slower moving or stagnant water.
Salt flats	Densely packed slat pans.
Salt marshes	Also known as coastal salt marshes or tidal salt marshes, is a coastal ecosystem I the upper coastal intertidal zone between the land ad open salt water or brackish water that is regularly flooded by the tides.
Sediment	Matter that settles to the bottom of a river or any other body of water.

Stressor	A physical, chemical or biological agent, environmental condition, external stimulus or an event which is observed as causing stress to an organism.
Toxicant	A toxicant is a chemical substance introduced into an environment and is known to be toxic.
Turbidity	The amount of cloudiness of water.
Turbulence	The flow of fluids which is characterized by disorderly changes in pressure and velocity of flow.

Author details

Charles Nyanga[1]*, Beatrice Njeri Obegi[2] and Loi To Thi Bich[3]

1 Zambia College of Agriculture, Mpika, Zambia

2 Kenya Marine and Fisheries Research Institute, Kenya

3 Nha Trang University, Vietnam

*Address all correspondence to: charles.nyanga@gmail.com

IntechOpen

References

[1] National Geographic. Delta. 2021. Available from: https://nationalgeogra phic.org/enclopedia/delta/ [Accessed: May 2, 2021]

[2] NASA. How a Delta Forms Where River Meets Lake. The Jet Propulsion Laboratory, California Institute of Technology; 2014. Available from: https://jpl.nasa.gov/images/how-a-deltaforms-where-river-meets-lake [Accessed: August 24, 2021]

[3] Scheltinga D,Counihan R, Moss A, Cox M, Bennet J. User's Guide to Estuarine, Coastal and Marine Indicators for Regional NRM Monitoring, Report to DEH, MEN, ICAG, Coastal Zone CRC. 2004. Available from: https://ozcoasts.org.au/ managemet/emf-frame/user-guide-rm/ [Accessed: June 2, 2021]

[4] Piman T, Manish S. Case Study on Sediment in the Mekong River Basin: Current State and Future Trends, Project Report 2017–03. Stockholm Environment Institute; 2017. Available from: https://www.sei.org/publica tions/sediment-mekong-river/ [Accessed: June 12, 2021]

[5] Allison MA, Nittroouer CA, Ogston AS, Mullarney JC, Nguyen TT. Sediment and survival of the Mekong delta: A case study of decreased sediment supply and accelerating rates of relative Sea level rise. Oceanography. 2017;**30**(3):98-109. DOI: 10.5670/oceanog.2017.318 [Accessed: June 12, 2021]

[6] World Wide Fund for Nature (WWF). Wildlife of the Great Mekong. 1986. Available from: https://www.fasia.awsasse ts.panda.org/discovering_the_greater_ mekong/species [Accessed: June 9, 2021]

[7] Asokan PK. Mangrove and its Importance to Fisheries. 2012. Available from: https://core.ac.uk/download/pdf/ 33019864.pdf [Accessed: June 8, 2021]

[8] Baran E, Guerin E, Nasielski J. Fish, Sediment and Dams in the Mekong. Penang, Malaysia: WorldFish, Land and Ecosystems (WLE) 108pp; Available from: https://cgspace.cgiar.org2015 [Accessed: June 10, 2021]

[9] Shinkai Y, Truc DV, Sun D, Canh D, Kumagai Y. Arsenic and other metal contamination of groundwater in the Mekong River Delta, Vietnam. Journal of Health Sciences. 2007;**53**(3):344-346. DOI: 10.1248/jns.53.344. Available from: https://researchgate,net/publication/ 241759562_Arsenic_and _Other_Metal_ Contamniation_of_Groundwater_in_ the_Mekong_River_Delta_Vienam/ [Accessed: June 2, 2021]

[10] Mekong River Commission (MRC). Climate. 2021. Available from: https:// mrcmekong.org/about/mekong-basin/ climate/# [Accessed: June 1, 2021]

[11] University of Hull. River of Plastic – The Journey of Plastics along the Mekong and its Ultimate Fate in the World's Oceans. 2021. Available from: https://hull.ac.uk/work-with-us/ research/institutes/energy-and-environment-institute/our-work/river-of-plastic [Accessed: June 7, 2021]

[12] United Nations Environment Programme (UNEP). Plastic Pollution Threatens the Mekong: a wildlife wonderland, UNEP, Ecosystems and Biodiversity. 2021. Available from: https://unep.org/news-and-stories/ story/plastic-pollution-threatens-mekong-wildlife-wonderland [Accessed: June 11, 2021]

[13] Contributing Writer. How is a Delta Formed?. 2021. Available from: https:// sciencing.com/delta-formed-6643968. html [Accessed: May 2, 2021]

[14] Seybod H, Andrade JS Jr, Herman HJ. Modelling river delta formation. Proceedings of the National Academy of

Sciences of the United States of America (PNAS). 2007. DOI: 10.1073/pnas. o705265104

[15] Briney A. Geography of River Deltas. Thought Co.; 2020. Available from: https://thoughco.com/geography-of-river-delts-1435824 [Accessed: May 7, 2021]

[16] Galloway WE. Process Framework for Describing the Morphologic and Stratigraphic Evolution of Deltaic Depositional Systems. 1975. Available from: https://reserachgate.net/publica tion/287828682 [Accessed: June 18th, 2021]

[17] Elliot T. Deltas. Available from: h ttps://geoweb.uwyo.edu/geol2100/Delta s.pdf [Accessed August 28, 2021]

[18] Central Connecticut State University. Define Fluvial and Outline the Fluvial Processes: Erosion, Transportation and Deposition. 2021. Available from: https://web.ccsu.edu/ faculty/Kyem/geog272/chapter11/rive rs_landforms.htm [Accessed: June 4, 2021]

[19] National Park Service. Geology: River Systems and Fluvial Landforms. 2021. Available from: https://nps.gov/ subjects/geology/fluvial-landforms.htm [Accessed: June 2, 2021]

[20] SEPM. Fluvial- Dominated Deltas. 2021. Available from: https://sepmstrata .org/page.aspx?pageid=315 [Accessed: May 11,2021]

[21] Nienhuis J. Wave-Dominated River Deltas. 2020. Available from: https:// www.coastalwiki.org/wiki/Wave-dominted_river_deltas [Accessed: May 8, 2021]

[22] Universiteit Utrecht. River Deltas Depositional Processes. 2021. Available from: https://geo.uu.nl/fg/gruessink/ PPT_files/Lecture%204%20delta% 20Systems.pdf [Accessed: May 2, 2021]

[23] Ozcoasts. Wave-Dominated Deltas. Australian Online Coastal Information; 2021). Available from: https://ozcoasts. org.au/conceptual-diagrams/science-models/geographic/wdd [Accessed: May 8, 2021]

[24] Goodbred SL, Saito Y. Tide-dominated deltas. In: Davis RA, Dalrymple RW, editors. Principles of Tide Sedimentology. 2021. pp. 129-149. DOI: 1007/978-94-007-0123-6_7. Available from: https://semanticscholar. org/paper/Tide-Dominated_Deltas-Goodbred_Saito/ [Accessed: May 20, 2021]

[25] InteGrate. Coastal Processes, Hazards, and Society: Deltas in Crisis. 2021. Available from: https://www.ed ucation.psu.edu/earth107/node/1024 [Accessed: May 7, 2021]

[26] InteGrate. Coastal Processes, Hazards, and Society: Delta Morphologies and Driving Processes. 2021. Available from: https://www.ed ucation.psu.edu/earth107/node/1023 [Accessed May 7, 2021]

[27] WWF. A Biological Treasure Trove. 2021. Available from: https://www.grea termekong.panda.org/discovering_ the_greater_mekong/ [Accessed: August 27, 2021]

[28] Wikipedia. Mekong. 2021. Available from: https://www.eu.wikipedia.org/ wiki/Mekong [Accessed: August 28, 2021]

[29] Singapore Management University. The Mekong. 2021. Available from: https://researchguides.smu.edu.sg/cp_ myanmar [Accessed: August 29, 2021]

[30] Wallonie/Bruxelles au Vietnam. Feel Inspired: The Mekong River, Between Issues and Opportunities. 2018. Available from: https://www.wallonie vn/fr/actualities/mekong-river-betwee n-issues-and-opportunities-vn [Accessed: August 23, 2021]

[31] Berg M, Stengel C, Pham TK, Pham H, Sampson ML, Leng M, et al. Magnitude of arsenic pollution in the Mekong and Red River deltas – Cambodia and Vietnam. Science of the Total Environment. 2007;**72**(2–3):13-25. DOI: 10.1016/j/scitoenv.2006.09.010. Available from: https://pubmed.ncbi. nlm.gov.17081593/ [Accessed: June 18, 2021]

[32] Eyler B. Science Shows Chinese Dams Are Devastating the Mekong. FP Special Reports. 2020. Available from: https://foreignpolicy.com/202/04/22/sc ience-shows-chinese-dams-devastating-mekong-river [Accessed: August 23, 2021]

[33] Hoang L, Lauri H, Kumma M, Koponen T, van Vliet M, Supit I, et al. Mekong River flow and hydrological under extreme climate change. Hydrology and Earth System Science. 2016 20.3027-3041.105194/hess-20-3027-2016. [Accessed: August 24, 2021]

[34] Adamson P, Rutherford I, Peel M, Caular I. The hydrology of the Mekong River. 2009. DOI: 10.1016/B978-0-12-374026-7.00004-8 [Accessed: August 24, 2021]

[35] Kazem M. Challenges and Difficulties of Living in River Deltas. A Review of the Major River Deltas in Asia and Africa (Scientific Essay). Munich GRIN Verlag; 2015. Available from: h ttps://www.grin.com/document/3064 [Retrieved: August 23, 2021]

[36] Safra de Campos R et al. Where people live and move in deltas. In: Nichells R, Adger W, Hutton C, Hanson S, editors. Deltas in the Anthropocene. Cham: Palgrave Macmillan; 2020. DOI: 10.1007/978-3-030-23517-8_7 [Accessed: August 23, 2021]

[37] Day et al. Approaches to defining Deltaic sustainability in the 21st Century. Estuarine and Coastal Shelf Science. Sustainability for Future Coasts and Estuaries. 2016;**183**:275-291. DOI: 10.1016/j.ecss.2016.06.018

[38] Syvitski PMJ. Sinking Deltas. 2009. Available from: https://www.core.ac.uk/ display/4167699?utm_source=pdf& utm_medium=banner_campaign=pdf-decorative-v1 [Accessed: August 31, 2021]

[39] Dunn et al. Projections of declining fluvial sediment delivery to major Deltas Worldwide in response to climate change and anthropogenic stress. Environmental Research Letters. 2019; **14**(8):084034. DOI: 10.1088/1748-9326/ ab304e

[40] Knapman D, Zmud J, Ecola L, Mao Z, Crane K. Quality of Life Indicators and Policy Strategies to Advance Sustainability in the Pearl River Delta. Santa Monica, CA: RAND Corporation. Available from: https://www.rand.org/ pubs/research_reports/RR871.html; 2015 [Accessed: August 31, 2021]

[41] Szabo S et al. Population dynamics, Delta vulnerability and environmental change: Comparison of the Mekong, Ganger Brahmaputra and Amazon Delta regions. Sustainability Science. 2016, 2016;**11**:539-554. DOI: 10.1007/ s11625-016-0372-6

[42] Schmidt CW. Delta subsidence: An imminent threat to coastal populations. Environmental Health Perspectives. 2015;**123**(8):A204-A209. DOI: 10.1289/ ehp.123-A204

[43] ANI. China's Hydroelectric Dams on River Mekong Leaving Region Dry, Livelihoods of Millions Affected. 2021. Available from: https://www.aninews. in/news/world/asia/china's-hydroelec tric-dams-on-river-mekong-leaving-reg ion-dry-livelihoods-of-millions-affected [Accessed: August 31, 2021]

[44] Macau-Markar M. Thailand Challenges Laos Dam Building Spree on

Mekong River. NIKKEI Asia; 2021.
Available from: https://asia.nikkei.com/
Politics/International-relations/Thaila
nd-challenegs-Laos-dam-building-
spree-on-Mekong-River [Accessed:
August 31, 2021]

[45] Dang T, Ouillon S, Vinh G. Water
and suspended sediment budgets in the
lower Mekong from high-frequency
measurements (2009-2016). Water.
2018. DOI: 10.103390/w10070846

[46] Mekong River Commission. The
ISH 0306 Study. Development of
guidelines for hydropower
environmental impact mitigation and
risk management in the lower Mekong,
mainstream and tributaries. In: Volume
2 – Hydropower Risks and Impact
Mitigation, Key Hydropower Risks,
Impacts and Vulnerability and General
Mitigation Options for Lower Mekong,
Final Version. 2018. Available from:
https://www.mrcmekong.org/assets/
Uploads/ISH0306-Volume-2-Final-Ma
nual2.pdf [Accessed: August 31, 2021]